Core Microbiome

Core Microbiome

Improving Crop Quality and Productivity

Edited by

Javid A. Parray
Department of Higher Education
Government Degree College
Eidgah, Srinagar, India

Nowsheen Shameem
Cluster University Srinagar
Jammu and Kashmir, India

Elsayed Fathi Abd-Allah
King Saud University
Riyadh, Saudi Arabia

Mohammad Yaseen Mir
University of Kashmir
Srinagar, India

WILEY

This edition first published 2022

Registered Office(s)
John Wiley & Sons, Inc., 111 River Street, Hoboken, NJ 07030, USA
John Wiley & Sons Ltd, The Atrium, Southern Gate, Chichester, West Sussex, PO19 8SQ, UK

Editorial Office
The Atrium, Southern Gate, Chichester, West Sussex, PO19 8SQ, UK

For details of our global editorial offices, customer services, and more information about Wiley products visit us at www.wiley.com.

Wiley also publishes its books in a variety of electronic formats and by print-on-demand. Some content that appears in standard print versions of this book may not be available in other formats.

Library of Congress Cataloging-in-Publication Data
Names: Parray, Javid A., author.
Title: Core microbiome : improving crop quality and productivity / edited by Javid A. Parray, Govt Sheikh Ul Alam Memorial College, Jammu and Kashmir, India, Nowsheen Shameem, Cluster University Srinagar, Jammu and Kashmir, India, Elsayed Fathi Abd-Allah, King Saud University, Riyadh, Saudi Arabia, Mohammad Yaseen Mir, University of Kashmir, Srinagar, India.
Description: Hoboken, NJ: John Wiley & Sons, 2022. | Includes bibliographical references and index.
Identifiers: LCCN 2021053026 (print) | LCCN 2021053027 (ebook) | ISBN 9781119830764 (hardback) | ISBN 9781119830788 (pdf) | ISBN 9781119830771 (epub) | ISBN 9781119830795 (ebook)
Subjects: LCSH: Soil microbiology. | Plant-microbe relationships. | Agricultural microbiology.
Classification: LCC QR111. P333 2022 (print) | LCC QR111 (ebook) | DDC 579/.1757--dc23/eng/20211221
LC record available at https://lccn.loc.gov/2021053026
LC ebook record available at https://lccn.loc.gov/2021053027

Cover image: © Thichaa/Shutterstock
Cover design by Wiley

Set in 9.5/12.5pt STIXTwoText by Integra Software Services Pvt. Ltd, Pondicherry, India

10 9 8 7 6 5 4 3 2 1

M 03M000030000005332 050522

Contents

List of Contributors

Elsayed Fathi Abd-Allah
Plant Production Department, College of Food and Agricultural Sciences, King Saud University, Riyadh, Saudi Arabia

Khaloud Mohammed Alarjani
Botany and Microbiology Department, College of Science, King Saud University, Riyadh, Saudi Arabia

Khalid S. Almaary
Botany and Microbiology Department, College of Science, King Saud University, Riyadh, Saudi Arabia

Khalid F. Almutariri
Plant Production Department, College of Food and Agricultural Sciences, King Saud University, Riyadh, Saudi Arabia

Shahnaz Anjum
Department of Biomedical Engineering, Sathyabama Institute of Science and Technology deemed to be University, Rajiv Gandhi, Salai, Chennai, Tamil Nadu, India

Himanshu R. Barot
Sardarkrushinagar Dantiwada Agricultural University, Sardarkrushinagar, Banaskantha, Gujarat, India

Shabir Ahmad Bhat
Center of Research for Development, University of Kashmir, Srinagar, Jammu and Kashmir, India

Songul Çanakcı-Gulengul
Department of Biology, Fırat University, Elazıg, Turkey

Kamal Choudhury
Department of Botany, SBMS College, Sualkuchi, Assam, India

Shabeer Ahmad Dar
Center of Research for Development, University of Kashmir, Srinagar, Jammu and Kashmir, India

Pranjal Pratim Das
Department of Biotechnology, Darrang College, Tezpur, Assam, India

Turki M. Dawoud
Botany and Microbiology Department, College of Science, King Saud University, Riyadh, Saudi Arabia

Rupak Dey
Department of Earth and Environmental Science, KSKV Kachchh University, Mundra Road, Bhuj, Gujarat, India

Bisma Farooq
Department of Biomedical Engineering, Sathyabama Institute of Science and Technology deemed to be University, Rajiv Gandhi, Salai, Chennai, Tamil Nadu, India

Madiha Farooq
Department of Biomedical Engineering, Sathyabama Institute of Science and Technology deemed to be University, Rajiv Gandhi, Salai, Chennai, Tamil Nadu, India

Bahman Fazeli-Nasab
Research Department of Agronomy and Plant Breeding, Agricultural Research Institute, University of Zabol, Zabol, Iran

Aparna Gunjal
Department of Microbiology, Dr D.Y. Patil, Arts, Commerce and Science College, Pimpri, Pune, Maharashtra

Abeer Hashem
Botany and Microbiology Department, College of Science, King Saud University, Riyadh, Saudi Arabia

Mycology and Plant Disease Survey Department, Plant Pathology Research Institute, ARC, Giza, Egypt

Afrozah Hassan
Department of Botany, University of Kashmir, Srinagar, Jammu and Kashmir, India

Nida Idrees
Toxicology Laboratory, Department of Zoology, Government Raza P.G. College, Rampur, Uttar Pradesh, India

Azra N. Kamili
Center of Research for Development, University of Kashmir, Srinagar, Jammu and Kashmir, India

Fadime Karabulut
Department of Biology, Fırat University, Elazıg, Turkey

Tulasi Korra
Department of Mycology and Plant Pathology, Institute of Agricultural Sciences, Banaras Hindu University, Varanasi, India

Aswini Krishnan
Division of Microbiology, ICAR-IARI, New Delhi, India

Ashwani Kumar
Metagenomics and Secretomics Research Laboratory, Department of Botany, Dr. Harisingh Gour University (A Central University), Sagar, Madhya Pradesh, India

Kailash Chand Kumawat
Department of Soil Science, Punjab Agricultural University, Ludhiana, Punjab, India

Nafiseh Mahdinezhad
Department of Agronomy and Plant Breeding, Faculty of Agricultural, University of Zabol, Zabol, Iran

Manoj Kumar.V
Department of Mycology and Plant Pathology, Institute of Agricultural Sciences, Banaras Hindu University, Varanasi, India

Dipal B. Minipara
Anand Agricultural University, Anand, Gujarat, India

Mohammad Yaseen Mir
Center of Research for Development, University of Kashmir, Srinagar, Jammu and Kashmir, India

Anima Nanda
Department of Biomedical Engineering, Sathyabama Institute of Science and Technology deemed to be University, Rajiv Gandhi, Salai, Chennai, Tamil Nadu, India

Irshad Ahmad Nawchoo
Department of Botany, University of Kashmir, Srinagar, Jammu and Kashmir, India

Bijaya Kumar Nayak
Kanchi Mamunivar Government Institute for Post Graduate Studies and Reserach, Department of Botany, Puducherry, India

Asma Nazir
Department of Biomedical Engineering, Sathyabama Institute of Science and Technology deemed to be University, Rajiv Gandhi, Salai, Chennai Tamil Nadu, India

Khushboo Pachhigar
Veer Narmad South Gujarat University, Surat, Gujarat, India

Javid A. Parray
Department of Higher Education, Government Degree College, Eidgah, Srinagar, India

Gulzar Ahmed Rather
Department of Biomedical Engineering, Sathyabama Institute of Science and Technology deemed to be University, Rajiv Gandhi Salai, Chennai, Tamil Nadu, India

Robeena Sarah
Toxicology Laboratory, Department of Zoology, Government Raza P.G. College, Rampur, Uttar Pradesh, India

Rupak Kumar Sarma
Department of Botany, Sadiya College, Chapakhowa, Sadiya, Assam, India

Ranjith Sellappan
Department of Agricultural Microbiology, Tamil Nadu Agricultural University, Coimbatore, Tamil Nadu, India

Nosheen Shameem
Department of Environmental Science, Cluster University Srinagar, Jammu and Kashmir, India

Barkha Sharma
Department of Microbiology, College of Basic Sciences & Humanities, G. B. Pant University of Agriculture & Technology, Pantnagar, Uttarakhand, India

Sandeep Sharma
Department of Soil Science, Punjab Agricultural University, Ludhiana, Punjab, India

Seema B. Sharma
Department of Earth and Environmental Science, KSKV Kachchh University, Mundra Road, Bhuj, Gujarat, India

Sushil K. Sharma
ICAR-National Bureau of Agriculturally Important Microorganisms, Kushmaur, Maunath Bhanjan, Uttar Pradesh, India

Niraj Singh
Department of Microbiology, Royal Global University, Guwahati, Assam, India

Baby Tabassum
Toxicology Laboratory, Department of Zoology, Government Raza P.G. College, Rampur, Uttar Pradesh, India

Kalaiselvi Thangavel
Department of Agricultural Microbiology, Tamil Nadu Agricultural University, Coimbatore, Tamil Nadu, India

Uday Kumar Thera
Department of Mycology and Plant Pathology, Institute of Agricultural Sciences, Banaras Hindu University, Varanasi, India

Thirunarayanan P
Department of Mycology and Plant Pathology, Institute of Agricultural Sciences, Banaras Hindu University, Varanasi, India

Shalini Tiwari
Department of Soil Science, Punjab Agricultural University, Ludhiana, Punjab, India

Preface

An overabundance of microbial cells is present in every gram of soil, and microbes are found extensively in plant and animal tissues. The mechanisms governed by microorganisms in the regulation of physiological processes of their hosts have been extensively studied in light of recent findings on microbiomes. In plants, the components of these microbiomes may form distinct communities, such as those inhabiting the plant rhizosphere, the endosphere, and the phyllosphere. In each of these niches, the "microbial tissue" is established and responds to specific selective pressures. Although there is no clear picture of the overall role of the plant microbiome, there is substantial evidence that these communities are involved in disease control, enhance nutrient acquisition, and affect stress tolerance. This book first summarizes features of microbial communities that compose the plant microbiome and further presents a series of studies describing the underpinning factors that shape the phylogenetic and functional plant-associated communities. We advocate that understanding the mechanisms by which plants select and interact with their microbiomes may directly affect plant development and health and further lead to the establishment of novel microbes. Despite being one of the simplest metazoans, corals harbor some of the most highly diverse and abundant microbial communities. Differentiating core, symbiotic bacteria from this diverse host-associated consortium is essential for characterizing the functional contributions of bacteria but has not been possible yet. Here, we describe the coral core microbiome and demonstrate clear phylogenetic and functional divisions between the micro-scale, niche habitats within the coral host. In doing so, we discover seven distinct bacterial phylotypes that are universal to the core microbiome of coral species, separated by thousands of kilometers of oceans. The two most abundant phylotypes are co-localized specifically with the corals' endosymbiotic algae and symbiont-containing host cells. These bacterial symbioses likely facilitate the success of the dinoflagellate endosymbiosis with corals in diverse environmental regimes. This book primarily focuses on selecting positive and effective interactive core microbiome that are both phenotypically and genotypically very adaptive and sustainable, which further improve crop quality and productivity vis-à-vis sustainable agriculture. The bioengineering concept for rhizosphere improvement has also been discussed in one of the chapters. The book also highlights the structure, characterization, and biotechnological application of aquatic core microbiomes.

Javid A. Parray
PhD, PDF

1

A Review of Endophytic Microbiota of Medicinal Plants and Their Antimicrobial Properties

Robeena Sarah, Nida Idrees, and Baby Tabassum

Toxicology Laboratory, Department of Zoology, Govt. Raza P.G. College, Rampur 244901, Uttar Pradesh, India

1.1 Introduction

Medicinal plants have been used in many parts of the world for thousands of years in traditional treatments for numerous diseases. In rural areas of developing countries, they are still used as a primary source of drugs [1]. About 80% of developing countries use conventional medicines for general health care [2]. Natural products can be obtained from medicinal plants that have proven to be a rich source of biologically active compounds; many of them are used to develop novel chemicals for the pharmaceutical industry. With regard to disease-causing microorganisms, the increasing resistance to therapeutic agents currently in use, such as antibiotics and antiviral agents, has led to renewed interest in exploring novel anti-infective compounds. As approximately 450,000 plant species are available worldwide, of which only one per cent has been phytochemically analyzed, the prospects of locating new bioactive compounds are tremendously positive.

Medicinal plants are essentially considered complex and dynamic when used in systems for remedial therapy. Hence, their chemical composition depends upon several factors, such as botanical species, genetically determined chemotypes, anatomically a part of the plant (e.g., seed, flower, root, and leaf), storage, sun exposure, humidity, kind of ground, time of harvesting, and ecological area. Moreover, biogenic factors, such as the fungal and bacterial endophytes related to diverse parts of the plant, can influence their chemical composition. In recent years, the research and study of the multiple interactions occurring between endophytes and medicinal plants have modernized our knowledge of plant biology, with entirely unexpected and remarkable application perspectives: the probability of modulating, amplifying, or interfering within the biosynthesis of phytoconstituents (e.g., terpenes, polyphenols, and alkamides), but also to engineer the synthesis of latest molecules directly, for instance with antibiotic activity.

1.2 Antimicrobial Properties of Medicinal Plants with Particular Reference to Neem (*Azadirachtaindica*)

There are many reports on conventional medicinal plants and natural products to treat several diseases. Many plant-based medicines utilized in traditional therapeutic systems as agents in the treatment of infectious diseases are recorded in pharmacopoeia, a variety of

Core Microbiome: Improving Crop Quality and Productivity, First Edition. Edited by Javid A. Parray, Nowsheen Shameem, Elsayed Fathi Abd-Allah, and Mohammad Yaseen Mir.

which have been investigated recently for their efficiency against disease-causing microorganisms. The general antimicrobial activities of medicinal plants and their products, such as essential oils, were analyzed earlier [3,4].

Moreover, medicinal plants also can play an elementary function against rising antibiotic resistance both directly for their antimicrobial activities (e.g., antibacterial, antiviral, antifungal, and antiparasitic ones) and indirectly by reducing resistance against antibiotics.

There are numerous medicinal plants with well-known antimicrobial activity, including traditional Chinese or Ayurvedic medicine. Among medicinal plants adopted in experimental studies, a classic example is the *Azadirachta indica* (Figure 1.1) commonly known as neem, an evergreen tree of the sub-tropics and tropics, indigenous to the Indian subcontinent, with a recognized ethnomedicinal value and significance in agriculture and also in the pharmaceutical industry [5]. Every part of the neem plant, such as leaves, fruits, seeds, bark, and roots, contains compounds with proven antioxidant, anti-inflammatory [6], antidiabetic, antibacterial, antigingivitic, antifungal, anticancer, antiviral, antiulcer, neuroprotective, antipyretic, hepatoprotective, nephroprotective, and wound-healing properties. It possesses incredible potential within the field of medicine, pest management, and environmental protection. Neem may be a natural source of eco-friendly herbal pesticides, insecticides, and agro-based chemicals [7]. The impact of mouthwash containing neem on plaque and gingivitis was analyzed in a clinical study, indicating that it may help maintain mouth hygiene and suggest an enhanced effect on preventing oral diseases since it is both cost-effective and easily accessible [8]. Approximately 400 compounds are isolated from various parts of neem until now, and significant bioactive secondary metabolites and relatively 30 compounds are isolated from endophytes of neem [5]. Endophytic fungi of neem produce a wide range of secondary bioactive metabolites with potential compounds, namely, melanin, antimicrobial, antioxidant, anti-inflammatory, insecticides, nematicides, etc. [5].

Neem imposes a check on microbial growth and, in the breakdown of cell membrane capability, exhibits antimicrobial properties [9]. Neem nanoemulsion displayed antibacterial activity by disturbing the integrity of the bacterial cell membrane against strains of a bacterial pathogen. Different parts of the plant reveal antimicrobial properties against a wide range of pathogenic microbes [10]. Neem extract has shown antimicrobial effects against *Streptococcus mutans* and *S. faecalis* [11]. A new vaginal contraceptive, NIM-76, obtained from neem oil, has shown an inhibitory effect on the growth of various pathogenic microorganisms, including fungi, bacteria, and viruses [12]. The antibacterial property of neem seed oil in vitro against 14 strains of pathogenic bacteria has been recently assessed [13].

1.3 Current Trends on Bioactive Metabolites from Endophytic Microbiota of Medicinal Plants

A significant development in the understanding of plant-associated microbiota has been prompted by massive DNA sequencing technology in the past decade. Such close connection of microorganisms with plants has a broad contribution to plant nutrient absorption,

Figure 1.1 Medicinally important parts of the *Azadirachta indica* (neem) tree showing flowers; fruits; twigs; bark and leaves.

Table 1.1 Some Endophytic fungi obtained from different parts of neem tree. [54]

Genus (former name) of Endophytic fungi	Parts of neem tree	References
Trichoderma	Root, seed, leaf, bark, flower	[51,53]
Verticillium	Root	[51,53]
Humicola	Flower	[51,53]
Chloridium	Root	[51]
Nigrospora	Root, seed, leaf, bark	[50,51,53]
Scytalidium, *Penicillium*	Root	[51]
Aspergillus	Root, seed, leaf, bark, flower	[52]
Alternaria, *Drechslera*	Flower	[50,51]
Phoma	Leaf	[53]
Periconia	Bark	[50,53]
Stenella	Bark	[50]
Cercinella	Root	[51]

growth, stress tolerance, and health status, and secondary metabolite production [14]. Metal hyper-accumulating plants are known to shelter a co-evolved microbiota [15,16], and an essential role has been recognized in microbial-assisted phytoremediation [17]. Medicinal plants are well-known to shelter endophytes that are potentially concerned with the biosynthesis of phytoconstituents and can synthesize bioactive compounds.

The composition of biologically active components of medicinal plants differs broadly depending on their soil type, plant species, and their association with microorganisms [18,19]. These plant-linked microbial communities and their physiological functions are also strongly influenced by bioactive secondary metabolites synthesized by medicinal plants [20–22]. Moreover, for specific character and activities, including growth, nutrient acquirement, induced systemic resistance, and tolerance to abiotic stress factors, plants depend on their microbiome [23–27]. Although many medicinal plants are well-studied with relation to their microbiome, phytochemical constituents, and pharmacological properties, the physiological interactions between host and microorganisms remain inadequately understood [28]. The plant-linked microbiome consists of different microbial communities existing within the endosphere, shoots, and roots [27,29]. The rhizosphere of many plants is known to be a potential source for selecting good microorganisms that may affect plant health and is well-studied [21,30,31]. Hence, a valuable understanding of the microbial ecology of plant-associated bacteria may provide by considering the response of microbial communities to adaptation within the physiochemical atmosphere of the rhizosphere. The abundance of antagonistic bacteria within the rhizosphere of medicinal plants like sweet false chamomile (*Matricaria chamomilla*), *Solanum distichum*, and common marigold (*Calendula officinalis*) was observed [32]. The root-linked bacteria of *Ajuga bracteosa* displayed a broad range of plant growth-promoting activities by producing siderophores (iron-chelating compounds) and indole ethanoic acid and reveal antioxidant activity [33]. The vast and considerable collection of endophytic fungi from neem may represent a distinct source of the useful bioactive antimicrobial and insecticidal compounds [table 1.1] associated with *Azadirachta indica* such as the azadirachtins and related tetranortriterpenoids [50]. Recently, endophytic microbes have been on the verge of increased investigation due to their close interaction with the host [34]; it is assumed that the phytochemical constituents of plants are related, however, directly or indirectly, to endophytic microorganisms and their associations with host plants [20,35].

1.4 Plant Growth-Promoting Rhizobacteria (PGPR): Biological Management of Plant Pathogens

In plant pathology, "biocontrol" is often considered weapon to manage plant diseases. It is the use of antimicrobial compounds produced by microbial antagonists. Microorganisms that control the causative agents of plant diseases are collectively known as "biological control agents" (BCA) [36] with microorganisms, chemical mediators, and natural substances [37]. Endophytes are the microbial antagonists, mainly fungal and bacterial, that play an essential role in controlling disease. Both of them act differently to effect biological control, from the struggle at the niche level (colonization) during the making of antimicrobial compounds to bringing on the host resistance response [38–42], thus aiding in enhancing plant development along with health.

The constricted zone of soil particularly inclined by the root system is the rhizosphere [43]. This zone is loaded with nutrients due to the abundance of plant secretions, like amino acids and sugars, compared with most of the earth, providing a good source of energy and nutrients for bacteria [44]. Various microorganisms colonize the rhizosphere, and therefore, the bacteria settling in this environment are called rhizobacteria [45]. Plant-linked bacteria are often categorized within three groups, i.e. beneficial, harmful, and neutral groups, based on their effects on plant growth [43]. Beneficial free-living soil bacteria are usually characterized as PGPR [46]. Bacteria of various genera are identified as PGPR, of which the most important groups are Bacillus and Pseudomonas spp. [47]. PGPR affects plant growth directly or indirectly. The direct promotion of plant growth by PGPR involves providing the plant with a compound synthesized by the bacterium, for example, plant hormones, or promoting the uptake of certain nutrients from the environment [48]. The indirect progress of plant growth happens when PGPR prevents or reduces the harmful effects of one or more phytopathogenic organisms. This can be done by inducing resistance to pathogens or producing antagonistic substances [49].

1.5 Conclusion

The rhizosphere of many plants is known to be a potential source for selecting good microorganisms that may affect plant health. PGPR offer an excellent substitute for environmental-friendly biological control of plant-pathogen and amelioration of the cropping systems into which they can be most beneficially applied. The role of medicinal plants and their endophytic microbiota in producing plant antimicrobial compounds is concerned with plant health and remedial applications. The production of antimicrobial bioactive compounds by endophytes is currently receiving urgent concern due to the emergence of multidrug-resistant pathogens. Indeed, the study of medicinal plants especially neem and their endophyte interaction is a flourishing prospect to develop sustainable methods to restrain human and plant pathogens. Therefore, more work on endophytes of medicinal plants for their antioxidant activities can impose a strong effect on the search for novel bioactive compounds.

References

1 Chitme, H.R., Chandra, R., and Kaushik, S. (2003). Studies on antidiarrheal activity of *Calotropis gigantea* R. Br. in experimental animals. *Journal of Pharmacy & Pharmaceutical Sciences* 7: 70–75.

2 Kim, H.S. (2005). Do not put too much value on conventional medicines. *Journal of Ethnopharmacology* 100 (1–2): 37–39.

3 Cowan, M.M. (1999). Plant products as antimicrobial agents. *Clinical Microbiology Reviews* 12 (4): 564–582.

4 Kalemba, D. and Kunicka, A. (2003). Antibacterial and antifungal properties of essential oils. *Current Medicinal Chemistry* 10 (10): 813–829.

5 Kharwar, R.N., Sharma, V.K., Mishra, A., Kumar, J., Singh, D.K., Verma, S.K., Gond, S.K., Kumar, A., Kaushik, N., and Revuru, B. (2020). Harnessing the phytotherapeutic treasure troves of the ancient medicinal plant *Azadirachta indica* (Neem) and associated endophytic microorganisms. *Planta Medica* 86: 906–940.

6 Chattopadhyay, R.R., Chattopadhyay, R.N., and Maitra, S.N. (1993). Possible mechanism of anti-inflammatory activity of *Azadirachta indica* leaf extract. *Indian Journal of pharmacology* 25: 99–100.

7 Brahmachari, G. (2004). Neem-an omnipotent plant: A retrospection. *Chem Biochem* 5: 408–421.

8 Jalaluddin, M., Rajasekaran, U.B., Paul, S., Dhanya, R.S., Sudeep, C.B., and Adarsh, V.J. (2017). Comparative evaluation of neem mouthwash on plaque and gingivitis: A double-blind crossover study. *The Journal of Contemporary Dental Practice* 18: 567–571.

9 Sarah, R., Tabassum, B., Idrees, N., and Hussain, M.K. (2019). Bio-active Compounds Isolated from Neem Tree and Their Applications. In: *Natural Bio-active Compounds*, (eds.M.Akhtar, M.Swamy and U.Sinniah). Singapore: Springer. https://doi.org/10.1007/978-981-13-7154-7_1

10 Jerobin, J., Makwana, P., Kumar, R.S.S., Sundaramoorthy, R., Mukharjee, A., and Chandrasekaran, N. (2015). Antibacterial activity of neem nanoemulsion and its toxicity assessment on human lymphocytes in vitro. *International Journal of Nanomedicine* 10 (77): 86.

11 Almas, K. (1999). The antimicrobial effects of extracts of *Azadirachta indica* (Neem) and *Salvadora persica* (Arak) chewing sticks. *Indian Journal of Dental Research.* 10 (1): 23–26.

12 Sai Ram, M. et al. (2000). Effect of Kombucha tea on Chromate(VI)-induced oxidative stress in albino rats. *Journal of Ethnopharmacology* 71 (1–2): 235-40.

13 Baswa, M., Rath, C.C., Dash, S.K., and Mishra, R.K. (2001). Antibacterial activity of Karanj (*Pongamia pinnata*) and Neem (*Azadirachta indica*) seed oil: a preliminary report. *Microbios* 105 (412): 183–189.

14 Compant, S., Samad, A., Faist, H., and Sessitsch, A. (2019). A review on the plant microbiome: Ecology, functions, and emerging trends in microbial application. *Journal of Advanced Research* 19: 29–37.

15 Mengoni, A., Pini, F., Huang, L.N., Shu, W.S., and Bazzicalupo, M. (2009). Plant-by-plant variations of bacterial communities associated with leaves of the nickel hyperaccumulator *Alyssum bertolonii* desv. *Microbial Ecology* 58: 660–667.

16 Mengoni, A., Schat, H., and Vangronsveld, J. (2010). Plants as extreme environments? Ni-resistant bacteria and Ni-hyperaccumulators of serpentine flora. *Plant and Soil* 331: 5–16.

17 Thijs, S., Sillen, W., Rineau, F., Weyens, N., and Vangronsveld, J. (2016). Towards an enhanced understanding of plant-microbiome interactions to improve phytoremediation: Engineering the metaorganism. *Frontiers in Microbiology* 7: 341.

18 Zhao, J., Chan, T., Mou, Y., and Zhou, L. (2011). Plant-derived bioactive compounds produced by endophytic fungi. *Mini Reviews in Medicinal Chemistry* 11: 159–168. DOI: 10.2174/138955711794519492

19 Morsy, N.M. (2014). Phytochemical analysis of biologically active constituents of medicinal plants. *Main Group Chemistry* 13: 7–21. DOI: 10.3233/MGC-130117

20 Qi, X., Wang, E., Xing, M., Zhao, W., and Chen, X. (2012). Rhizosphere and non-rhizosphere bacterial community composition of the wild medicinal plant *Rumex patientia*. *World Journal of Microbiology and Biotechnology* 28: 2257–2265. DOI: 10.1007/s11274-012-1033-2

21 Philippot, L., Raaijmakers, J.M., Lemanceau, P., and van der Putten, W.H. (2013). Going back to the roots: The microbial ecology of the rhizosphere. *Nature Reviews Microbiology* 11: 789–799. DOI: 10.1038/nrmicro3109

22 Chaparro, J.M., Badri, D.V., and Vivanco, J.M. (2014). Rhizosphere microbiome assemblage is affected by plant development. *The ISME Journal* 8: 790–803. DOI: 10.1038/ismej.2013.196

23 Egamberdieva, D., Berg, G., Lindstrom, K., and Rasanen, L. (2010). Root colonizing Pseudomonas spp. improve growth and symbiosis performance of fodder Galega (*Galega orientalis* LAM) grown in potting soil. *European Journal of Soil Biology* 46: 269–272. DOI: 10.1016/j.ejsobi.2010.01.005

24 Egamberdieva, D., Kucharova, Z., Davranov, K., Berg, G., Makarova, N., Azarova, T. et al. (2011). Bacteria able to control foot and root rot and to promote the growth of cucumber in salinated soils. *Biology and Fertility of Soils* 47: 197–205. DOI: 10.1007/s00374-010-0523-3

25 Malfanova, N., Kamilova, F., Validov, S., Shcherbakov, A., Chebotar, V., Tikhonovich, I.et al. (2011). Characterization of *Bacillus subtilis* HC8, a novel plant-beneficial endophytic strain from giant hogweed. *Microbial Biotechnology* 4: 523–532. DOI: 10.1111/j.1751-7915.2011.00253.x

26 Sessitsch, A., Kuffner, M., Kidd, P., Vangronsveld, J., Wenzel, W., Fallmann, K. et al. (2013). The role of plant-associated bacteria in the mobilization and phytoextraction of trace elements in contaminated soils. *Soil Biology and Biochemistry* 60: 182–194. DOI: 10.1016/j.soilbio.2013.01.012

27 Berg, G., Grube, M., Schloter, M., and Smalla, K. (2014). Unraveling the plant microbiome: Looking back and future perspectives. *Frontiers in Microbiology* 5: 148. DOI: 10.3389/fmicb.2014.00148

28 Köberl, M., Schmidt, R., Ramadan, E.M., Bauer, R., and Berg, G. (2014). The microbiome of medicinal plants: Diversity and importance for plant growth, quality and health. *Frontiers in Microbiology* 4: 400. DOI: 10.3389/fmicb.2013.00400

29 Beneduzi, A., Ambrosini, A., and Passaglia, L.M.P. (2012). Plant growth-promoting rhizobacteria (PGPR): Their potential as antagonists and biocontrol agents. *Genetics and Molecular Biology* 35: 1044–1051. DOI: 10.1590/S1415-47572012000600020

30 Weller, D.M., Raaijmakers, J.M., McSpadden Gardner, B.B., and Thomashow, L.S. (2002). Microbial populations responsible for specific soil suppressiveness to plant pathogens. *Annual Review of Phytopathology* 40: 308–348. DOI: 10.1146/annurev.Phyto.40.030402.110010

31 Berendsen, R.L., Pieterse, C.M.J., and Bakker, P.A.H.M. (2012). The rhizosphere microbiome and plant health. *Trends in Plant Science* 17: 478–486. DOI: 10.1016/j.plants.2012.04.001

32 Köberl, M., Ramadan, E.M., Adam, M., Cardinale, M., Hallmann, J., Heuer, H. et al. (2013). Bacillus and Streptomyces were selected as broad-spectrum antagonists against soilborne pathogens from arid areas in Egypt. *FEMS microbiology letters* 342: 168–178. DOI: 10.1111/1574-6968.12089

33 Kumar, G., Kanaujia, N., and Bafana, A. (2012). Functional and phylogenetic diversity of root-associated bacteria of *Ajuga bracteosa* in Kangra valley. *Microbiological Research* 167: 220–225. DOI: 10.1016/j.micres.2011.09.001

34 Hardoim, P.R., van Overbeek, L.S., Berg, G., Pirttilä, A.M., Compante, S., Campisano, A. et al. (2015). The hidden World within plants: Ecological and evolutionary considerations for defining functioning of microbial endophytes. *Microbiology and Molecular Biology Reviews* 79: 293–320. DOI: 10.1128/MMBR.00050-14

35 Chandra, S. (2012). Endophytic fungi: Novel sources of anticancer lead molecules. *Applied Microbiology and Biotechnology* 95: 47–59. DOI: 10.1007/s00253-012-4128-7

36 Pal, K.K. and McSpadden Gardener, B. (2006). *Biological Control of Plant Pathogens*. Plant Health Instr.

37 Lecomte, C., Alabouvette, C., Edel-Hermann, V., Robert, F., and Steinberg, C. (2016). Biological control of ornamental plant diseases caused by *Fusarium oxysporum*: A review. *Biological Control* 101: 17–30.

38 Eljounaidi, K., Lee, S.K., and Bae, H. (2016). Bacterial endophytes as potential biocontrol agents of vascular wilt diseases—Review and future prospects. *Biological Control* 103: 62–68.

39 Latz, M.A.C., Jensen, B., Collinge, D.B., and Jørgensen, H.J.L. (2018). Endophytic fungi as biocontrol agents: Elucidating mechanisms in disease suppression. *Plant Ecology & Diversity* 11: 555–567.

40 De Silva, N.I., Brooks, S., Lumyong, S., and Hyde, K.D. (2019). Use of endophytes as biocontrol agents. *Fungal Biology Reviews* 33: 133–148.

41 Morales-Cedeño, L.R., delCarmenOrozco-Mosqueda, M., Loeza-Lara, P.D., Parra-Cota, F.I., de losSantos-Villalobos, S., and Santoyo, G. (2021). Plant growth-promoting bacterial endophytes as biocontrol agents of pre-and post-harvest diseases: Fundamentals, methods of application and future perspectives. *Microbiological Research* 242: 126612.

42 Santoyo, G., delOrozco-mosqueda, M.C., and Govindappa, M. (2012). Mechanisms of biocontrol and plant growth-promoting activity in soil bacterial species of Bacillus and Pseudomonas: A review. *Biocontrol Science and Technology* 22: 855–872.

43 Dobbelaere, S., Vanderleyden, J., and Okon, Y. (2003). Plant growth-promoting effects of diazotrophs in the rhizosphere. *Critical Reviews in Plant Sciences* 22: 107–149.

44 Gray, E.J. and Smith, D.L. (2005). Intracellular and extracellular PGPR: Commonalities and distinctions in the plant-bacterium signaling processes. *Soil Biology and Biochemistry* 37: 395–412.

45 Schroth, M.N. and Hancock, J.G. (1982). Disease-suppressive soil and root-colonizing bacteria. *Science* 216: 1376–1381.

46 Kloepper, J.W., Lifshitz, R., and Zablotowicz, R.M. (1989). Free-living bacterial inocula for enhancing crop productivity. *Trends in Biotechnology* 7: 39–43.

47 Podile, A.R. and Kishore, G.K. (2006). Plant growth-promoting rhizobacteria. In: *Plant-Associated Bacteria*, (ed. S.S.Gnanamanickam), 195–230. Netherlands: Springer.

48 Glick, B.R. (1995). The enhancement of plant growth by free-living bacteria. *Canadian Journal of Microbiology* 41: 109–117.

49 Castronovo, L.M., Vassallo, A., Mengoni, A., Miceli, E., Bogani, P., Firenzuoli, F., Fani, R., and Maggini, V. (2021). Medicinal plants and their bacterial microbiota: A review on antimicrobial compounds production for plant and human health. *Pathogens* 10 (106). https://doi.org/10.3390/pathogens10020106

50 Verma, V.C., Gond, S.K., Kumar, A., Kharwar, R.N., and Strobel, G.(2007). The endophytic mycoflora of bark, leaf, and stem tissues of *Azadirachta indica* A. Juss (Neem) from Varanasi (India). *Microbial Ecology* 54: 119–125.

51 Verma, V.C., Gond, S.K., Kumar, A., Kharwar, R.N., Boulanger, L.A. and Strobel, G.A. (2011). Endophytic Fungal Flora from Roots and Fruits of an Indian Neem Plant *Azadirachta indica* A. Juss., and Impact of Culture Media on their Isolation. *Indian Journal of Microbiology* 51: 469–476.

52 James, T.Y., Kauff, F., Schoch, C.L., Matheny, P.B., Hofstetter, V., Cox, C.J., Celio, G., Gueidan, C., Fraker, E., and Miadlikowska, J. et al. (2006). Reconstructing the early evolution of fungi using a six-gene phylogeny. *Nature* 443: 818–822.

53 Maharachchikumbura, S.S., Hyde, K.D., Jones, E.G., McKenzie, E.H., Huang, S.K., Abdel-Wahab, M.A., Daranagama, D.A., Dayarathne, M., D'souza, M.J., and Goonasekara, I.D. et al. (2015). Towards a natural classification and backbone tree for Sordariomycetes. *Fungal Diversity* 72: 199–301.

54 Chutulo, E.C. and Chalannavar, R.K. (2018). Endophytic Mycoflora and Their Bioactive Compounds from Azadirachta Indica: A Comprehensive Review. *Journal of Fungi* 4 (2): 42.

2

Plant Microbiome

A Key to Managing Plant Diseases

Dipal B. Minipara[1], Khushboo Pachhigar[2], and Himanshu R. Barot[3]

[1]*Anand Agricultural University, Anand, Gujarat, India*
[2]*Veer Narmad South Gujarat University, Surat, Gujarat, India*
[3]*Sardarkrushinagar Dantiwada Agricultural University, Sardarkrushinagar, Banaskantha, Gujarat, India*

2.1 Introduction

Microbes are fundamental for a balanced life on Earth, and understanding their function is crucial. Culture-based microbial techniques and next-generation sequencing have added benefits. Metagenomics studies have allowed studying the microbial genome and its function, which are not culturable but have an essential role in the ecosystem. Microbiome study enables the exploration of the genome of all microorganisms, including symbiont and pathogens. The microbial community varies with host health and environmental factors that shape the microbiome. The concept of association of microbiome with plant health and disease state received focus over the past decades. However, it is critical to identify prominent microbiota associated with biotic stress in variable environmental conditions on the field, as most studies were conducted under greenhouse conditions. Core microbiota, a host-associated persistence microbial community carrying functional genes, are critically important for plant health. Therefore, identifying host-associated core microbiota and their response toward biotic stress is essential to understanding disease management and improving productivity.

Plant microbiome can be categorized at the interface where plant and microbes can communicate, which include plant–microbe interaction at rhizosphere, phyllosphere, and endosphere (Figure 2.1) [1]. Plant microbiome alters as plant species or cultivar changes; it also depends on plant developmental stages, disease conditions, and geographical locations [2].

2.2 Plant–Microbe Interaction in the Rhizosphere

"Rhizosphere" is derived from "rhiza" or root and "sphere" or the field of influence in ancient Greek. The rhizosphere is the soil environment surrounding plant roots, and it is a crucial underground region for plant–microbe interaction. The rhizosphere comprises 10-fold microbes more than that in the surrounding soil, and the rhizosphere microbiome may consist of fungi, bacteria, archaea, actinomycetes, and viruses [3]. Roots utilize photosynthates to produce exudates used by rhizosphere microbiota to exchange nutrients

Core Microbiome: Improving Crop Quality and Productivity, First Edition. Edited by
Javid A. Parray, Nowsheen Shameem, Elsayed Fathi Abd-Allah, and Mohammad Yaseen Mir.

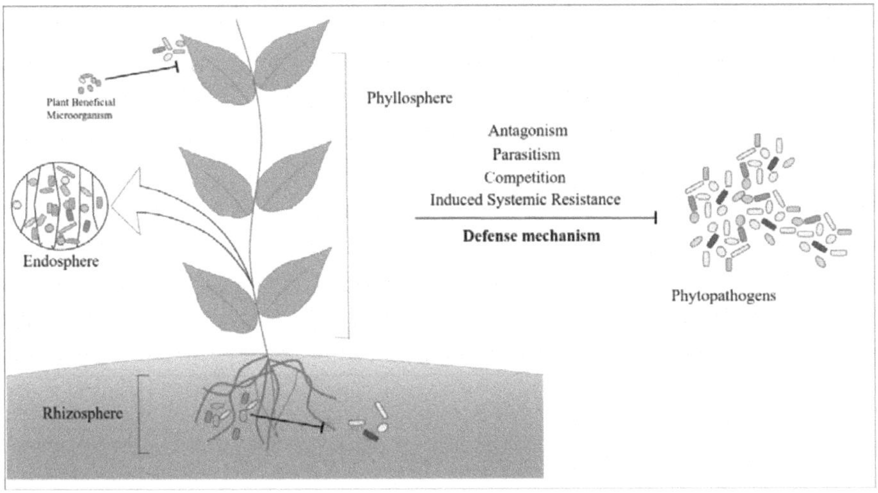

Figure 2.1 Plant–microbe interaction in rhizosphere, endosphere, and phyllosphere and biocontrol mechanism acquired by beneficial microorganism.

and water [4]. Root exudates also function as signaling molecules, attractants, stimulants, or repellents to rhizosphere microbes. Bioactive compounds of exudates vary with plant species that define microbial community [5]. Root secretion influences soil pH, soil structure, oxygen obtainability, and influences plant–microbe interaction. A study on secondary metabolite benzoxazinoids showed it's defensive role in cereals, alters rhizosphere fungal and bacterial microbiota, and influences plant growth [6]. Bacteria produce protein and polysaccharides, which allow them to attach with roots as biofilm aggregates [7]. Symbiotic association evolved millions of years ago, and most living plants form a symbiotic relationship with arbuscular mycorrhizal fungi (AMF) [8]. Upon sensation of strigolactone from plant roots, AMF initiates hyphal branching to colonize roots and releases lipochito oligosaccharides, promoting plant growth by increasing root surface area [1]. Almost all soilborne fungi are necrotrophic, and they do not require the living cell to acquire nutrients.

2.2.1 Microbial Population in the Rhizosphere

The most enriched microbiota in the rhizosphere are bacteria, which influence the plant rhizosphere significantly. Next-generation sequencing (NGS) has provided insight into microbial diversity and composition. Internal transcribed spacer (ITS), 16s rRNA sequencing of fungi and bacteria, respectively [9,10] and shotgun metagenomic sequencing [11] have unravelled root-associated microbiome. The relative abundance of taxonomic marker genes has provided more information on phyla *Proteobacteria*, *Bacteroidetes*, and *Actinobacteria*, which were augmented in the rhizosphere [12]. Predominant genera in the rhizosphere are *Pseudomonas, Bacillus, Rhizobia, Azotobacter, Mycobacterium, Flavobacter, Cellulomonas, Agrobacterium*, and *Micrococcus* [13]. So far, various gram-positive and aerobic spore-forming bacteria are inferior because of the low availability of oxygen [14]. For fungal communities, phyla *Ascomycota, Basidiomycota* followed by *Zygomycota, Glomeromycota* dominate in the rhizosphere of *Arabidopsis thaliana*. Plant-friendly microorganisms comprise nitrogen-fixing bacteria *viz., Azotobacter, cyanobacteria, clostridium,* plant growth-promoting rhizobacteria, fungi, and endo ectomycorrhizal fungi [13].

2.2.2 Biocontrol Mechanism in the Rhizosphere

To grow a healthy plant, it is necessary to examine soil niches surrounding the root area to detect pathogens and enhance advantageous microorganisms. The rhizosphere is the battlefield where phytopathogens acquire parasitic relations with the plant. Biocontrol agents have a mechanism that includes rivalry for space and nutrients, antagonistic activity, and hyperparasitism. The level of decomposition of organic matter influences microbial communities and biocontrol activities [14]. The introduction of beneficial microorganisms can also change microbial community structure in the rhizosphere.

2.2.2.1 Competition

Some strains of *Pseudomonas fluorescens* act as biocontrol by competing for nutrients and root surface colonization aggressively. *Pseudomonas sp.* RU47 colonizes the rhizosphere with sufficient density and compete with *Rhizoctonia solani* and minimizes the severity of black potato scurf and lettuce bottom rot in diluvial sand, alluvial loam, and loess loam soils [15]. The competition can also be for micro-nutrients, specifically iron, associated with high-affinity chelators called siderophores secreted by microorganisms. An experiment with transgenic tobacco overexpressing ferritin showed that fluorescent *Pseudomonads* could survive in a less iron-containing environment and have an antagonistic activity to the pathogen [16].

2.2.2.2 Parasitism

Parasitism as a biocontrol mechanism is mainly associated with fungal biocontrol agents. Biocontrol agents parasitize the pathogen by looping around fungal hyphae and derive nutrition from the pathogen. Some beneficial microorganism secretes hydrolytic enzymes such as chitinase and cellulase [17]. *Trichoderma* and *Gliocladium* parasitize on *Rhizoctonia, Sclerotinia, Verticillium,* and *Gaeumannomyces* and cause cell damage to the pathogen [18]. *Firmicute* bacteria *Pasteuria penetrans* also control *Meloidogyne* nematode, and its activity is stimulated by two beneficial rhizospheric bacteria [19].

2.2.2.3 Antagonism

Antagonism is mediated by antibiosis in which antimicrobial compounds, lytic enzymes, or effector molecules are produced by biocontrol agents toxic to pathogens. Antibiotics compounds, *viz.*, phenazines, pyrrolnitrin, lipopeptides, hydrogen cyanide, and 2,4-diacetylphloroglucinol, produced by specific biocontrol agents are well characterized [20]. Certain Bacillus species make antifungal lipopeptide iturin A [21], antimicrobial and antiviral cyclic lipodecapeptide fengycin [22], and various biosurfactants [23]. *Bacillus subtilis* RB14 produces antibiotic iturin, and surfactant was found to control *Rhizoctonia solani,* which causes damping-off disease of tomato seedlings [24]. Several strains can also induce plant defense-related genes by biocontrol metabolites or lipopolysaccharides and flagella and can control phytopathogen [25]. *Streptomyces pactum*, a biocontrol agent, increases the indigenous *P. koreensis* GS population in ginseng rhizoplane by expressing chemotaxis and flagellar biosynthesis-related genes and antagonizes soilborne pathogens [26].

Extracellular lytic enzymes such as cellulase, protease, and chitinase produced by certain antagonistic microorganisms are common. Other lytic enzymes target phytotoxins like fusaric acid produced by *Fusarium oxysporum* [23]. Another mechanism acquired by *Pseudomonas* is the type III secretion system that targets oomycetes [27]. According to Vacheron et al. [28] plant roots preferentially shaped *Pseudomonads* had one to five plant-beneficial properties [28].

2.2.2.4 Induced Systemic Resistance (ISR)

In the theory of plant immune system, the initial defense system activated by plants is pathogen-associated molecular patterns (PAMP), PAMP-triggered immunity (PTI), in which pattern recognition receptors (PRRs) recognize bacterial flagella or fungal chitin. The subsequent defense after initial defense is effector-triggered immunity (ETI), in which the nucleotide binding leucine-rich repeat (NB-LRR) receptor recognizes effector molecules of a pathogen. These two lines of defense often trigger induced resistance in unexposed parts of the plant by pathogens, and the mechanism is termed "systemic acquired resistance" (SAR) [29].

Plant-microorganism are evolved to gather according to different taxa and soil types but nowadays focus has been on functional microbiota, which provides fitness to halobiont. Induced systematic resistance (ISR) is a well-studied mechanism and first described for *Pseudomonas* in which certain beneficial bacteria, *viz.*, *Pseudomonas*, *Bacillus*, and *Serratia* strains, and nonpathogenic fungi, *viz.*, *Trichoderma*, *F. oxysporum*, and *Piriformospora* isolated strains, and the symbiotic *mycorrhiza* species in the rhizosphere prime immune system in the unexposed parts of plants to fight against infectious agents [30]. Seed-grown maize plant exposed to *Trichoderma harzianum* strain T22 had minimized symptoms of anthracnose, which was explained by root colonization by *Trichoderma* strain-ISR [31].

2.2.3 Plant Disease Management

Every natural soil can suppress pathogens to a certain level; this phenomenon is called *general disease suppression*, and it depends on the total microbial biomass. Specific suppression comes into the picture when an individual or group of microorganisms causes soil to suppress disease [32]. Some soil can retain its condition by suppressing activity for a long time, while some could develop after several years of monoculture. Moreover, the establishment of disease suppression in soil was described for several diseases. In an experiment conducted on a tomato plant, a susceptible cultivar was treated with rhizosphere microbiota of resistant cultivar, which suppressed disease symptoms.

Further analysis showed that *flavobacterium* TRM1 could antagonize *R. solanacearum* and inhibited the progress of bacterial wilt in tomatoes [33]. A root disease termed as "take-all" of *Triticum aestivum* caused by *Gaeumannomyces graminis var. tritici* was suppressed through several years of monoculture in soil by disease-suppressive microorganisms. This phenomenon is called "take-all decline," which is associated with the antagonistic activity of fluorescent *Pseudomonas* spp. by the antifungal compound 2,4-diacetylphloroglucinol (DAPG) [32]. Soil metagenomics study of potato plants associated with potato common scab disease revealed a positive correlation between pathogenic *Streptomyces* and scab severity but negative correlation with *Geobacillus*, *Curtobacterium* unclassified *Geodermatophilaceae* [34]. Some commercial products such as "Mycostop," containing *Streptomyces griseoviridis* strain, which can suppress root rot and wilt diseases by occupying the rhizosphere are also available [35], "BlightBan A506" containing *P. Fluorescens* can control fire blight frost damage in fruits caused by *Erwinia amylovora* and "Epic Kodiak" to target *Rhizoctonia solani* by *Bacillus subtilis* [36]. A microbial consortium containing four isolates, *Serratia marcescens isolate 59*, *Pseudomonas fluorescens* 57, *Rahnella aquatilis* 36, and *Bacillus amyloliquefaciens* 63 was able to increase soil disease suppression ability against *Fusarium spp.* in chickpea rhizosphere, which causes wilt and root rot [37].

Various management practices in combination or isolation are being used to suppress diseases, including intercropping, organic amendments, crop rotation, and tillage management. Several years of monoculture of one variety in the same field causes replanting disease. Intercropping changes root exudates that alter the rhizospheric microbiome. An experiment with intraspecific intercropping on *Radix pseudostellariae* plant increased beneficial *Nitrosomonadales*, *Nitrospirales*, *Pseudomonadales*, and decreased pathogenic *Aspergillus* and *Talaromyces* [38]. *Atractylodes lancea*, a medicinal plant, suppressed *Fusarium oxysporum* mediated root rot disease in peanut while used as inter-crop [39] and while intercropping with aerobic rice, conquered *Fusarium oxysporum* mediated watermelon wilt [40]. Intercropping between maize and soybean can cause phenolic acid-mediated inhibition of *Phytophthora sojae*, responsible for Phytophthora blight of soybean [41]. Reduced conventional tillage practices increase organic matter, which prevents *C. sativus* from germinating and changes nutrient availability, which influences pathogen survival. Crop rotation changes root exudates and compounds released from crop residue decomposition, which affects the rhizosphere microbial community. Fungicide compounds released from canola residues breakdown procedure can diminish the extremity of common root rot in cereals [42]. Animal manures and organic composts also influence plant pathogens. Providing beneficial microbiota from compost to conductive soil remains the critical strategy for increasing suppression against soilborne pathogens. Five compost–peat mixtures were used to control *Pythium ultimum*, *Rhizoctonia solani*, and *Sclerotinia minor* – *Lepidium sativum* to manage dumping off diseases [43]. Combining effort with selective biological control agents, organic amendments, and suppressive disease compost can achieve natural disease control against soilborne pathogens. A positive balance is obligatory between phytopathogens and beneficial microorganisms to obtain optimal plant growth and health.

2.3 Plant–Microbe Interaction in the Endosphere

Plants develop an association with their surrounding ecosystem to thrive in their natural environment. The most common type of association is the plant–microbe association, where the indigenous microbes help plant survival in biotic and abiotic stress. The plant endosphere consists of complex microbial communities whose function ranges from mutualism to pathogenicity. These microorganisms colonize at least a part of their life inside the plant interior and are termed "endophytes" [44]. While living near plant hosts, endophytic bacteria exchange for a consistent nutrient supply exerts a beneficial effect [45]. For establishing an asymptomatic association with the host plant, the endophyte must avoid triggering the plant's defense system, which is achieved by maintaining low cell densities. Colonization of endophytes involves competition in the plant rhizosphere for space and nutrients, which is assisted by the production of polysaccharides and motility. Once they are established in the rhizosphere and rhizoplane, they make their way in by producing adhesion molecules and ultimately gain entry into the root by an active (low levels of cell-wall degrading enzymes) or passive (through cracks in roots) process [46]. After entering the roots, they migrate to above-ground plant tissue through the plant transpiration stream. Movement through intercellular spaces requires cell-wall degrading enzymes. However, migration through the xylem element occurs through perforated plates [47].

2.3.1 Microbial Population in the Endosphere

Numerous studies have already characterized endophytes from a plethora of plant hosts and different plant compartments above and below the ground. Endophytes are present ubiquitously in most of the plant, either actively or latently colonizing the plant tissue. Generally, the plant endosphere is enriched with members of *Proteobacteria* [48] and to a lesser amount with *Actinobacteria*, *Firmicutes*, and *Bacteroidetes* [49]. Other classes that are less commonly found include *Acidobacteria, Chloroflexi, Verrucomicrobia*, and *Planctomycetes* [48]. Endophytic genera of bacteria commonly found to belong to *Microbacterium*, *Burkholderia*, *Micrococcus, Bacillus, Pseudomonas*, and *Pantoea*, where *Bacillus* and *Pseudomonas* spp. dominate [50]. Fungal endophytes commonly found belong to *Ascomycota*, and a few belong to *Basidiomycota, Zygomycota*, and *Oomycota*. Common genera that are reported are *Trichoderma, Fusarium*, and *Aureobasidium* [51]. However, the dominance of the phyla varies depending on the plant host species.

2.3.2 Biocontrol Mechanism in the Endosphere

Endophytes produce various substances that directly help enhance the growth of the host plant and discourage phytopathogens and plant pests' survival. These metabolites could be antibiotics, siderophores, hydrolytic enzymes, volatile organic compounds (VOCs), and toxins [52]. Antimicrobial activity has been reported for endophytic *Pseudomonas putida* (PpBP25) in black pepper with aggressive action against plant pathogen *Phytophthora capsici* and *Radopholus similis* [52]. The most common genera with antagonistic activity against phytopathogens include *Bacillus, Enterobacter, Actinobacteria, Pseudomonas, Serratia*, and *Paenibacillus* [53,54].

2.3.2.1 Competition

There is a race between endophytes and phytopathogens to prevent host tissue colonization [55]. They colonize either systemically or locally and act by inhabiting locations available for the pathogens and lurking for nutrients that are available for the proper functioning of the plant [56]. There are not many reports on how nutrient management/uptake by beneficial microbiomes is related. Still, some reports confirm the crosstalk between Fe-deficient/nutrient starvation and resistance elicited by microorganisms. For instance, *Herbaspirillum seropedicae* Z67, a nitrogen-fixing endophyte, is dependent on iron for its vital cellular processes and produces serobactins (siderophores) to fulfill its iron requirement [57]. Some studies have provided valuable insights into the iron competition between endophyte *Streptomyces sporocinereus* OsiSh-2 and a pathogen of rice *Magnaporthe oryzae* [58]. The study indicated that *M. oryzae* is dependent more on iron supplementation than OsiSh-2 for growth and follows different strategies to acquire iron. Many genes involved in siderophore synthesis and iron uptake were present in endophytes than in pathogens, and OsiSh-2 has an added advantage for capturing iron over *M. oryzae* [58].

2.3.2.2 Parasitism

Parasitism is another mechanism that the endophytes use to defend their plant host. Generally, it is observed when there is an interaction between bacteria and fungi or fungi and fungi. Endophytes fight the pathogens directly and produce lyase that helps in destroying the pathogen's cell wall. For instance, out of all the endophytes isolated from poplar, the *Burkholderia cepacia* complex could efficiently control *Cytospora chrysosperma*, *Phomopsis macrospora*, and *Fusicoccum aesculi* causative agent of poplar canker. The interaction of endophytic cotton bacteria *Bacillus halotolerans* (Y6) hinders spore germination *in vitro* along with mycelial growth of *Verticillium dahlia* pathogen. Mousa et al. (2016) observed that the endophyte swarmed toward the root-invading pathogen and completely covered *F. graminearum* hyphae by forming a biofilm that acts as a physical barrier that obstructs the passage and entraps the pathogen and is eliminated then. Thus, the bacteria construct a microhabitat of their own to invade the pathogen. Furthermore, the isolated strain can increase root hair proliferation and create a barrier to block and destroy the invaders [59].

2.3.2.3 Antagonism

Endophytes and phytopathogens live in a similar niche; they release a range of bioactive compounds to suppress or impede pathogens' average growth and activity. Production of antibiotics, metabolites with antifungal properties, and production of volatile compounds such as hydrogen cyanide (HCN) by endophytes are directed mainly toward inhibiting plant pathogens [60]. Several compounds with antimicrobial properties have been purified and identified from endophytes, including peptides, terpenoids, steroids, alkaloids, phenols, quinones, flavonoids, and polyketides [61]. When different microbial species are near each other, endophytes or the host plants obstruct the growth of invading microbes, which is evident from the secretion of bioactive metabolites [62]. *Phomopsis cassia*, an endophyte from *Cassia spectabilis*, synthesized compounds similar to cadinene sesquiterpenes and 3,11,12-trihydroxycadalene, which showed activity against fungi like *Cladosporium cladsporioides* and *C. sphaerospermum* [63]. Similarly, the antagonistic potential of five endophytic *Bacillus* sp. isolated from *Solanum sp.* was screened against *Fusarium oxysporum*.

It was reported that the endophyte was capable of limiting pathogen sporulation and mycelial growth by the production of extracellular metabolites such as chitinases and lipopeptide antibiotics [64]. Effective inhibition of phytopathogenic nematodes [65,66], oomycetes, and fungi [67–71] by bacterial species have been reported through the production of VOCs. Thus, VOCs have been shown to have enormous potential in biocontrol.

2.3.2.4 Induced Systemic Resistance

The ISR and SAR primes plant defense against pathogenic microbes and herbivore insects by protecting plants from future attacks. Plants exhibit resistance by several molecular defense mechanisms such as ethylene/jasmonic/salicylic acid (ET/JA/SA) signaling pathway, callose deposition, regulation of polyamines uptake, accumulation of phytoalexins, producing reactive oxygen species, and gene expression that codes for pathogenesis-related (PR) proteins [46,69]. Defense-related genes were expressed in *Arabidopsis*, which simultaneously activated both defense pathways triggered by *Bacillus cereus* AR156 [70]. This increased biomass of plants and a simultaneous reduction in pathogen density and disease severity was observed. Moreover, an up-regulation of both of the defense pathways in *Arabidopsis thaliana was observed* by Conn et al. [71]. Inoculation of *Actinobacteria* protected against infection from both fungus *Fusarium oxysporum* and bacteria *Erwinia carotovora*. However, the defense pathways primed were different for both types of pathogens. The resistance to *F. oxysporum* involved the SA pathway, while the jasmonic acid pathway provided resistance to *E. carotovora*. Thus, the bacteria used two different ways, which helped in conferring resistance to two different pathogens [71].

2.3.3 Plant Disease Management

Disease management involves strategies and methods for manipulating the antagonistic populations to hinder pathogen survival and plant growth promotion. Endophytes are environmentally benign agents and thus are ideal candidates for efficiently promoting plant growth and act as bioactive candidates against parasitic nematodes. Bogner et al. (2016) showed that *Fusarium solani* and *F. oxysporum* complexes successfully reduced penetration and subsequent galling and reproduction of root-knot nematode, which can further be used for developing disease management systems in tomato [72]. In a study on activity against nematode, endophytic bacteria that belong to *Bacillus* spp., *Streptomyces* spp., and *Pseudomonas* spp. suppressed phytopathogenic burrowing nematode (*Meloidogyne javainica*) in banana roots with 70.7% biocontrol efficiency in sterile soil compared with control [73].

Furthermore, in another study, the endophytic *B. subtilis* strain E1 R-j was administered to determine its inhibitory effect on the causative agent of wheat stripe rust disease *Puccinia striiformis* [74]. *In-vitro* studies revealed an 84.1% inhibition rate when 10-fold dilution of fermentation liquid was used to treat bacterial cells. At the same time, in field trials, a significant decrease in disease severity was observed compared with non-treated plots. Moreover, a seven-month trial period of endophytic *Burkholderia cenocepacia* 869T2 against Fusarium wilt reduced the disease incidence to 3.4% compared to 24.5% of non-inoculated plants [75]. Similarly, pre-inoculation of peanut with beneficial root endophyte curbed root rot, which was evident from a decrease in the colonization by pathogens, cell death, and disease severity [76].

After a long period of continuous cropping system, a single type of root exudate can lead to an imbalance in the microbial diversity of the soil and can cause soil deterioration. In a greenhouse experiment, exudates from root endophyte *Phomopsis liquidambaris* improved significantly the rhizospheric abundance of bacteria. They promoted growth in peanuts by alleviating soil sickness [77]. Adopting a rational approach for disease management for sustainable agriculture has vast potential in the future. One of the benefits of using microorganisms is their role in declining agrochemicals, making agricultural practices more viable. An extensive field trial using inoculants containing an endophytic strain of *Rhizobium leguminosarum bv. Trifolii* enhanced grain yield in 19 tests out of 24, thus promoting rice production capacity and curtailing the necessity of chemical N-fertilizer inputs [78]. *Muscodor albus*, a novel endophytic isolate, is already registered and patented by Marrone Bio Innovations as a potent mycofumigating agent. Many studies have reported its potential as a biofumigant [79,84,80].

2.4 Plant–Microbe Interaction in the Phyllosphere

"Phyllosphere" is an ancient Greek word. "Phyllo" means leaf, and "sphere" refers to the field of influence. The term "Phyllosphere" referred to the above-ground part of a plant and was coined by Last and Ruinen in the 1950s in "Analogy to the Rhizosphere" [81,82]. The phyllosphere is associated with the largest biosphere–atmosphere interphase on Earth and is dominated by aerial plant parts, such as leaves, flowers, fruits, and others. The phyllosphere microbiota interacts with host and other microbes to colonize outer plant surfaces in addition to biotic, abiotic, and anthropogenic factors. Most biomes reach the phyllosphere via soil, wind, water, or insect transmission [83]. Leaf microbiotas first interacts with the outer cuticular layer, and the leaf micro-environment determines microbial communities [84]. Stomatal opening and wounds on the leaves surface are the primary paths for microbiota entering the inner phyllosphere [85]. Leaf exudates, such as terpenes, benzenoids, methanol, sugar, amino acids, and organic acids, also shape phyllosphere microbial assembly [86].

2.4.1 Microbial Population in the Phyllosphere

Microbial populations associated with aerial parts of plants have been known as epiphytes [87]. *Proteobacteria, Actinobacteria, Bacteriodetes*, and *Firmicutes* are consistent microbial phylotypes [88]. 16S rRNA sequencing has provided information about important phyllosphere colonizing bacteria, including diazotrophic and methylotrophic bacteria *viz.*, *Beijerinckia* and *Methylobacterium* that consume nitrogen and methanol compounds, respectively [89]. *Methylobacterium, Sphingomonas*, and *Pseudomonas* are the most dominant genera in the phyllosphere microbial diversity, which possess specific adaptation factors to the phyllosphere [90,91]. For instance, the *Sphingomonas spp.* manage to survive in a nutrient-deficient environment by metabolizing many carbon sources [90]. Similarly, *Pseudomonas spp.* can move at auspicious sites by using flagellar motility [92]. The most enriched fungal communities in the phyllosphere are *Ascomycota* and *basidiomycetous* yeasts, and the most common genera are *Cladosporium, Aureobasidium*, and *Taphrina* [85].

2.4.2 Biocontrol Mechanism in the Phyllosphere

The plant immune system plays a crucial role in controlling phyllosphere diversity [93]. Phyllosphere microbiota has adapted against extreme environmental conditions along with biocontrol activity that contains pathogenic microorganisms. A large amount of the inhibitory strains carries gene clusters encoding metabolites with prophylactic effects. The *Brevibacillus sp.* was reported to produce several antimicrobial compounds, such as marthiapeptide A, streptocidin D, and an unusual lysophospholipid that acts on gram-negative bacteria [94].

2.4.2.1 Competition

Leaf exudates are carbon sources for both resident microbiota and pathogens, which allow plant–microbe and microbe–microbe interaction. Competition for carbon source determines the phyllosphere community; for example, methylotrophic yeast *Candida boidinii* in *Arabidopsis thaliana* and *Methylobacterium extorquens* in the rice phyllosphere can metabolize methanol compounds emitted by plant leaves and survive on the leaves [95,96].

2.4.2.2 Parasitism

Parasitism acts by producing antagonistic agents that can parasite other fungi and feed on them. Biocontrol agents either destroy fungal propagules or cause partial obstruction and lysis of pathogen structure by producing hydrolytic enzymes. The host plant also acquires a defense mechanism by delivering chitinases and β-1,3-glucanase, which act on pathogen cell walls [97]. Under field conditions, strawberry plants are protected by *Bacillus subtilis* against *Bacillus cinerea* by the secreting extracellular enzymes *viz.*, chitinase and proteases [98]. Moreover, five species of *Tilletiopsis* from the phyllosphere have antagonistic activity against powdery mildew by producing chitinase and endo- and exo- β-1,3-glucanase [99].

2.4.2.3 Antagonism

Microorganisms with antagonistic activity suppress the growth of phytopathogens and inhibits the subsequent development and activity of infectious agents. The richness of bacterial species and negative interactions positively promote the growth and development of the plant [100]. *Moesziomyces albugensis* carries potent antagonistic activity against the oomycete pathogen *Albugo laibachii* in the phyllosphere of wild *A. thaliana* [101]. *Trichoderma* strains with the antagonistic potential of secreting cell-wall hydrolyzing enzymes such as chitinases, proteases, and β-glucanases can degrade the cell wall of pathogens [102].

Pseudomonas strains directly inhibit wide variety of pathogens (i.e., *P. syringae and B. cinerea*) in the laboratory as well as infield [103,104]. *Pseudomonas* spp. produced phenazines, a group of heterocyclic secondary metabolites that antagonize phytopathogens [105]. The *Pseudomonas* spp. is also reported to produce many other potential biocontrol metabolites, i.e., siderophores, 4-hydroxy-2-alkylquinolines, cyclic lipopeptides, and rhamnolipids, which are a class of biosurfactants [106]. *Xanthomonas oryzae pv. oryzae* is responsible for bacterial blight, which is the most destructive and serious disease-affecting rice. Among bacteria, *actinomycetes* are gram-positive and produce 70% of the compounds with biological effects. A total of eight strains of *actinomycetes* are reported to have shown a positive effect on controlling *Xanthomonas in vitro* and reducing disease severity of bacterial leaf blight. STG-15 can control bacterial leaf blight incidence and is identified as *Nonomuraea* sp [107].

2.4.2.4 Induced Systemic Resistance

The Plant immune system responds the same way toward both pathogenic and nonpathogenic microbes. However, response toward nonpathogenic microbes remains mild as they lack virulence factors. The ISR mechanism is well characterized in the rhizosphere and endosphere. *Sphingomonas melonis* fr1, a phyllosphere bacteria, suppressed disease development in *Arabidopsis thaliana* by inducing camalexin production [108]. Camalexin is a tryptophan-derived alkaloid that interferes with the membrane integrity of bacteria and fungi [109]. Foliar application of *Bacillus amyloliquefasciens* 5B6 decreased the relative abundance of cucumber mosaic virus by activating SA and ethylene signaling pathway in three years of field trials [110].

2.4.3 Plant Disease Management

Pathogens and pests are responsible for 15–30% of crop yield losses worldwide [111]. We depend on chemical-based crop production and protection strategies using chemical fertilizers, pesticides, herbicides, and fungicides for a sufficient and steady yield. Biological chemical-free agriculture is gaining more and more support ecologically. For disease control, ecological control has been the desirable policy in which nonpathogenic microorganisms are applied to the foliar parts of the plant to affect disease suppression [112]. Microorganisms of the phyllosphere take over a large assortment of adaptation and biocontrol factors, which permit them to adopt the phyllosphere condition and discourage pathogen growth, thus assisting in plant health [113].

The genera of *Pantoea* and *Sphingomonas* have a vital contribution to wildfire disease inhibition [114]. *Bacillus amyloliquefaciens* as a biocontrol agent to mandarin fruit suppresses *Penicillium digitatum* infection by 77% [115]. *Trichoderma atroviridae* can be correlated with reduced infection of *Botrytis cinerea* in strawberries by 88% [116]. Black leaf streak disease in bananas caused by *Pseudocercospora musae* is restrained by *Bacillus subtilis* B106 up to 72% [117]. The biocontrol agent *Trichoderma koningii* controls up to 93% of *Phytophthora cactorum* pathogen that causes collar rot in apples [118]. Similarly, infection caused by *Phytophthora medium* in rubber trees can be controlled by using biocontrol agents, such as *Alcaligenes sp.* EIL-2 [119].

2.5 Conclusion and Prospects

We studied the plant microbiome, aiming to elucidate on the new and improved tools for developing products that can enhance disease management. Our first step was to explore and discover prospective products that can invade phytopathogens and help in maintaining the overall health of plants. Plant microbiota are very useful in improving plant growth as they produce secondary metabolites that confer resistance to a wide variety of pathogens. As a whole, metabolites secreted by the plant microbiome will offer an immense contribution in opening up new opportunities to fight challenges in the environment and agriculture. Finally, the advances in metagenomics combined with NGS techniques will uncover new pools of defense metabolites by novel microbes in the rhizosphere, endosphere, and phyllosphere.

References

1 Kumar, V., Prasad, R., Kumar, M., and Choudhary, D.K., (eds). (2019). *Microbiome in Plant Health and Disease: Challenges and Opportunities*. Springer. Aug 10.

2 Hammonds, K., Trivedi, P., Garg, A., Janitz, C., Grinyer, J., Holford, P., Botha, F.C., Anderson, I.C., and Singh, B.K. (2018). Field study reveals core plant microbiota and the relative importance of their drivers. *Environmental Microbiology* Jan 20 (1): 124–140.

3 Mendes, R., Garbeva, P., and Raaijmakers, J.M. (2013). The rhizosphere microbiome: Significance of plant beneficial, plant pathogenic, and human pathogenic microorganisms. *FEMS Microbiology Reviews* Sep 1 37 (5): 634–663.

4 Tkacz, A. and Poole, P. (2015). Role of root microbiota in plant productivity. *Journal of Experimental Botany* Apr 1 66 (8): 2167–2175.

5 Baetz, U. and Martinoia, E. (2014). Root exudates: The hidden part of plant defense. *Trends in Plant Science* Feb 1 19 (2): 90–98.

6 Hu, L., Robert, C.A., Cadot, S., Zhang, X., Ye, M., Li, B., Manzo, D., Chervet, N., Steinger, T., Van Der Heijden, M.G., and Schlaeppi, K. (2018). Root exudate metabolites drive plant-soil feedbacks on growth and defense by shaping the rhizosphere microbiota. *Nature Communications* Jul 16 9 (1): 1–3.

7 Tkacz, A. and Poole, P. (2021). The plant microbiome: The dark and dirty secrets of plant growth. *Plants, People, Planet* Mar 3 (2): 124–129.

8 Parniske, M. (2008). Arbuscular mycorrhiza: The mother of plant root endosymbioses. *Nature Reviews Microbiology* Oct 6 (10): 763–775.

9 Claesson, M.J., Wang, Q., O'Sullivan, O., Greene-Diniz, R., Cole, J.R., Ross, R.P., and O'Toole, P.W. (2010). Comparison of two next-generation sequencing technologies for resolving highly complex microbiota composition using tandem variable 16S rRNA gene regions. *Nucleic Acids Research* Dec 1 38 (22): e200.

10 Schoch, C.L., Seifert, K.A., Huhndorf, S., Robert, V., Spouge, J.L., Levesque, C.A., and Chen, W. (2012). Fungal Barcoding Consortium. Nuclear ribosomal internal transcribed spacer (ITS) region as a universal DNA barcode marker for Fungi. *Proceedings of the National Academy of Sciences* Apr 17 109 (16): 6241–6246.

11 Sessitsch, A., Hardoim, P., Döring, J., Weilharter, A., Krause, A., Woyke, T., Mitter, B., Hauberg-Lotte, L., Friedrich, F., Rahalkar, M., and Hurek, T. (2012). Functional characteristics of an endophyte community colonizing rice roots as revealed by metagenomic analysis. *Molecular Plant–Microbe Interactions* Jan 25 (1): 28–36.

12 Wang, X., Wang, M., Xie, X., Guo, S., Zhou, Y., Zhang, X., Yu, N., and Wang, E. (2020). An amplification-selection model for quantified rhizosphere microbiota assembly. *Science Bulletin* Mar 7 65 (12): 983–986.

13 Prashar, P., Kapoor, N., and Sachdeva, S. (2014). Rhizosphere: Its structure, bacterial diversity and significance. *Reviews in Environmental Science and Bio/Technology* Mar 13 (1): 63–77.

14 Garbeva, P.V., Van Veen, J.A., and Van Elsas, J.D. (2004). Microbial diversity in soil: Selection of microbial populations by plant and soil type and implications for disease suppressiveness. *Annual Review of Phytopathology* Sep 8 42: 243–270.

15 Schreiter, S., Babin, D., Smalla, K., and Grosch, R. (2018). Rhizosphere competence and biocontrol effect of *Pseudomonas sp.* RU47 Independent from Plant Species and Soil Type at the Field Scale. *Frontiers in Microbiology* Feb 1 9: 97.

16 Robin, A., Mazurier, S., Mougel, C., Vansuyt, G., Corberand, T., Meyer, J.M., and Lemanceau, P. (2007). Diversity of root-associated fluorescent *Pseudomonads* as affected by ferritin overexpression in tobacco. *Environmental Microbiology* Jul 9 (7): 1724–1737.

17 Di Pietro, A., Lorito, M., Hayes, C.K., Broadway, R.M., and Harman, G.E. (1993). Endochitinase from *Gliocladium virens*: Isolation, characterization, and synergistic antifungal activity in combination with gliotoxin. Phytopathology.

18 Harman, G.E., Petzoldt, R., Comis, A., and Chen, J. (2004). Interactions between *Trichoderma harzianum* strain T22 and maize inbred line Mo17 and effects of these interactions on diseases caused by *Pythium ultimum* and *Colletotrichum graminicola. Phytopathology* Feb 94 (2): 147–153.

19 Amadou, M., Duponnois, R., and Marseille, T. (1999). Beneficial effects of *Enterobacter cloacae* and *Pseudomonas mendocina* for biocontrol of *Meloidogyne incognita* with the endospore-forming bacterium *Pasteuria penetrans. Nematology* Jan 1 1 (1): 95–101.

20 Haas, D. and Keel, C. (2003). Regulation of antibiotic production in root-colonizing *Pseudomonas spp.* and relevance for biological control of plant disease. *Annual Review of Phytopathology* Sep 41 (1): 117–153.

21 Tsuge, K., Akiyama, T., and Shoda, M. (2001). Cloning, sequencing, and characterization of the iturin A operon. *Journal of Bacteriology* Nov 1 183 (21): 6265–6273.

22 Steller, S., Vollenbroich, D., Leenders, F., Stein, T., Conrad, B., Hofemeister, J., Jacques, P., Thonart, P., and Vater, J. (1999). Structural and functional organization of the fengycin synthetase multienzyme system from *Bacillus subtilis* b213 and A1/3. *Chemistry & Biology* Jan 1 6 (1): 31–41.

23 Raaijmakers, J.M., Paulitz, T.C., Steinberg, C., Alabouvette, C., and Moënne-Loccoz, Y. (2009). The rhizosphere: A playground and battlefield for soilborne pathogens and beneficial microorganisms. *Plant and Soil* Aug 321 (1): 341–361.

24 Asaka, O. and Shoda, M. (1996). Biocontrol of *Rhizoctonia solani* damping-off of tomato with *Bacillus subtilis* RB14. *Applied and Environmental Microbiology* Nov 1 62 (11): 4081–4085.

25 Couillerot, O., Prigent-Combaret, C., Caballero-Mellado, J., and Moënne-Loccoz, Y. (2009). *Pseudomonas fluorescens* and closely-related fluorescent *Pseudomonads* as biocontrol agents of soilborne phytopathogens. *Letters in Applied Microbiology* May 48 (5): 505–512.

26 Guo, Q., Shi, M., Chen, L., Zhou, J., Zhang, L., Li, Y., Xue, Q., and Lai, H. (2020). The biocontrol agent *Streptomyces pactum* increases *Pseudomonas koreensis* populations in the rhizosphere by enhancing chemotaxis and biofilm formation. *Soil Biology & Biochemistry* May 1 144: 107755.

27 Rezzonico, F., Binder, C., Défago, G., and Moënne-Loccoz, Y. (2005). The type III secretion system of biocontrol *Pseudomonas fluorescens* KD targets the phytopathogenic Chromista *Pythium ultimum* and promotes cucumber protection. *Molecular Plant–Microbe Interactions* Sep 18 (9): 991–1001.

28 Vacheron, J., Moënne-Loccoz, Y., Dubost, A., Gonçalves-Martins, M., Müller, D., and Fluorescent, P.-C.-C. (2016). *Pseudomonas* strains with only few plant-beneficial properties are favored in the maize rhizosphere. *Frontiers in Plant Science* Aug 25 7: 1212.

29 Pieterse, C.M., Zamioudis, C., Berendsen, R.L., Weller, D.M., Van Wees, S.C., and Bakker, P.A. (2014). Induced systemic resistance by beneficial microbes. *Annual Review of Phytopathology* Aug 4 52.

30 Pascale, A., Proietti, S., Pantelides, I.S., and Stringlis, I.A. (2020). Modulation of the root microbiome by plant molecules: The basis for targeted disease suppression and plant growth promotion. *Frontiers in Plant Science* Jan 24 (10): 1741.

31 Harman, G.E., Björkman, T., Ondik, K., and Shoresh, M. (2008). Changing paradigms on the mode of action and uses of *Trichoderma spp.* for biocontrol. *Outlooks on Pest Management* Feb 1 19 (1): 24.

32 Weller, D.M., Raaijmakers, J.M., Gardener, B.B., and Thomashow, L.S. (2002). Microbial populations responsible for specific soil suppressiveness to plant pathogens. *Annual Review of Phytopathology* Sep 40 (1): 309–348.

33 Kwak, M.J., Kong, H.G., Choi, K., Kwon, S.K., Song, J.Y., Lee, J., Lee, P.A., Choi, S.Y., Seo, M., Lee, H.J., and Jung, E.J. (2018). Author correction: Rhizosphere microbiome structure alters to enable wilt resistance in tomato (*Nat. Biotechnology* 36 (11): (1100–1116). 10.1038/nbt. 4232). *Nature Biotechnology* 2018 Nov 9 36 (11): 1117.

34 Shi, W., Li, M., Wei, G., Tian, R., Li, C., Wang, B., Lin, R., Shi, C., Chi, X., Zhou, B., and Gao, Z. (2019). The occurrence of potato common scab correlates with the community composition and function of the geocaulosphere soil microbiome. *Microbiome* Dec 7 (1): 1–8.

35 Minuto, A., Spadaro, D., Garibaldi, A., and Gullino, M.L. (2006). Control of soilborne pathogens of tomato using a commercial formulation of *Streptomyces griseoviridis* and solarization. *Crop Protection* May 1 25 (5): 468–475.

36 Sindhu, S.S., Rakshiya, Y.S., and Sahu, G. (2009). Biological control of soilborne plant pathogens with rhizosphere bacteria. *Pest Technology* 3 (1): 10–21.

37 Palmieri, D., Vitullo, D., De Curtis, F., and Lima, G. (2017). A microbial consortium in the rhizosphere as a new biocontrol approach against fusarium decline of chickpea. *Plant and Soil* Mar 1 412 (1-2): 425–439.

38 Wu, H., Lin, M., Rensing, C., Qin, X., Zhang, S., Chen, J., Wu, L., Zhao, Y., Lin, S., and Lin, W. (2020). Plant-mediated rhizospheric interactions in intraspecific intercropping alleviate the replanting disease of *Radix pseudostellariae*. *Plant and Soil* Sep 454 (1): 411–430.

39 Li, X., De Boer, W., Ding, C., Zhang, T., and Wang, X. (2018). Suppression of soilborne Fusarium pathogens of peanut by intercropping with the medicinal herb *Atractylodes lancea*. *Soil Biology & Biochemistry* Jan 1 116: 120–130.

40 Ren, L., Su, S., Yang, X., Xu, Y., Huang, Q., and Shen, Q. (2008). Intercropping with aerobic rice suppressed Fusarium wilt in watermelon. *Soil Biology & Biochemistry* Mar 1 40 (3): 834–844.

41 Zhang, H., Yang, Y., Mei, X., Li, Y., Wu, J., Li, Y., Wang, H., Huang, H., Yang, M., He, X., and Zhu, S. (2020). Phenolic acids released in maize rhizosphere during maize-soybean intercropping inhibit Phytophthora blight of soybean. *Frontiers in Plant Science* Jul 28 (11): 886.

42 Bailey, K.L. and Lazarovits, G. (2003). Suppressing soilborne diseases with residue management and organic amendments. *Soil and Tillage Research* Aug 1 72 (2): 169–180.

43 Pane, C., Spadaccini, R., Piccolo, A., Scala, F., and Bonanomi, G. (2011). Compost amendments enhance peat suppressiveness to *Pythium ultimum*, *Rhizoctonia solani* and *Sclerotinia minor*. *Biological Control* Feb 1 56 (2): 115–124.

44 Hardoim, P.R., van Overbeek, L.S., and van Elsas, J.D. (2008). Properties of bacterial endophytes and their proposed role in plant growth. *Trends in Microbiology* Oct 1 16 (10): 463–471.

45 Compant, S., Clément, C., and Sessitsch, A. (2010). Plant growth-promoting bacteria in the rhizo-and endosphere of plants: Their role, colonization, mechanisms involved and prospects for utilization. *Soil Biology & Biochemistry* May 1 42 (5): 669–678.

46 Compant, S., Duffy, B., Nowak, J., Clément, C., and Barka, E.A. (2005). Use of plant growth-promoting bacteria for biocontrol of plant diseases: Principles, mechanisms of action, and future prospects. *Applied and Environmental Microbiology* Sep 1 71 (9): 4951–4959.

47 Sapers, G.M., Gorny, J.R., and Yousef, A.E., (editors). (2005). *Microbiology of Fruits and Vegetables*. CRC Press. Aug 29.

48 Santoyo, G., Moreno-Hagelsieb, G., del Carmen Orozco-mosqueda, M., and Glick, B.R. (2016). Plant growth-promoting bacterial endophytes. *Microbiological Research* Feb 1 183: 92–99.

49 Reinhold-Hurek, B. and Hurek, T. (2011). Living inside plants: Bacterial endophytes. *Current Opinion in Plant Biology* Aug 1 14 (4): 435–443.

50 Chaturvedi, H., Singh, V., and Gupta, G. (2016). Potential of bacterial endophytes as plant growth-promoting factors. *Journal of Plant Pathology and Microbiology* 7 (9): 1–6.

51 Anjum, R., Afzal, M., Baber, R., Khan, M.A., Kanwal, W., Sajid, W., and Raheel, A. (2019). Endophytes: As potential biocontrol agent—review and future prospects. *The Journal of Agricultural Science* 11: 113.

52 Sheoran, N., Nadakkakath, A.V., Munjal, V., Kundu, A., Venugopal, V., Rajamma, S., Eapen, S.J., and Kumar, A. (2015). Genetic analysis of plant endophytic *Pseudomonas putida* BP25 and chemo-profiling of its antimicrobial volatile organic compounds. *Microbiological Research* Apr 173: 66–78.

53 Sessitsch, A., Reiter, B., and Berg, G. (2004). Endophytic bacterial communities of field-grown potato plants and their plant-growth-promoting and antagonistic abilities. *Canadian Journal of Microbiology* May 50: 239–249.

54 Lodewyckx, C., Vangronsveld, J., Porteous, F., Moore, E.R.B., Taghavi, S., Mezgeay, M., and Lelie, D.V.D. (2002). Endophytic bacteria and their potential applications. *Critical Reviews in Plant Sciences* Nov 21 (6): 583–606.

55 Martinuz, A., Schouten, A., and Sikora, R.A. (2012). Systemically induced resistance and microbial competitive exclusion: Implications on biological control. *Phytopathology* Mar 102 (3): 260–266.

56 Rodriguez, R.J., White, J.F., Jr, Arnold, A.E., and Redman, A.R. (2009). Fungal endophytes: Diversity and functional roles. *New Phytologist* Apr 182 (2): 314–330.

57 Trover, M.F., Scavone, P., Platero, R., de Souza, E.M., Fabiano, E., and Rusconi, F. (2018). *Herbaspirillum seropedicae* differentially expressed genes in response to iron availability. *Frontiers in Microbiology* Jul 3 (9): 1430.

58 Zeng, J., Xu, T., Cao, L., Tong, C., Zhang, X., Luo, D., Han, S., Pang, P., Fu, W., Yan, J., and Liu, X. (2018). The role of iron competition in the antagonistic action of the rice endophyte *Streptomyces sporocinereus* OsiSh-2 against the pathogen *Magnaporthe oryzae*. *Microbial Ecology* Nov 76 (4): 1021–1029.

59 Mousa, W.K., Shearer, C., Limay-Rios, V., Ettinger, C.L., Eisen, J.A., and Raizada, M.N. (2016). Root-hair endophyte stacking in finger millet creates a physicochemical barrier to trap the fungal pathogen *Fusarium graminearum*. *Nature Microbiology* Sep 26 1 (12): 1–2.

60 Arnold, A.E., Mejía, L.C., Kyllo, D., Rojas, E.I., Maynard, Z., Robbins, N., and Herre, E.A. (2003). Fungal endophytes limit pathogen damage in a tropical tree. *Proceedings of the National Academy of Sciences* Dec 23 100 (26): 15649–15654.

61 Mousa, W.K. and Raizada, M.N. (2013). The diversity of antimicrobial secondary metabolites produced by fungal endophytes: An interdisciplinary perspective. *Frontiers in Microbiology* Mar 27 4: 65.

62 Kusari, S., Hertweck, C., and Spiteller, M. (2012). Chemical ecology of endophytic fungi: Origins of secondary metabolites. *Chemistry & Biology* Jul 27 19 (7): 792–798.

63 Silva, G.H., Teles, H.L., Zanardi, L.M., Young, M.C., Eberlin, M.N., Hadad, R., Pfenning, L.H., Costa-Neto, C.M., Castro-Gamboa, I., Da Silva Bolzani, V., and Araújo, Â.R. (2006). Cadinane sesquiterpenoids of *Phomopsis cassiae*, an endophytic fungus associated with *Cassia spectabilis* (*Leguminosae*). *Phytochemistry* Sep 1 67 (17): 1964–1969.

64 Abdallah, R.A., Steel, C., Garagounis, C., Nefzi, A., Jabnoun-Khiareddine, H., Papadopoulou, K.K., and Daami-Remadi, M. (2017). Involvement of lipopeptide antibiotics and chitinase genes and induction of host defense in suppression of Fusarium wilt by endophytic *Bacillus* spp. in tomato. *Crop Protection* Sep 1 99: 45–58.

65 Munjal, V., Nadakkakath, A.V., Sheoran, N., Kundu, A., Venugopal, V., Subaharan, K., Rajamma, S., Eapen, S.J., and Kumar, A. (2016). Genotyping and identification of broad-spectrum antimicrobial volatiles in black pepper root endophytic biocontrol agent, *Bacillus megaterium* BP17. *Biological Control* Jan 1 92: 66–76.

66 Huang, Y., Xu, C., Ma, L., Zhang, K., Duan, C., and Mo, M. (2010). Characterisation of volatiles produced from *Bacillus megaterium* YFM3. 25 and their nematicidal activity against *Meloidogyne incognita*. *European Journal of Plant Pathology* Mar 126 (3): 417–422.

67 Athukorala, S.N., Fernando, W.D., Rashid, K.Y., and De Kievit, T. (2010). The role of volatile and non-volatile antibiotics produced by *Pseudomonas chlororaphis* strain PA23 in its root colonization and control of *Sclerotinia sclerotiorum*. *Biocontrol Science and Technology* Jan 1 20 (8): 875–890.

68 Kai, M., Haustein, M., Molina, F., Petri, A., Scholz, B., and Piechulla, B. (2009). Bacterial volatiles and their action potential. *Applied Microbiology and Biotechnology* Jan 81 (6): 1001–1012.

69 Pieterse, C.M., Van Der Does, D., Zamioudis, C., Leon-Reyes, A., and Van Wees, S.C. (2012). Hormonal modulation of plant immunity. *Annual Review of Cell and Developmental Biology* Nov 10 (28): 489–521.

70 Niu, D.D., Liu, H.X., Jiang, C.H., Wang, Y.P., Wang, Q.Y., Jin, H.L., and Guo, J.H. (2011). The plant growth-promoting rhizobacterium *Bacillus cereus* AR156 induces systemic resistance in *Arabidopsis thaliana* by simultaneously activating salicylate and jasmonate/ethylene-dependent signaling pathways. *Molecular Plant–Microbe Interactions* May 24 (5): 533–542.

71 Conn, V.M., Walker, A.R., and Franco, C.M. (2008). Endophytic actinobacteria induce defense pathways in *Arabidopsis thaliana*. *Molecular Plant–Microbe Interactions* Feb 21 (2): 208–218.

72 Bogner, C.W., Kariuki, G.M., Elashry, A., Sichtermann, G., Buch, A.K., Mishra, B., Thines, M., Grundler, F.M., and Schouten, A. (2016). Fungal root endophytes of tomato from Kenya and their nematode biocontrol potential. *Mycological Progress* Mar 1 15 (3): 30.

73 Su, L., Shen, Z., Ruan, Y., Tao, C., Chao, Y., Li, R., and Shen, Q. (2017). Isolation of antagonistic endophytes from banana roots against *Meloidogyne javanica* and their effects on soil nematode community. *Frontiers in Microbiology* Oct 26 (8): 2070.

74 Li, H., Zhao, J., Feng, H., Huang, L., and Kang, Z. (2013). Biological control of wheat stripe rust by an endophytic *Bacillus subtilis* strain E1R-j in greenhouse and field trials. *Crop Protection* Jan 1 43: 201–206.

75 Ho, Y.N., Chiang, H.M., Chao, C.P., Su, C.C., Hsu, H.F., Guo, C.T., Hsieh, J.L., and Huang, C.C. (2015). In planta biocontrol of soilborne Fusarium wilt of banana through a plant endophytic bacterium, *Burkholderia cenocepacia* 869T2. *Plant and Soil* Feb 387 (1): 295–306.

76 Sun, K., Xie, X.G., Lu, F., Zhang, F.M., Zhang, W., He, W., and Dai, C.C. (2021). Peanut preinoculation with a root endophyte induces plant resistance to soilborne pathogen *Fusarium oxysporum* via activation of salicylic acid-dependent signaling. *Plant and Soil* Jan 12: 1–6.

77 Xie, X.G., Zhao, Y.Y., Yang, Y., Lu, F., and Dai, C.C. (2020). Endophytic Fungus Alleviates Soil Sickness in Peanut Crops by Improving the Carbon Metabolism and Rhizosphere Bacterial Diversity. *Microbial Ecology* Jul 12: 1–3.

78 Yanni, Y.G. and Dazzo, F.B. (2010). Enhancement of rice production using endophytic strains of *Rhizobium leguminosarum* bv. *Trifolii in Extensive Field Inoculation Trials within the Egypt Nile Delta*. *Plant and Soil*. Nov 336 (1): 129–142.

79 Mercier, J. and Jiménez, J.I. (2004). Control of fungal decay of apples and peaches by the biofumigant fungus *Muscodor albus*. *Postharvest Biology and Technology* Jan 1 31 (1): 1–8.

80 Stinson, A.M., Zidack, N.K., Strobel, G.A., and Jacobsen, B.J. (2003). Mycofumigation with *Muscodor albus* and *Muscodor roseus* for control of seedling diseases of sugar beet and Verticillium wilt of eggplant. *Plant Disease* Nov 87 (11): 1349–1354.

81 Last, F.T. (1955). Seasonal incidence of *Sporobolomyces* on cereal leaves. *Transactions of the British Mycological Society* Sep 38 (3): 221–239.

82 Ruinen, J. (1956). Occurrence of Beijerinckia species in the 'phyllosphere'. *Nature* Feb 4 177: 220–221.

83 Maignan, L., DeForce, E.A., Chafee, M.E., Murat Eren, A., and Simmons, S.L. (2014). Ecological succession and stochastic variation in the assembly of *Arabidopsis thaliana* phyllosphere communities. *mBio* Jan 21 5 (1): 1–10.

84 Chaudhry, V., Runge, P., Sengupta, P., Doehlemann, G., Parker, J.E., and Kemen, E. (2021). Shaping the leaf microbiota: Plant–microbe–microbe interactions. *Journal of Experimental Botany* Jan 20 72 (1): 36–56.

85 Vacher, C., Hampe, A., Porté, A.J., Sauer, U., Compant, S., and Morris, C.E. (2016). The phyllosphere: Microbial jungle at the plant–climate interface. *Annual Review of Ecology, Evolution, and Systematics* Nov 1 47: 1–24.

86 Shakir, S., Zaidi, S.S., de Vries, F.T., and Mansoor, S. Plant genetic networks shaping phyllosphere microbial community. *Trends in Genetics*. 2020 Oct 6.

87 Carlos, A.R.P., Silvia, R., and Maria, M.Z. (2016). Microbial and functional diversity within the phyllosphere of *Espeletia* species in an Andean high-mountain ecosystem. *Applied and Environmental Microbiology* March 82 (6): 1807–1817.

88 Carlstrom, C.I., Field, C.M., Bortfeld-Miller, M., Müller, B., Sunagawa, S., and Vorholt, J.A. (2019). Synthetic microbiota reveal priority effects and keystone strains in the *Arabidopsis* phyllosphere. *Nature Ecology & Evolution* Oct 3 (10): 1445–1454.

89 Steven, W.K., Timothy, K.O., Holly, K.A., Stephen, P.H., Joseph, W., and Jessica, L.G. (2014). Relationships between phyllosphere bacterial communities and plant functional traits in a neotropical forest. *PNAS* sep 23 111 (38): 13715–13720.

90 Delmotte, N., Claudia, K., Samuel, C., Gerd, I., Bernd, R., Ralph, S., Christian, V.M., and Julia, A.V. (2009). Community proteogenomics reveals insights into the physiology of phyllosphere bacteria. *Proceedings of the National Academy of Sciences of the United States of America* Sep 22 106 (38): 16428–16433.

91 Vorholt, J.A. (2012). Microbial life in the phyllosphere. *Nature Reviews. Microbiology* Dec 10 (12): 828–840.

92 Haefele, D.M. and Lindow, S.E. (1987). Flagellar motility confers epiphytic fitness advantages upon *Pseudomonas syringae*. *Applied and Environmental Microbiology* Oct 1 53 (10): 2528–2533.

93 Tao, C., Kinya, N., Xiaolin, W., Reza, S., Jin, X., Lingya, Y., Bradley, C.P., Li, M., James, K., Yu, C., Li, Z., Nian, W., Ertao, W., Xiu-Fang, X., and Sheng, Y.H. (2020). A plant genetic network for preventing dysbiosis in the phyllosphere. *Nature* Apr 8 580 (7805): 653–657.

94 Blin, K., Shaw, S., Steinke, K., Villebro, R., Ziemer, N., Lee, S.Y., Medema, M.H., and Weber, T. (2019). antiSMASH 5.0: Updates to the secondary metabolite genome mining pipeline. *Nucleic Acids Research* Apr 29 47 (1): 81–87.

95 Kawaguchi, K., Yurimoto, H., Oku, M., and Sakai, Y. (2011). Yeast methylotrophy and autophagy in a methanol-oscillating environment on growing *Arabidopsis thaliana* leaves. *PLoS One* Sep 26 6 (9): e25257.

96 Knief, C., Delmotte, N., Chaffron, S., Stark, M., Innerebner, G., Wassmann, R., Von Mering, C., and Vorholt, J.A. (2012). Metaproteogenomic analysis of microbial communities in the phyllosphere and rhizosphere of rice. *The ISME Journal* Jul 6 (7): 1378–1390.

97 Boller, T. (1993). "Antimicrobial Functions of the Plant Hydrolases, Chitinase and ß-1,3-Glucanase," in *Mechanisms of Plant Defense Responses*. Springer 2: 391–400.

98 Essghaier, B., Hedi, A., Halaoui, M.R., Boudabous, A., and Sadfi-Zouaoui, N. (2012). *In vivo* and *in vitro* evaluation of antifungal activities from a halotolerant *Bacillus subtilis* strain J9. *African Journal of Microbiology* May 23 6 (19): 4073–4083.

99 Urquhart, E.J. and Punja, Z.K. (2002). Hydrolytic enzymes and antifungal compounds produced by *Tilletiopsis* species, phyllosphere yeasts that are antagonists of powdery mildew fungi. *Canadian Journal of Microbiology* Mar 1 48 (3): 219–229.

100 Muhammad, S., Nicole, M., Zahida, H.P., and Milton, B.T. (2017). Microbial interactions in the phyllosphere increase plant performance under herbivore biotic stress. *Frontiers in Microbiology* Jan 20 8 (41): 1–10.

101 Eitzen, K., Sengupta, P., Kroll, S., Kemen, E., and Doehlemann, G. (2020). An antagonistic driver of the microbial phyllosphere suppresses infection of *Arabidopsis thaliana* by the oomycete pathogen *Albugo laibachii* via a secreted hydrolase. *bioRxiv* Apr 21: 1–36.

102 Lopez-Mondejar, R., Ros, M., and Pascual, J.A. (2011). Mycoparasitism-related genes expression of *Trichoderma harzianum* isolates to evaluate their efficacy as biological control agent. *Biological Control* Jan 56 (1): 59–66.

103 Volks, B. and May, R. (2001). Biological control of *Pseudomonas syringae* PV. glycine by epiphytic bacteria under field conditions. *Microbial Ecology* Feb 41: 132–139.

104 Simionato, A.S., Navarro, M.O.P., de Jesus, M.L.A, Barazetti, A.R., da Silva, C.S., and Simoes, G.C. (2017). The effect of phenazine-1-carboxylic acid on mycelial growth of *Botrytis cinerea* produced by *Pseudomonas aeruginosa* LV strain. *Front Microbiology* Jun 14 8 (1102): 1–9.

105 Chin-A-Woeng, T.F.C. Bloemberg, G.V., and Lugtenberg, B.J.J. (2003). Phenazines and their role in biocontrol by *Pseudomonas* bacteria. *The New Phytologist* Mar 3 157: 503–523.

106 Yasmin, S., Hafeez, F.Y., Mirza, M.S., Rasul, M., Arshad, H.M.I., Zubair, M., and Iqbal, M. (2017). Biocontrol of Bacterial Leaf Blight of rice and profiling of secondary metabolites produced by rhizospheric *Pseudomonas aeruginosa* BRp3. *Front. Microbiology* Sep 26 8 (1895): 1–23.

107 Noor, A.I., Abdiad, A.N., and Aris, T.W. (2016). Rice phyllosphere actinomycetes as biocontrol agent of bacterial leaf blight disease on rice. *Asian Journal of Plant Pathology* Mar 15 10 (1-2): 1–8.

108 Vogel, C., Bodenhausen, N., Gruissem, W., and Vorholt, J.A. (2016). The Arabidopsis leaf transcriptome reveals distinct but also overlapping responses to colonization by phyllosphere commensals and pathogen infection with impact on plant health. *New Phytologist* Oct 212 (1): 192–207.

109 Ryffel, F., Helfrich, E.J., Kiefer, P., Peyriga, L., Portais, J.C., Piel, J., and Vorholt, J.A. (2016). Metabolic footprint of epiphytic bacteria on *Arabidopsis thaliana* leaves. *The ISME Journal* Mar 10 (3): 632–643.

110 Lee, G.H. and Ryu, C.M. (2016). Spraying of leaf-colonizing *Bacillus amyloliquefaciens* protects pepper from Cucumber mosaic virus. *Plant Disease* Oct 14 100 (10): 2099–2105.

111 Savary, S., Willocquet, L., Pethybridge, S.J., Esker, P., McRoberts, N., and Nelson, A. (2019). The global burden of pathogens and pests on major food crops. *Nature Ecology & Evolution* Mar 3 (3): 430–439.

112 Jeffrey, B.J., Gary, E.V., Fanny, B.I., Aleksa, O., Mine, H.W., Lee, E.J., Botond, B., Jason, C.H., and Timur, M. (2012). Considerations for using bacteriophages for plant disease control. *Bacteriophage*. Oct 1 2 (4): 208–214.

113 Marie, L., Wenke, S., Dieter, V., Tom, E., Babette, M., Els, P., Roeland, S., and Sarah, L. (2020). Modes of Action of Microbial Biocontrol in the Phyllosphere. *Frontiers in Microbiology* July 14 11 (1610): 1–18.

114 Qin, C., Tao, J., Liu, T., Liu, Y., Xiao, N., Li, T., Gu, Y., and Meng, D. (2019). Responses of phyllosphere microbiota and plant health to application of two different biocontrol agents. *AMB Express* Mar 28 9 (42): 1–13.

115 Hao, W.N., Li, H., Hu, M.Y., Yang, L., and Rizwan-ul-Haq, M. (2011). Integrated control of citrus green and blue mold and sour rot by *Bacillus amyloliquefaciens* in combination with tea saponin. *Postharvest Biology and Technology* 59: 316–323.

116 Card, S.D., Walter, M., Jaspers, M.V., Sztejnberg, A., and Stewart, A. (2009). Targeted selection of antagonistic microorganisms for control of *Botrytis cinerea* of strawberry in New Zealand. *Australas Plant Pathology* Mar 38 (2): 183–192.

117 Fu, G., Huang, S., Ye, Y., Wu, Y., Cen, Z., and Lin, S. (2010). Characterization of a bacterial biocontrol strain B106 and its efficacy on controlling banana leaf spot and post-harvest anthracnose diseases. *Biological Control* 55 (1): 1–10.

118 Alexander, B.J.R. and Stewart, A. (2001). Glasshouse screening for biological control agents of *Phytophthora cactorum* on apple (*Malus domestica*). *New Zealand Journal of Crop and Horticultural Science* 29 (3): 159–169.

119 Abraham, A., Philip, S., Jacob, C.K., and Jayachandran, K. (2013). Novel bacterial endophytes from *Hevea brasiliensis* as biocontrol agent against Phytophthora leaf fall disease. *BioControl* Oct 58 (5): 675–684.

120 Hoitink, H.A. and Boehm, M.J. (1999). Biocontrol within the context of soil microbial communities: A substrate-dependent phenomenon. *Annual Review of Phytopathology* Sep 37 (1): 427–446.

121 Fernando, W.D., Ramarathnam, R., Krishnamoorthy, A.S., and Savchuk, S.C. (2005). Identification and use of potential bacterial organic antifungal volatiles in biocontrol. *Soil Biology & Biochemistry* May 1 37 (5): 955–964.

122 Mercier, J. and Manker, D.C. (2005). Biocontrol of soilborne diseases and plant growth enhancement in greenhouse soilless mix by the volatile-producing fungus *Muscodor albus*. *Crop Protection* Apr 1 24 (4): 355–362.

123 Zheng, M., Shi, J., Shi, J., Wang, Q., and Li, Y. (2013). Antimicrobial effects of volatiles produced by two antagonistic Bacillus strains on the anthracnose pathogen in postharvest mangos. *Biological Control* May 1 65 (2): 200–206.

124 Wan, M., Li, G., Zhang, J., Jiang, D., and Huang, H.C. (2008). Effect of volatile substances of *Streptomyces platensis* F-1 on control of plant fungal diseases. *Biological Control* Sep 1 46 (3): 552–559.

3

Impact of Microbiomes to Counter Abiotic Stresses in Medicinal Plants- A Review

Abeer Hashem[1,2], Khaloud Mohammed Alarjani[1], Khalid F. Almutariri[3], Javid A. Parray[4], Sushil K. Sharma[5], Ashwani Kumar[6], Turki M. Dawoud[1], Khalid S. Almaary[1], Nosheen Shameem[7], and Elsayed Fathi Abd-Allah[3]

[1]Botany and Microbiology Department, College of Science, King Saud University, P.O. Box. 2460, Riyadh 11451, Saudi Arabia
[2]Mycology and Plant Disease Survey Department, Plant Pathology Research Institute, ARC, Giza, Egypt
[3]Plant Production Department, College of Food and Agricultural Sciences, King Saud University, P.O. Box. 2460, Riyadh 11451, Saudi Arabia
[4]Department of Environmental Science, Higher Education Department, Government Degree College, Eidgah, Srinagar, India
[5]ICAR-National Institute of Biotic Stress Management, Baronda, Raipur – 493 225, Chhattisgarh, INDIA
[6]Metagenomics and Secretomics Research Laboratory, Department of Botany, Dr. Harisingh Gour University (A Central University), Sagar – 470003, Madhya Pradesh, India
[7]Department of Environmental Science, Cluster University, Srinagar – 190001, Jammu & Kashmir, India
Corresponding author. Abeer Hashem, Botany and Microbiology Department, College of Science, King Saud University, P.O. Box. 2460, Riyadh 11451, Saudi Arabia.

3.1 Introduction

Medicinal plants are a major group of crops that are considered very valuable due to their role in helping people maintain good health. These medicinal plants are traditionally used as herbal formulations for the prevention and treatment of diseases (Williamson 2003). About 80% of the world's population depends on medicinal herbs directly or indirectly as reported by the World Health Organization (Kasilo and Trapsida 2010). These medicinal plants are associated with many beneficial bacteria. The microbes have a positive and antagonistic effect with other microbes in the soil. Moreover, the rhizosphere biome helps plants grow as well as manage pathogenic microorganisms residing in the rhizosphere ruthlessly to break the protective shield and overwhelm the innate plant defense mechanism (Hashem et al. 2017; Dubey et al. 2019).

Bioactive phytochemicals were the focus of previous medicinal plant research, however, because to the realization that a considerable number of phyto-therapeutic chemicals are really created by related microorganisms or through contact with their host, the focus is currently moving. The host rhizosphere, a narrow zone of the soil contains more than 30,000 prokaryotic species and a large number of microbial cells in a one-gram root (Berendsen 2012). These are attracted by rhizosphere plant root secretion and deposits into the soil. Among all microbes, *Bacillus* and *Pseudomonas* are well-known genera that are used to produce organic biofertilizers and improve soil fertility and biogeochemical cycles for better growth of the entire plant (Freitas et al. 2004). The tolerance and secondary metabolites mechanism for commercial medicinal plants has not been investigated.

Core Microbiome: Improving Crop Quality and Productivity, First Edition. Edited by Javid A. Parray, Nowsheen Shameem, Elsayed Fathi Abd-Allah, and Mohammad Yaseen Mir.

So, there is a need to determine the response of medicinal plants under stress conditions, as salt stress reduced the quality of essential oils and secondary metabolites in medicinal plants. Both biotic and abiotic stresses are significant constraints in agriculture production, which is affected by several factors such as hormonal and nutritional imbalance, ion toxicity, physiological disorders, and susceptibility to disease (Dubey et al. 2018; Malla et al. 2018).

In this study, we discuss plants and microbes and their interaction with the microbiomes and implications of abiotic stresses toward medicinal plant response; physiological, biological, and molecular mechanisms; and microbe-mediated stress-alleviation processes. Less information is available for the molecular mechanism of metabolites in medicinal plants. Modern omics approaches, i.e. quantitative trait loci, functional genomics techniques, proteomics, and mass spectrometry, are used to understand plant–microbe interactions to optimize under abiotic stresses (Kumar and Dubey 2020).

3.2 Structure and Function of Microbiota

Plant rhizosphere contains microorganisms including bacteria, algae, fungi, actinomycetes, and protozoa, among which bacteria are important contributors in the rhizosphere (Kaymak 2010). *Actinomycetes* belong to filamentous bacteria (gram positive) with a high guanine–cytosine ratio in their DNA, and they are good sources of natural compounds used in the pharmaceutical industry (Golinska et al. 2015). Endophytic actinomycetes showed enhanced growth and decreased disease infestation through the production of bioactive secondary metabolites. Some important bioactive compounds that are released by host plants and their therapeutic properties along with microorganisms are presented in Table 3.1. Moreover, they alter plant physiology by stimulating the acquired resistance in medicinal plants. Specific antibiotics produced by genera of *Penicillium*, *Trichoderma*, *Aspergillus*, and *Actinomycetes* may enhance producer diligence when competitors or predators are present in soil habitats. Moreover, *Azospirillum*, *Gluconacetobacter*, *Pseudomonas*, and *Rhizobium*, *Bacillus*, and *Paenibacillus* and gram-positive genera are also included in this circle. Plant–microbe mutualism produces diverse biochemical mechanism that improves bacterial health and ultimately plant growth (Bulgarelli et al. 2013). It also enables plants to maintain microbiota in the soil. The concentration of the bacteria found around the roots of plants is greater than the rest of the soil. It might also be due to the presence of nutrients including sugars, amino acids, and organic acids as well as other small molecules from plant root exudates that fix one-third of carbon by the plant (Badri et al. 2009). However, it mostly depends on soil type and host genotype interaction. Both the type and number of bacteria found in different soils are influenced by soil conditions. It was also influenced by soil temperature, moisture, chemicals as well as the concentration of the salts, and the number and types of crop plants present in the soil. Moreover, the effect of specific bacterium on the plant may change with the change of conditions. For example, a bacterium facilitates plant growth either by fixing phosphorus compounds or by fixing nitrogen that benefits only plants (Kumar et al. 2015, 2016). Plants with medicinal properties, their use, and secondary metabolites production under abiotic stress condition are presented in Table 3.1.

Table 3.1 Plants with Medicinal Properties Under Abiotic Stress, Their Use, and Their Secondary Metabolites.

Medicinal plants	Uses	Secondary metabolites	Abiotic stresses	References
Pluchea lanceolata	Bronchitis, dyspepsia, and rheumatoid arthritis	Quercetin	Heavy metals	Kumar et al. (2004)
Dioscorea bulbifera	Antispasmodic, analgesic, aphrodisiac, and diuretic	Diosgenin	Metal	Narula et al. (2005)
Catharanthus roseus	Cancer and diabetes mellitus	Vinblastine	Salinity	Jaleel et al. (2009)
Jatropha curcas	Skin diseases and rheumatism, piles	Curcin	Salinity	Gao et al. (2008)
Orthosiphon stamineus	Antiallergenic, antihypertensive, anti-inflammatory, and nephritis	Polyphenols	Salinity	Ting et al. (2009)
Thymus maroccanus Ball	Antitussive, antiseptic, antispasmodic, and antihelmintic	Thymol	Salinity	Belaqziz et al. (2009)
Matricaria chamomilla	Mucositis and irritable bowel syndrome	Umbelliferone	$CuCl_2$	Eliasova et al. (2004)
Bacopa monnieri	Antioxidant, epilepsy, and asthma	Bacoside	Salinity and Drought	Debnath et al. (2011)
Olea europaea L.	Antipruritic, antiseptic, astringent, and cholagogue	Oleosides	Salinity	Rejskova et al. (2007)
Populus euphratica	Anodyne, anti-inflammatory, febrifuge, and vermifuge	Gallic acid	Salinity	Zhang et al. (2004)
Matricaria chamomilla	Sore stomach, irritable bowel syndrome, and oral mucositis	Apigenin	Salinity	Razmjoo et al. (2008)
Ziziphora clinopodioides	Antibacterial and sedative stomachache	Leucoanthocyanins	Salinity and Defoliation	Koocheki et al. (2008)
Dioscorea dregeana	Sedative and anti-inflammatory	Paclobutrazol	Smoke, temperature	Kulkarni et al. (2007)
Datura innoxia	Anodyne, antispasmodic, hallucinogenic, hypnotic, and narcotic	Scopolamine	Light, dark, and HCHO deprivation	Laszlo et al. (2004)

However, the composition of the secondary metabolite, which is common in the rhizosphere belongs to these bacteria, such as actinobacteria, bacteroidetes, firmicutes, and proteobacteria. Microbial communities provide protection directly or indirectly from the pathogen in the leaf regions while root microbiota provides an additional host function by getting nutrients from the soil. Other functions of the microbiota associated with these medicinal plants such as phytohormone production, nutrient solubilization, and nitrogen metabolism are also important (Kumar et al. 2015).

3.3 Physiological and Molecular Response of Plant and Microbiota against Stress

Salinity reduces mineral nutrients transport into the plants that retard growth and yield of medicinal plant as well as their active components. Some studies showed that the composition of essential oil of marjoram (*Origanum majorana*), coriander (*Coriandrum sativum*), and peppermint (*Mentha piperita*) have changed due to these stresses. Moreover, salinity and sodicity affect shoots and root yield of citronella, lemongrass, and vetiver plants (Naeem et al. 2013). Seed germination showed a significant reduction at 200 mM NaCl stress in *Thymus maroccanus* (Belaqziz et al. 2009), sweet marjoram (Ali et al. 2007), *Ocimum basilicum* (Miceli et al. 2003), and *Ocimum* (Chowdhary et al. 2018). Moreover, seedling growth has also been affected in *Thymus maroccanus* (Belaqziz et al. 2009), basil (Ramin 2006), *marjoram*, and *chamomile* under salt stress (Ali et al. 2007). Salt stress has affected the transport of reserved foods, reduced cell division, and increased hypocotyls (Said-Al Ahl and Omer 2011). The growth characteristics such as leaf area, the number of leaves, and accumulation of biomass have been reduced in a variety of medicinal plants. Defense mechanisms adopted by plants toward abiotic stresses are presented in Figure 3.1. However, salt stress also has effects on geranium (El-Leithy et al. 2010) and sage (Ben Taarit et al. 2009). High salinity at 200 mM NaCl concentration was responsible for the increase in ascorbic acid contents while glutathione contents were reduced at 25 mM salinity level in *Catharanthus roseus* (L.) (Jaleel et al. 2008). Enzymatic activity such as SOD activity was increased at 50 mM NaCl, while it was decreased at a higher 200 mM level. Moreover, peroxidase activity at 25 mM NaCl was insignificant while it was significant at 50 mM salinity level (Jaleel et al. 2008). Drought affects the morphology, physiology, and biochemistry of medicinal plants, resulting in a significant loss in yield, though the effects of drought on many crop plants were investigated while little was reported about the biosynthesis and accumulation of oil in medicinal plants under drought conditions (Farahani et al. 2009). Drought at the vegetative stage reduced plant height and leaf area of mint (Abbaszadeh et al. 2009). Furthermore, *Calendula* plants use less water due to reduction in plant height (Rahmani et al. 2011) while dry matter reduced at the vegetative stage was recorded in balm. The oil quality of rosemary (*Rosmarinus officinalis* L.) is affected due to drought while thyme oil production increases (Aziz et al. 2008). Medicinal plants such as *Catharanthus roseus* cultivation under drought enhances its proline metabolism, osmoregulation, and defense system (Jaleel et al. 2008). Drought stress response depends

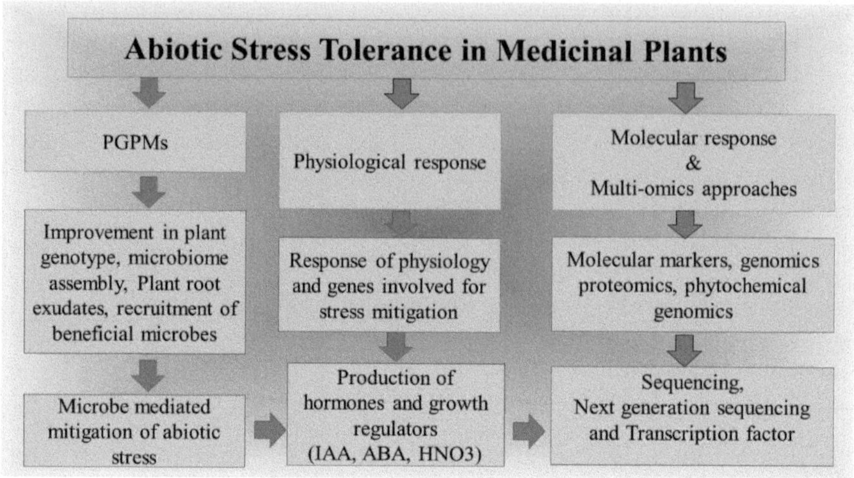

Figure 3.1 Diverse abiotic stresses and their response toward medicinal plants and the defense mechanisms adopted by plants.

on cultivar and plant density in medicinal plants. Proper amount of nutrients might increase resistance against stress and improve productivity when crops are subject to drought stress (Cakmak 2005).

Another type of stress called heat shock, which is normally brought on by an increase by 10–15°C above ambient temperature also influences plant growth and metabolism. On the other hand, the highest photosynthetic rate was determined at 30°C. Besides other symptoms, the prominent symptoms are inhibition in photosynthesis at high-temperature stress (Camejo et al. 2005). Temperature stress also affects several changes at the cellular level and metabolism. High and low temperatures incite the biochemical and physiological changes in plants. These stresses lead to lipid peroxidation and membrane disruption. Moreover, chlorophyll and protein degradation were also recorded due to high-temperature stress. High-temperature stress showed a negative effect on photosynthesis as well as reduced various enzyme activities in rice (Mohammad and Tarpley 2009). Moreover, RUBISCO activity was also reduced due to the inhibition of photosynthetic activity in *Festuca arundinacea* under high-temperature stress. High-temperature stress ranges between 25°C and 40°C encouraged not only the activities of reactive oxidative species (ROS) scavenging enzymes but also increased the activity of lipoxygenase and the content of 3,4-methylenedioxyamphetamine (MDA) and cysteine as well as that of protein and non-protein thiol in the leaf and root of *Phalaenopsis* (Ali et al. 2005). Low temperature also acts as a growth-limiting factor to limit the productivity and distribution of crop plants. When a plant is subjected to low temperature, ROS produced repeatedly and may disrupt normal metabolism through oxidative damage to lipids, nucleic acid, and protein. Low-temperature stress affects the physiological and biochemical characteristics of Crofton weed (*Eupatorium adenophorum*) (Ma et al. 2020). As a result, MDA, total soluble protein contents, soluble sugar, chlorophyll changes, and fluctuation in SOD activity increase (Li et al. 2008). Exposure of sweet potato (*Ipomoea batatas* L.) to chilling stress reduced transpiration rate and stomatal conductance while enhancing peroxidase activity and electrolyte leakage (Islam et al. 2009).

3.3.1 Effect of Plant Genotype on Rhizosphere Microbiome Assembly

A large number of bacteria associated with the rhizosphere of plants produce phyto-therapeutic compounds that increase the growth of medicinal plants. Rhizosphere microbes of *Matricaria chamomilla* L., and *Calendula officinalis* L growing naturally in desert ecosystem showed abundant occurrence of bacteria (gram positive), which is very important for pathogens suppression. A large difference in endophytic bacterial diversity in medicinal plants, i.e., *Lamiophlomis rotate, Codonopsis pilosula*, and *Ephedra sinica* was recorded by Li et al. (2008). In contrast, different varieties of microbial groups are present in the soil. About 50 strains of 47 genera *Stigmatella* (1), *Archangium* (8), *Myxococcus* (18), *Corallococcus* (11), Cystobacter (7), *Pyxidicoccus* (1), with dominant genera of *Corallococcus* and *Myxococcus*. The incessant cropping of *Rahmannia glutinossa* decreases the productivity of some medicinal plants due to plant–microbe interaction activity in the rhizosphere (Qi et al. 2009). Different medicinal plant–microbe interactions in the rhizosphere is presented in Table 3.2.

3.4 Microbe-Mediated Mitigation of Abiotic Stresses

Crop plants need to manage external environment pressure exerted by edaphic factors with intrinsic biological mechanisms, as a result of which growth, development, and productivity fail. Microorganisms are the natural populations of complex environments that have various metabolic abilities to mitigate abiotic stresses. Plant–microbe interaction comprises a complex mechanism within the plant cellular system that modulates local and systemic mechanisms in plants to offer defense under adverse conditions (Meena et al. 2017). Many microbes have genetic and metabolic abilities to mitigate abiotic stresses in plants (Gopalakrishnan et al. 2015). The most important rhizosphere inhabitants comprise *Pseudomonas* (Sorty et al. 2016), *Azotobacter* (Sahoo et al. 2014), *Pantoea* (Sorty et al. 2016), *Enterobacter* (Nadeem et al. 2014), *Burkholderia* (Ait Barka et al. 2006), *Trichoderma* (Ahmad et al. 2015), and cyanobacteria (Singh et al. 2005), which have been used for plant growth to alleviate multiple abiotic stresses.

A sequence of plant-protective mechanisms was acquired during evolution to combat adverse environmental situations (Yolcu et al. 2012). These processes cause re-programming in the cells to enable repetitive bio-physico-chemical processes regardless of the external situation. Most of the time, plants tend to decrease the burden of environmental stresses with the help of the microbiome residing in the soil (Ngumbi and Kloepper 2016). The application of phytohormones supplementation has been reported to improve plant growth and development as well as metabolic activity under stressful conditions. Root microbes are very important for managing target metabolism and induced host tolerance against abiotic stress in medicinal plants (Egamberdieva et al. 2017).

In addition to the stress-tolerant microorganism's application selection, screening might be a good option to overcome stress limitation in crop plants. Indian mustard (*Brassica juncea*) oil contents were improved in NaCl-affected soil by *Trichoderma harzianum* application. It improves essential nutrients, triggers antioxidants accumulation and osmolytes while decreases Na^+ uptake (Ahmad et al. 2015). *Pseudomonas sp.* and *Acinetobacter sp* were reported to enhance the production of indole acetic acid (IAA) and 1-aminocyclopropane-1-carboxylic acid deaminase (ACC) in barley and oats in salt-affected soil.

Table 3.2 Rhizosphere Microbes Interactions in Medicinal Plants.

Plant species	Microorganism	References
Angelica sinensis	*Terriglobus saanensis, Mucilaginibacter polysacchareus, Mucilaginibacter myungsuensis, Mucilaginibacter ximonensis*	Whang et al. (2014)
Matricaria chamomilla, Calendula officinal, Solanum distichum	Bacillus sp.	Koeberl et al. (2013)
Rumex patientia	*Proteobacterium Bacteroidetes, Acidobacteria, Gemmatimonadetes, Verrucomicrobia, Planctomycetes, Actinobacteria, Firmicutes, Chloroflexi*	Qi et al. (2012)
Atractylodes lancea	Gram-negative bacteria	Dai et al. (2013)
Origanum vulgare	Pseudomonas, Stenotrophomonas	Bafana (2013)
Typhonium giganteum	*Kribbella flavida, Kribbella karoonensis, Kribbella alba*	Xu et al. (2012)
Ginseng plants	Actinomycetes	Zhang et al. (2013)
Ajuga bracteosa	*Pseudomonas*	Kumar et al. (2012)
Nerium indicum	*Pontibacter*	Kumar et al. (2012)
Nerium indicum	*Pontibacter*	Raichand et al. (2011)
Fritillaria thunbergii	*Proteobacteria* *Acidobacteria* *Actinobacteria* *Bacteroidetes*	Shi et al. (2011)
Astragalus membranaceus	*Geodermatophilus obscurus*	Zhang et al. (2011)
Phytolacca acinose	*Aspergillus fumigatus*	Guo et al. (2010)
Agathosma betulina	*Cryptococcus laurentii*	Cloete et al. (2010)
Ocimum sanctum, Coleus forskohlii, Catharanthus roseus, Aloe vera	*Azospirillum, Azotobacter, Pseudomonas*	Karthikeyan et al. (2008)
Annona squamosa, Eclipta alba, Cassia auriculata	*Bacillus, Pseudomonas, Enterobacter, Corynebacterium, Micrococcus, Serratia*	Tamilarasi et al. (2008)

3.5 Plant Root Exudates and the Recruitment of Beneficial Microbes

Root exudates serve as indicators that start the symbiotic relation between arbuscular mycorrhizal fungi and rhizobia. Root exudates contain ions such as oxygen, water, an inorganic acid, and (H^+) but mainly consist of carbon compounds (Bais et al. 2004). Lower-weight organic compounds (amino acid, organic acid, phenols, sugars, secondary metabolites) and high-weight organic compounds (proteins, polysaccharides) are present in the soil (Badri et al. 2009). Microorganisms established a sensory system called chemotaxis, which guides these components secreted from roots to deliver essential nutrition and energy for survival under stressful conditions. Moreover, Gao et al. (2011) observed that plants might enhance the degradation by the exudation of enzymes, such as phenol oxidase, laccase, and peroxidase through the root. As a result, the oxidation of various hydrocarbons degrades them into intermediate products. The combined application of arbuscular mycorrhizal fungi and septate endophyte was studied in 36 medicinal plant species. The level of abundance of arbuscular mycorrhizal fungi in the roots varied from 2.5% (*Helianthus tuberosus*) to 77.9% (*Convallaria majalis*) (Zubek and Błaszkowski 2009).

3.5.1 Multi-omics Approaches Used to Mitigate Abiotic Stresses in Medicinal Plants

Multi-omics approaches would help in dissecting the mechanism of combating stress by these plant–microbe interactions (Dubey et al. 2020). Plant–microbe interaction improves metabolic activity and resistance against stresses (Nguyen et al. 1997). For this purpose, molecular, biochemical, and physiological aspects are important to understand complex processes. At the same time, we need to determine the process of stress mitigation in medicinal plants for higher productivity (Meena et al. 2017). Stress alleviation is an important and challenging issue and gets results by altering crop genetics and it requires extensive breeding efforts (Grainger and Rajcan 2014). Low genetic diversity and temperature fluctuation make breeding efforts more challenging (Manavalan et al. 2009). Molecular marker breeding is effective to speed up tolerance in cultivars. For this purpose, the availability of molecular markers that are tightly linked with genomic loci that are responsible for tolerance is important for selection. With the new advancement of DNA sequencing and metabolic profiling technology, a new era of phytochemical genomics has emerged. In this new technology, the biosynthetic mechanism, regulation, function, and evolution of phytochemicals are determined by the integration of genomics and omics technology, such as metabolomics, proteomics, and transcriptomics (Saito 2013). Endophytic actinomycetes are the source of producing natural products with antifungal activity (Passari et al. 2015). Transcriptome data of *Catharanthus roseus* recognized genes involved in monoterpenoid indole alkaloid biosynthesis. Furthermore, these gene products have been established to be cytochrome P450 hydroxylating *aspidosperma* alkaloids at 19 positions

(Giddings et al. 2011) and iridoid synthase reductively converting 10-oxogeranial scaffold (Geu-Flores et al. 2012). Morphine biosynthesis genes are identified by the omics technique in opium poppy (*Papaver somniferum*) (Hagel and Facchini 2010). Multi-omics approaches such as genomics, transcriptomic, proteomics, metabolic pathway, and phenomics integrative studies are compulsory to study plant and microbes and their interaction with environment, and generated data could be helpful in mitigating abiotic stresses. It will also be helpful to report microbial groups for their proper working and functions in the soil.

3.5.1.1 Genomics

An analytical database (http://metnetdb.org/mpmr_public/) regarding transcriptome and metabolic data for 14 medicinal plants are available for gene function. In this way, the possible microbial production of glycyrrhetinic acid was recorded. Yamazaki et al. (2013) reported the differential transcriptome analysis along with metabolic profiling, and identified candidate's genes involved in the biosynthesis pathway of alkaloids and anthraquinones. Transcriptomics of glandular trichomes, of *Cannabis sativa flowers, which are the site for cannabinoid biosynthesis, recognized the enzyme involved in cannabinoid biosynthesis, i.e., polyketide synthase, which is responsible for olivetolic acid (and acyl-activating enzyme).* Integrated transcriptomics and genetic techniques proposed the presence of a cluster of the gene responsible for alkaloid synthesis in the opium poppy genome (Winzer et al. 2012). Apx1, gene encoding cytosolic ascorbate peroxidase 1 has the characteristics of tolerance to drought and heat stress in *Arabidopsis*. Three genes of ectoine (compatible osmolyte for salt tolerance) required for biosynthesis were cloned and transferred to tobacco plant *Nicotiana tabacum* L. bright yellow 2 cv. (Koussevitzky et al. 2007). These three genes are used to enhance tolerance to hyperosmotic shock by the accumulation of ectoine as a result of which normal growth occurs under harsh conditions (Nakayama et al. 2000). Identification and selection of these genes are important processes for introduction into new cultivar to increase the resistance against drought and saline stress. The transcription factor, *MsCBF3*, could be expressed for the induction of stress tolerance in sensitive genotypes (Purdy et al. 2013). Non-targeted metabolomics analysis results coincide with linking genes function with metabolites, resulting in mutants containing glycosyltransferase and methyltransferase. The genes responsible for the biosynthetic pathway could be characterized and integrated for metabolic profiling, helpful in understanding the stress mechanism in medicinal plants.

3.5.1.2 Phytochemical Genomics in Medicinal Plants

Integrated approaches were followed for phytochemical genomics in medicinal plants and their correlation among corresponding elements, such as transcriptome and metabolome, and after which, their successful application in medicinal plants produced bioactive specialized metabolites. For this purpose, next-generation sequencing triggers transcriptome study for an analysis of the co-response of the compound gene that is activated during the stress response. *Catharanthus roseus* transcriptome co-expression results predicted that specific genes are linked to monoterpenoid indole biosynthesis (Giddings et al. 2011) as a result of converting oxogeranial to iridoid scaffold (Geu-Flores et al. 2012). As part of the chemical expression profiling for quinolizidine alkaloids in *Lupinus angustifolius*, gene encoding lysine decarboxylase, the first step of the alkaloid biosynthesis was characterized. Moreover, the molecular basis for the catalytic property of the enzyme was elucidated by site-directed mutagenesis and protein modeling (Winzer et al. 2012).

3.5.2 Proteomics

In studying the expression of plant stress response, protein profiling is important as it directly reflects phenotypic traits. It was also helpful in determining the physiological and metabolic pathway and plant–microbe protein interaction. Comparative analysis of non-stressed stress and target proteins associated with plant–microbe interaction can help identify target proteins and networks (Karthik et al. 2014). The differences in the metabolic pathway among microbes make them more responsive toward stress conditions. The meta-proteome analysis and resolution technique are relatively difficult. The extraction and analysis of meta-proteome data could provide significant output and give a better response among organisms toward stresses. The *haloarchea* and *halobacteria* are getting much attention due to their ability to thrive under high salinity stress. Under *in situ* conditions, culturing of these organisms is used to persuade effective metabolites to impart tolerance against stress. GC-MS/MS protein profiling was used for *Araucaria angustifolia* at the embryogenic stage to figure out large-scale protein identification against cold-tolerance. About 106 differentially expressed protein data were collected from the blocked type and stress response varieties (Dos Santos et al. 2016). The ginsenosides protein samples were taken from hairy roots of ginseng and proteome analysis predicted that about 20% of selected proteins originated from energy metabolism and stress response (Nam et al. 2005). Milky sap isolated from *C. majus* L. produced different types of protein signals such as a nucleic acid-binding protein that could be favorable under stress conditions (Nawrot et al. 2013). Moreover, *aldolase*, *G3P dehydrogenase*, and enolase were also identified as isotypes during the analysis (Nam et al. 2005).

3.6 Medicinal Plants: Plant- and Microbe-Derived Ingredients

Plants contain many active biological compounds, and plant-based medicine has been a part of traditional health care in most parts of the world, the most important being traditional Chinese medicine, phyto-therapeutic knowledge from the Mayans, Australian aboriginal medicine, and many other cultures that provide a huge spectrum for natural remedies exploited as therapeutic agents. Up till now, more than 5000 plant species biological compounds have been used against many diseases, providing the basis for the discovery of modern drugs, i.e. anticancer agents (Miller et al. 2010). These bioactive phytochemical compounds shift toward phyto-therapeutic agents produced by specific microbes during plant–host interaction (Compant et al. 2010).

Chinese medicinal plants such as (*Ainsliaea henryi* Diels, *Potentilla discolor* Bge, *Juncus effusus* L. var. *decipiens* Buchen, and *Rhizoma arisaematis*) showed specific actino-bacterial community and remarkable colonization with actinobacteria (Zhao et al. 2012). Moreover, they also showed antimicrobial and anticancer characteristics (Zhao et al. 2011). The levels of metabolites changed under stress conditions and produced hemiterpenoid compounds responsible for mitigating stresses. Ward et al. (2011) reported the isoprenoid pathway in *Arabidopsis* in leaves, which was diverted to hemiterpenoids. Such compounds are induced by root wounding or oxidative stress while being weakly induced by potassium deficiency. However, other stresses such as osmotic, cold, or saline stress did not induce compounds production. Nutrient deficiency stress has an impact on metabolism as well as hemiterpenoid glycosides production in leaves as compared with *scopolin* and conifer productions in roots. Sulfur deficiency in crops also affects secondary metabolites production and results in a reduction in the synthesis of glucosinolates (Kusano et al. 2011) and induction of anthocyanin biosynthesis. Glucuronosyl–diacylglycerol reportedly reduced stress metabolites under phosphorus-deficient stress (Okazaki et al. 2013), which could be used as *lipidomics* function genomics (Okazaki et al. 2013). Modern research predicted that a significant number of natural products are produced by microbes and supposed that endo-microbiome (plant–microbe) is involved directly or indirectly in producing phytochemicals. On the other hand, the only subset of potential microbial strains was qualified as phyto-therapeutic properties (Chandra et al. 2012) while their contribution toward known bioactivity of medicinal plants is not clear yet. Bioactive compounds released by the host plants and the microorganisms that produce those compounds with therapeutic properties are shown in Table 3.3.

3.7 Biological Control and Plant Improvement

Plant–microbe interaction proves to have increased plant growth directly by stimulating nutrient availability or hormonal production or indirectly by suppressing plant pathogens. Many rhizobacteria strains, such as *Bacillus* group, *Taphylococcus*, and *Burkholderia cepacian*, promote plant growth. These strains also produced volatile organic compounds (VOCs) that enhance growth (Bitas et al. 2013). Biosurfactants, VOCs, antibiotics, and enzymes are used to suppress the growth of pathogens (Berg et al. 2010). Plant extract with antifungal activity is used to suppress pathogens and increase the growth of medicinal plants. Medicinal plants effectively control the development of diseases caused by *Macrophomina phaseolina*. It is also reported that medicinal plants exhibited varying degrees of inhibition activity against *M. phaseolina*. Different concentrations of *Cleome viscose*, *Hyptis sueolences*, *Grewia arborea*, *Avicennia officinalis*, *Ocimum*, *Tephrosia villosa*, *Peltophorum pterophorus*, *Sesbanian grandiflora*, and *Terminalia chebula* showed high activity with strong antifungal activity against *M. phaseolina*, while chloroform extracts showed very low activity (Bobbarala et al. 2009).

Biological compounds produced by medicinal plants depend on microbes, plant interaction, plant and microbe species, and soil types (Morsy 2014). These metabolites produced by medicinal plants have a strong effect on biological and physiological pathways and help them withstand harsh environments. *Solanum distichum*, *Matricaria chamomilla*,

Table 3.3 Bioactive compounds that are released by host plants and the microorganisms involved in the production of therapeutic properties.

Bioactive compound	Therapeutic properties	Host plant	Producing microorganism	References
Cochliodinol	Antibacterial, antimycotic, anticancer	*Salvia officinalis*	*Chaetomium* sp.	Debbab et al. (2009)
Botryorhodines	Antimycotic, anticancer	*Bidens pilosa*	*Botryosphaeria rhodina*	Abdou et al. (2010)
Phomol	Antiphlogistic, antibacterial	*Erythrina crista-galli*	*Phomopsis* sp.	Weber et al. (2004)
Podophyllotoxin	Anticancer, antiphlogistic	*Podophyllum hexandrum, Juniperus communis*	*Alternaria* sp., *Aspergillus fumigatus*	Yang et al. (2003)
Camptothecin	Anticancer, antiviral (HIV)	*Nothapodytes foetida, Camptotheca acuminate*	*Entrophospora infrequens, Fusarium solani*	Puri et al. (2005)
Maytansine	Anticancer	*Putterlickia verrucosa*	*Actinosynnema pretiosum*	Wings et al. (2013)
Rohitukine	Antiphlogistic, anticancer	*Dysoxylum binectariferum*	*Fusarium proliferatum*	Mohana Kumara et al. (2012)
Munumbicins	Antibacterial, antimycotic	*Kennedia nigriscans*	*Streptomyces* sp.	Castillo et al. (2002)
Kakadumycins	Antibacterial, antiplasmodial	*Grevillea pteridifolia*	*Streptomyces* sp.	Castillo et al. (2003)
Coronamycins	Antimycotic, antiplasmodial	*Monstera* sp.	*Streptomyces* sp.	Ezra et al. (2004)

and *Calendula officinalis* showed an antagonistic effect in the bacterial rhizosphere (Köberl et al. 2013). *Ajuga bracteosa* is a medicinal plant used to cultivate diverse root-associated bacteria for plant growth promotion. About 123 different morphological bacteria were isolated from soil and roots. The results predicted in the majority of isolates showed plant growth promotion due to the production of IAA and siderophores (Kumar et al. 2012). Plant growth-promoting rhizobacteria activities include the production of IAA, contributing to phytohormone-regulating plant growth, while siderophores can solubilize iron and make it available for plants. Moreover, the chitinase enzyme that can lyse pathogenic fungal cells inhibits pathogens. Phytochemicals produced by plants affect plant–microbe interaction (Qi et al. 2009). Greater antimicrobial activity was observed in *Hypericum* plant extract that protects endophytes as compared with *Ziziphora* that showed lesser antimicrobial characteristics. Moreover, methanol extract exhibited strong antibacterial activity against *Bacillus subtilis*, *E. facecalis*, and *S. aureus*. *Chelidonim majus* L. and much potential against fungal pathogens (Baker and Satish 2013).

3.8 Management Strategies to Alleviate Abiotic Stress in Medicinal Plants

Various bacterial genera can promote plant growth and are known as plant growth-promoting rhizobacteria (PGPR). Moreover, scavenging of ROS, the compatible solutes, and accumulation of antioxidant metabolites are induced signals for important transcriptional modulation activities under heat stress (Wahid et al. 2007). Most endophytic *actinomycetes* belong to the genus *Streptomyces*, which is responsible for about 80% of biologically active compounds. *Actinomycetes* protect medicinal plants by reducing the growth of fungal pathogens with chitinases acting as inhibitors of fungal growth. The use of endophytic actinomycetes for controlling *S. rolfsii* offer an eco-friendly environment and disease management (Nagpure et al. 2014). IAA has a role in stem elongation and root growth. The IAA value is high in the rhizosphere, where rhizosphere bacteria produce auxin as secondary metabolites because of rich supplies of root exudates. It also triggers the propagation of lateral roots that increase the nutrients-absorbing area, which leads to better assimilation of water and nutrients from the surroundings.

Actinobacteria also produce IAA, cytokinin's and gibberellic acid. *Aspergillus* and *Streptomyces* produce a high level of IAA, which leads to the production of hematophagous actinomycetes (Ruanpanun et al. 2010). The application of rhizobacteria improves plant growth as well as soil microbial activity (Karami et al. 2010). With nourishment from microbes in the stress environment and adaptable response at the phenotypic level, it is always appropriate to implement microbe-derived natural products that can perform a good job to alleviate stress regardless of environmental conditions.

3.9 Conclusion and Future Consensus

Medicinal plants have been gaining much importance in developing countries due to the re-emerging health concerns from the rising costs of prescription drugs for maintaining personal health. A WHO report stated that about 80% of people rely on herbs. The quality of medicinal plant products is affected by biotic and abiotic stresses. Microbes in the soil improve medicinal plant growth and quality, which explains the increasing research interest in medicinal plants and their interaction with rhizosphere microbes for crop improvement, withstanding abiotic stresses. PGPR are a great option to improve the quantity and quality of medicines and plants. However, it mainly depends upon the selection and identification of bacteria, streptomycetes, fungi, and microbes and their interaction with plants that could help improve growth and withstand harsh environmental conditions as well as high-quality secondary plant metabolites production. Plants react to different stresses, and microbe-based mitigation approaches necessitate studies for gaining a deeper insight into the molecular, physiological, biochemical, and ultra-structural composition. These studies require different approaches, i.e. genomics, metagenomics, and molecular proteomics for a clear understanding of behind-the-scenes plant–microbe interactions. As the medicinal product value and requirement keep increasing day by day, we strongly recommend greater attention to keener studies on the selection of microbes, their characterization, and plant–microbe compatibility when introduced into the new environment to mitigate abiotic stress in medicinal plants.

References

Abbaszadeh, B., Valadabadi, H.A.F.S.A., and Moaveni, P. (2009). Investigation of variations of the morphological values and flowering shoot yield in different mint species at Iran. *Journal of Horticulture and Forestry 1* (7): 109–112.

Abdou, R., Scherlach, K., Dahse, H.-M., Sattler, I., and Hertweck, C. (2010). Botryorhodines A–D, antifungal and cytotoxic depsidones from *Botryosphaeria rhodina*, an endophyte of the medicinal plant *Bidens Pilosa*. *Phytochemistry 71* (1): 110–116.

Ahmad, P., Hashem, A., Abd-Allah, E.F., Alqarawi, A.A., John, R., Egamberdieva, D., and Gucel, S. (2015). Role of *Trichoderma harzianum* in mitigating NaCl stress in Indian mustard (*Brassica juncea L*) through antioxidative defense system. *Frontiers in Plant Science 6*: 868.

Ait Barka, E., Nowak, J., and Clément, C. (2006). Enhancement of chilling resistance of inoculated grapevine plantlets with a plant growth-promoting rhizobacterium, *Burkholderia phytofirmans* strain PsJN. *Applied and Environmental Microbiology 72* (11): 7246–7252.

Ali, M.B., Hahn, E.-J., and Paek, K.-Y. (2005). Effects of temperature on oxidative stress defense systems, lipid peroxidation and lipoxygenase activity in *Phalaenopsis*. *Plant Physiology and Biochemistry 43* (3): 213–223.

Ali, R.M., Abbas, H.M., and Kamal, R.K. (2007). The effects of treatment with polyamines on dry matter, oil and flavonoid contents in salinity stressed *chamomile* and *Sweet Marjoram*. *Plant Soil and Environment 53* (12): 529.

Aziz, E.E. and Hendawy, S.F. (2008). Effect of soil type and irrigation intervals on plant growth, essential oil yield and constituents of *Thymus vulgaris* plant.

Badri, D.V., Weir, T.L., Daniel, V.D.L., and Vivanco, J.M. (2009). Rhizosphere chemical dialogues: Plant–microbe interactions. *Current Opinion in Biotechnology 20* (6): 642–650.

Bafana, A. (2013). Diversity and metabolic potential of culturable root-associated bacteria from *Origanum vulgare* in sub-Himalayan region. *World Journal of Microbiology and Biotechnology 29* (1): 63–74.

Bais, H.P., Park, S.-W., Weir, T.L., Callaway, R.M., and Vivanco, J.M. (2004). How plants communicate using the underground information superhighway. *Trends in Plant Science 9* (1): 26–32.

Baker, S. and Satish, S. (2013). Antimicrobial evaluation of fluorescent pseudomonas sp inhabiting medicinal plant *Annona squamosa* L. *Journal of Pure and Applied Microbiology 7* (2): 1027–1033.

Belaqziz, R., Romane, A., and Abbad, A. (2009). Salt stress effects on germination, growth and essential oil content of an endemic thyme species in Morocco (*Thymus maroccanus Ball.*). *Journal of Applied Sciences Research* (July): 858–863.

Berendsen, R.L., Pieterse, C.M., and Bakker, P.A. (2012). The rhizosphere microbiome and plant health. *Trends in Plant Science 17*: 478–486. doi: 10.1016/j.tplants.2012.04.001.

Bulgarelli, D., Schlaeppi, K., Spaepen, S., Themaat, V., Loren, E.V., and Schulze-Lefert, P. (2013). Structure and functions of the bacterial microbiota of plants. *Annual Review of Plant Biology 64*: 807–838.

Cakmak, I. (2005). The role of potassium in alleviating detrimental effects of abiotic stresses in plants. *Journal of Plant Nutrition and Soil Science 168* (4): 521–530.

Camejo, D., Rodríguez, P., Morales, M.A., Dell'Amico, J.M., Torrecillas, A., and Alarcón, J.J. (2005). High temperature effects on photosynthetic activity of two tomato cultivars with different heat susceptibility. *Journal of Plant Physiology 162* (3): 281–289.

Castillo, U.F., Strobel, G.A., Ford, E.J., Hess, W.M., Porter, H., Jensen, J.B., and Teplow, D.B. (2002). Munumbicins, wide-spectrum antibiotics produced by *Streptomyces* NRRL 30562, endophytic on *Kennedia nigriscansaa*. The GenBank accession number for the sequence determined in this work is AY127079. *Microbiology 148* (9): 2675–2685.

Castillo, U., Harper, J.K., Strobel, G.A., Sears, J., Alesi, K., Ford, E., and Ge, H. (2003). Kakadumycins, novel antibiotics from Streptomyces sp. NRRL 30566, an endophyte of *Grevillea pteridifolia*. *FEMS Microbiology Letters 224* (2): 183–190.

Chandra, D., Kim, J.G., and Kim, Y.P. (2012). Changes in microbial population and quality of microgreens treated with different sanitizers and packaging films. *Horticulture, Environment, and Biotechnology 53* (1): 32–40.

Chowdhary, K., Kumar, A., Sharma, S., Pathak, R., and Jangir, M. (2018). *Ocimum sp.*: Source of biorational pesticides. *Industrial Crops and Products 122*: 686–701.

Cloete, K.J., Przybylowicz, W.J., Mesjasz-przybylowicz, J.O.L.A.N.T.A., Barnabas, A.D., Valentine, A.J., and Botha, A. (2010). Micro-particle-induced X-ray emission mapping of elemental distribution in roots of a Mediterranean-type sclerophyll, *Agathosma betulina* (Berg.) Pillans, colonized by *Cryptococcus laurentii*. *Plant, Cell & Environment 33* (6): 1005–1015.

Compant, S., Van Der Heijden, M.G., and Sessitsch, A. (2010). Climate change effects on beneficial plant–microorganism interactions. *FEMS Microbiology Ecology 73* (2): 197–214.

Dai, C.-C., Chen, Y., Wang, X.-X., and Li, P.-D. (2013). Effects of intercropping of peanut with the medicinal plant *Atractylodes lancea* on soil microecology and peanut yield in subtropical China. *Agroforestry Systems 87* (2): 417–426.

Debbab, A., Aly, A.H., Edrada-Ebel, R., Müller, W.E.G., Mosaddak, M., Hakikj, A., and Proksch, P. (2009). Bioactive secondary metabolites from the endophytic fungus *Chaetomium sp.* isolated from Salvia officinalis growing in Morocco. *Biotechnology, Agronomy, Society and Environment (BASE) 13* (2): 229–234.

Debnath, M., Pandey, M., and Bisen, P.S. (2011). An omics approach to understand the plant abiotic stress. *Omics: A Journal of Integrative Biology 15* (11): 739–762.

Dos Santos, A.L.W., Elbl, P., Navarro, B.V., de Oliveira, L.F., Salvato, F., Balbuena, T.S., and Floh, E.I.S. (2016). Quantitative proteomic analysis of *Araucaria angustifolia* (Bertol.) Kuntze cell lines with contrasting embryogenic potential. *Journal of Proteomics 130*: 180–189.

Dubey, A., Malla, M.A., Kumar, A., Dayanandan, S., and Khan, M.L. (2020). Plants endophytes: Unveiling hidden agenda for bioprospecting toward sustainable agriculture. *Critical Reviews in Biotechnology 40* (8): 1210–1231.

Dubey, D., Kothapalli, N., McKeon, A., Flanagan, E.P., Lennon, V.A., Klein, C.J., and Tillema, J.-M. (2018). Predictors of neural-specific autoantibodies and immunotherapy response in patients with cognitive dysfunction. *Journal of Neuroimmunology 323*: 62–72.

Egamberdieva, D., Wirth, S.J., Alqarawi, A.A., Abd_Allah, E.F., and Hashem, A. (2017). Phytohormones and beneficial microbes: Essential components for plants to balance stress and fitness. *Frontiers in Microbiology 8*: 2104.

Eliašová, A., Repčák, M., and Pastírová, A. (2004). Quantitative changes of secondary metabolites of *Matricaria chamomilla* by abiotic stress. *Zeitschrift Für Naturforschung C 59* (7-8): 543–548.

El-Leithy, E.S., Shaker, D.S., Ghorab, M.K., and Abdel-Rashid, R.S. (2010). Evaluation of mucoadhesive hydrogels loaded with diclofenac sodium–chitosan microspheres for rectal administration. *Aaps Pharmscitech 11* (4): 1695–1702.

Ezra, D., Castillo, U.F., Strobel, G.A., Hess, W.M., Porter, H., Jensen, J.B., and Maranta, M. (2004). Coronamycins, peptide antibiotics produced by a verticillate Streptomyces sp. (MSU-2110) endophytic on Monstera sp. *Microbiology 150* (4): 785–793.

Farahani, H.A., Valadabadi, S.A., Daneshian, J., and Khalvati, M.A. (2009). Evaluation changing of essential oil of balm (*Melissa officinalis* L.) under water deficit stress conditions. *Journal of Medicinal Plants Research 3* (5): 329–333.

Freitas, H., Prasad, M.N.V., and Pratas, J. (2004). Plant community tolerant to trace elements growing on the degraded soils of Sao Domingos mine in the southeast of Portugal: Environmental implications. *Environment International 30* (1): 65–72.

Gao, D.-W., Fu, Y., Tao, Y., Li, X.-X., Xing, M., Gao, X.-H., and Ren, N.-Q. (2011). Linking microbial community structure to membrane biofouling associated with varying dissolved oxygen concentrations. *Bioresource Technology 102* (10): 5626–5633.

Gao, S., Ouyang, C., Wang, S., Xu, Y., Tang, L., and Chen, F. (2008). Effects of salt stress on growth, antioxidant enzyme and phenylalanine ammonia-lyase activities in *Jatropha curcas* L. seedlings. *Plant, Soil and Environment 54* (9): 374–381.

Geu-Flores, F., Sherden, N.H., Courdavault, V., Burlat, V., Glenn, W.S., Wu, C., and O'Connor, S.E. (2012). An alternative route to cyclic terpenes by reductive cyclization in iridoid biosynthesis. *Nature 492* (7427): 138–142.

Giddings, L.-A., Liscombe, D.K., Hamilton, J.P., Childs, K.L., DellaPenna, D., Buell, C.R., and O'Connor, S.E. (2011). A stereoselective hydroxylation step of alkaloid biosynthesis by a unique cytochrome P450 in Catharanthus roseus. *Journal of Biological Chemistry 286* (19): 16751–16757.

Golinska, P., Wypij, M., Agarkar, G., Rathod, D., Dahm, H., and Rai, M. (2015). Endophytic actinobacteria of medicinal plants: Diversity and bioactivity. *Antonie Van Leeuwenhoek 108* (2): 267–289.

Gopalakrishnan, S., Sathya, A., Vijayabharathi, R., Varshney, R.K., Gowda, C.L.L., and Krishnamurthy, L. (2015). Plant growth promoting rhizobia: Challenges and opportunities. *3 Biotech 5* (4): 355–377.

Govindasamy, V., Senthilkumar, M., Magheshwaran, V., Kumar, U., Bose, P., Sharma, V., and Annapurna, K. (2010). *Bacillus* and *Paenibacillus* spp.: Potential PGPR for sustainable agriculture. In: *Plant Growth and Health Promoting Bacteria*. 333–364, Springer

Grainger, C.M. and Rajcan, I. (2014). Characterization of the genetic changes in a multi-generational pedigree of an elite Canadian soybean cultivar. *Theoretical and Applied Genetics 127* (1): 211–229.

Guo, D., Chen, J., Du, X., and Han, B. (2010). Screening of molluscicidal strain against *Oncomelania hupensis* from the rhizosphere of medicinal plant *Phytolacca acinosa Roxb. Pharmacognosy Magazine 6* (23): 159.

Hashem, A., Abd_Allah, E.F., Alqarawi, A.A., Radhakrishnan, R., and Kumar, A. (2017). Plant defense approach of *Bacillus subtilis* (BERA 71) against *Macrophomina phaseolina* (Tassi) Goid in mung bean. *Journal of Plant Interactions 12* (1): 390–401.

Islam, S., Izekor, E., and Garner, J.O. (2009). Development of selection procedures of Sweetpotato genotypes according to chilling tolerance. *Journal of Food, Agriculture & Environment 7* (2): 329–331.

Jaleel, C.A., Gopi, R., Manivannan, P., and Panneerselvam, R. (2008). Soil salinity alters the morphology in *Catharanthus roseus* and its effects on endogenous mineral constituents. *EurAsian Journal of BioSciences 2* (1): 18–25.

Karami, O.R.A., Khodaverdi, M., and Ali, A.F. (2010). Antibacterial effect of effective compounds of *Satureja hortensis* and *Thymus vulgaris* essential oils against Erwinia amylovora.

Karthik, L., Kumar, G., Keswani, T., Bhattacharyya, A., Chandar, S.S., and Bhaskara Rao, K.V. (2014). Protease inhibitors from marine actinobacteria as a potential source for antimalarial compound. *PloS One 9* (3): e90972.

Karthikeyan, B., Jaleel, C.A., Lakshmanan, G.M.A., and Deiveekasundaram, M. (2008). Studies on rhizosphere microbial diversity of some commercially important medicinal plants. *Colloids and Surfaces B: Biointerfaces 62* (1): 143–145.

Kasilo, O.M.J. and Trapsida, J.-M. (2010). Regulation of traditional medicine in the WHO African region. *Afr. Health Monit. (Online)*, 25–31.

Kato, T., HINoo, H.I.R.O.S.H.I., and Shoji, J.U.N.I.C.H.I. (1978). The structure of tridecaptin A studies on antibiotics from the genus *Bacillus*. XXIV. *The Journal of Antibiotics 31* (7): 652–661.

Kaymak, H.C. (2010). Potential of PGPR in agricultural innovations. In: *Plant Growth and Health Promoting Bacteria*. 45–79.

Köberl, M., Schmidt, R., Ramadan, E.M., Bauer, R., and Berg, G. (2013). The microbiome of medicinal plants: Diversity and importance for plant growth, quality and health. *Frontiers in Microbiology 4*: 400.

Koocheki, A., Nassiri-Mahallati, M., and Azizi, G. (2008). Effect of drought, salinity, and defoliation on growth characteristics of some medicinal plants of Iran. *Journal of Herbs, Spices & Medicinal Plants 14* (1–2): 37–53.

Koussevitzky, S., Suzuki, N., Huntington, S., Armijo, L., Sha, W., Cortes, D., and Mittler, R. (2008). Ascorbate peroxidase 1 plays a key role in the response of *Arabidopsis thaliana* to stress combination. *Journal of Biological Chemistry 283* (49): 34197–34203.

Kulkarni, M.G., Street, R.A., and Van Staden, J. (2007). Germination and seedling growth requirements for propagation of *Dioscorea dregeana* (Kunth) Dur. and Schinz—A tuberous medicinal plant. *South African Journal of Botany 73* (1): 131–137.

Kumar, A. and Dubey, A. (2020). Rhizosphere microbiome: Engineering bacterial competitiveness for enhancing crop production. *Journal of Advanced Research 24*: 337–352.

Kumar, M., Tomar, R.S., Lade, H., and Paul, D. (2016). Methylotrophic bacteria in sustainable agriculture. *World Journal of Microbiology and Biotechnology 32* (7): 1–9.

Kumar, S., Narula, A., Sharma, M.P., and Srivastava, P.S. (2004). In vitro propagation of *Pluchea lanceolata*, a medicinal plant, and effect of heavy metals and different aminopurines on quercet in content. *Vitro Cellular & Developmental Biology-Plant 40* (2): 171–176.

Kumar, P., Dubey, R.C., and Maheshwari, D.K. (2012). *Bacillus* strains isolated from rhizosphere showed plant growth promoting and antagonistic activity against phytopathogens. *Microbiological Research 167* (8): 493–499.

Kumar, V., Singh, S., Singh, J., and Upadhyay, N. (2015). Potential of plant growth promoting traits by bacteria isolated from heavy metal contaminated soils. *Bulletin of Environmental Contamination and Toxicology 94* (6): 807–814.

Kumara, P.M., Zuehlke, S., Priti, V., Ramesha, B.T., Shweta, S., Ravikanth, G., and Shaanker, R.U. (2012). *Fusarium proliferatum*, an endophytic fungus from *Dysoxylum binectariferum Hook.* f, produces rohitukine, a chromane alkaloid possessing anti-cancer activity. *Antonie Van Leeuwenhoek 101* (2): 323–329.

Kusano, M., Tohge, T., Fukushima, A., Kobayashi, M., Hayashi, N., Otsuki, H., and Matsuda, F. (2011). Metabolomics reveals comprehensive reprogramming involving two independent metabolic responses of Arabidopsis to UV-B light. *The Plant Journal 67* (2): 354–369.

Laszlo, I., Szoke, E., and Tyihak, E. (1998). Relationship between abiotic stress and formaldehyde concentration in tissue culture of *Datura Innoxia Mill. Plant Growth Regulation 25* (3): 195–199.

Li, H., Qiang, S., and Qian, Y. (2008). Physiological response of different croftonweed (*Eupatorium adenophorum*) populations to low temperature. *Weed Science 56* (2): 196–202.

Ma, J., Feng, X., Yang, X., Cao, Y., Zhao, W., and Sun, L. (2020). The leaf extract of crofton weed (*Eupatorium adenophorum*) inhibits primary root growth by inducing cell death in maize root border cells. *Plant Diversity 42* (3): 174–180.

Malla, B.A., Maqbool, I., Lone, S.A., Ahmad, S., Bhat, A.M., Para, I.A., and Dar, P.S. (2018). An overview on Neurocysticercosis. *Journal of Pharmacognosy and Phytochemistry 7* (6): 303–312.

Manavalan, L.P., Guttikonda, S.K., Phan Tran, L.-S., and Nguyen, H.T. (2009). Physiological and molecular approaches to improve drought resistance in soybean. *Plant and Cell Physiology 50* (7): 1260–1276.

Meena, K.K., Sorty, A.M., Bitla, U.M., Choudhary, K., Gupta, P., Pareek, A., and Gupta, V.K. (2017). Abiotic stress responses and microbe-mediated mitigation in plants: The omics strategies. *Frontiers in Plant Science 8*: 172.

Miceli, A., Moncada, A., Vetrano, F., and D'Anna, F. (2003). First results on yield and quality response of basil (Ocimum basilicum L.) grown in a floating system. Paper presented at the International Symposium on Managing Greenhouse Crops in Saline Environment 609. https://www.ishs.org/ishs-article/609_57

Mohammad, A.R. and Tarpley, L. (2009). Instrumentation enabling study of plant physiological response to elevated night temperature. *Plant Methods 5*: 7–16.

Nadeem, S.M., Ahmad, M., Zahir, Z.A., Javaid, A., and Ashraf, M. (2014). The role of mycorrhizae and plant growth promoting rhizobacteria (PGPR) in improving crop productivity under stressful environments. *Biotechnology Advances 32* (2): 429–448.

Naeem, M., Khan, M.N., and Khan, M.M.A. (2013). Adverse effects of abiotic stresses on medicinal and aromatic plants and their alleviation by calcium. In: *Plant Acclimation to Environmental Stress*. 101–146, Springer.

Nagpure, A., Choudhary, B., and Gupta, R.K. (2014). Mycolytic enzymes produced by *Streptomyces violaceusniger* and their role in antagonism towards wood-rotting fungi. *Journal of Basic Microbiology 54* (5): 397–407.

Narula, A., Kumar, S., and Srivastava, P.S. (2005). Abiotic metal stress enhances diosgenin yield in *Dioscorea bulbifera* L. cultures. *Plant Cell Reports 24* (4): 250–254.

Nam, M.H., Kim, S.I., Liu, J.R., Yang, D.C., Lim, Y.P., Kwon, K.-H., and Park, Y.M. (2005). Proteomic analysis of Korean ginseng *(Panax ginseng CA Meyer). Journal of Chromatography B 815* (1-2): 147–155.

Nawrot, R., Tomaszewski, Ł., Czerwoniec, A., and Goździcka-Józefiak, A. (2013). Identification of a coding sequence and structure modeling of a glycine-rich RNA-binding protein (CmGRP1) from *Chelidonium majus* L. *Plant Molecular Biology Reporter 31* (2): 470–476.

Ngumbi, E. and Kloepper, J. (2016). Bacterial-mediated drought tolerance: Current and future prospects. *Applied Soil Ecology 105*: 109–125.

Nguyen, H.T., Babu, R.C., and Blum, A. (1997). Breeding for drought resistance in rice: Physiology and molecular genetics considerations. *Crop Science 37* (5): 1426–1434.

Okazaki, Y., Kamide, Y., Hirai, M.Y., and Saito, K. (2013). Plant lipidomics based on hydrophilic interaction chromatography coupled to ion trap time-of-flight mass spectrometry. *Metabolomics 9* (1): 121–131.

Patwardhan, B., Warude, D., Pushpangadan, P., and Bhatt, N. (2005). Ayurveda and traditional Chinese medicine: A comparative overview. *Evidence-Based Complementary and Alternative Medicine 2*: 465–473.

Puri, S.C., Verma, V., Amna, T., Qazi, G.N., and Spiteller, M. (2005). An endophytic fungus from *Nothapodytes f oetida* that produces Camptothecin. *Journal of Natural Products 68* (12): 1717–1719.

Passari, A.K., Mishra, V.K., Saikia, R., Gupta, V.K., and Singh, B.P. (2015). Isolation, abundance and phylogenetic affiliation of endophytic actinomycetes associated with medicinal plants and screening for their in vitro antimicrobial biosynthetic potential. *Frontiers in Microbiology 6*: 273.

Purdy, S.J., Maddison, A.L., Jones, L.E., Webster, R.J., Andralojc, J., Donnison, I., and Clifton-Brown, J. (2013). Characterization of chilling-shock responses in four genotypes of Miscanthus reveals the superior tolerance of M.× giganteus compared with *M. sinensis* and *M. sacchariflorus. Annals of Botany 111* (5): 999–1013.

Qi, J., Yao, H., Ma, X., Zhou, L., and Li, X. (2009). Soil microbial community composition and diversity in the rhizosphere of a Chinese medicinal plant. *Communications in Soil Science and Plant Analysis 40* (9–10): 1462–1482.

Rahmani, N., Daneshian, J., Farahani, H.A., and Taherkhani, T. (2011). Evaluation of nitrogenous fertilizer influence on oil variations of Calendula (*Calendula officinalis* L.) under drought stress conditions. *Journal of Medicinal Plants Research 5* (5): 696–701.

Ramin, A.A. (2006). Effects of salinity and temperature on germination and seedling establishment of sweet basil (*Ocimum basilicum* L.). *Journal of Herbs, Spices & Medicinal Plants 11* (4): 81–90.

Razmjoo, K.H.O.R.S.H.I.D., Heydarizadeh, P.A.R.I.S.A., and Sabzalian, M.O.H.A.M.M.A.D.R. (2008). Effect of salinity and drought stresses on growth parameters and essential oil content of Matricaria Chamomile. *International Journal of Agriculture And Biology 10* (4): 451–454.

Rejšková, A., Patková, L., Stodůlková, E., and Lipavská, H. (2007). The effect of abiotic stresses on carbohydrate status of olive shoots (*Olea europaea* L.) under in vitro conditions. *Journal of Plant Physiology 164* (2): 174–184.

Ruanpanun, P., Tangchitsomkid, N., Hyde, K.D., and Lumyong, S. (2010). Actinomycetes and fungi isolated from plant-parasitic nematode infested soils: Screening of the effective biocontrol potential, indole-3-acetic acid and siderophore production. *World Journal of Microbiology and Biotechnology 26* (9): 1569–1578.

Sahoo, R.K., Ansari, M.W., Dangar, T.K., Mohanty, S., and Tuteja, N. (2014). Phenotypic and molecular characterisation of efficient nitrogen-fixing *Azotobacter strains* from rice fields for crop improvement. *Protoplasma 251* (3): 511–523.

Said-Al Ahl, H. and Omer, E.A. (2011). Medicinal and aromatic plants production under salt stress. A review. *Herba Polonica 57* (1): 72–87.

Saito, K. (2013). Phytochemical genomics—a new trend. *Current Opinion in Plant Biology 16* (3): 373–380.

Shi, J.-Y., Yuan, X.-F., Lin, H.-R., Yang, Y.-Q., and Li, Z.-Y. (2011). Differences in soil properties and bacterial communities between the rhizosphere and bulk soil and among different production areas of the medicinal plant *Fritillaria thunbergii*. *International Journal of Molecular Sciences 12* (6): 3770–3785.

Singh, S., Kate, B.N., and Banerjee, U.C. (2005). Bioactive compounds from cyanobacteria and microalgae: An overview. *Critical Reviews in Biotechnology 25* (3): 73–95.

Sorty, A.M., Meena, K.K., Choudhary, K., Bitla, U.M., Minhas, P.S., and Krishnani, K.K. (2016). Effect of plant growth promoting bacteria associated with halophytic weed (*Psoralea corylifolia L*) on germination and seedling growth of wheat under saline conditions. *Applied Biochemistry and Biotechnology 180* (5): 872–882.

Taarit, M.B., Msaada, K., Hosni, K., and Marzouk, B. (2010). Changes in fatty acid and essential oil composition of sage (*Salvia officinalis* L.) leaves under NaCl stress. *Food Chemistry 119* (3): 951–956.

Tamilarasi, S., Nanthakumar, K., Karthikeyan, K., and Lakshmanaperumalsamy, P. (2007). Diversity of root associated microorganisms of selected medicinal plants and influence of rhizomicroorganisms on the antimicrobial property of *Coriandrum Sativum*. *Journal of Environmental Biology 29* (1): 127.

Ting, A.S.Y., Tan, L.M., and Ling, A.P.K. (2009). In vitro assessment of tolerance of *Orthosiphon stamineus* to induced water and salinity stress. *Asian Journal of Plant Sciences 8* (3): 206–211.

Vishwakarma, K., Kumar, N., Shandilya, C., Mohapatra, S., Bhayana, S., and Varma, A. (2020). Revisiting plant–microbe interactions and microbial consortia application for enhancing sustainable agriculture: A review. *Frontiers in Microbiology 11*: 3195.

Wahid, A., Gelani, S., Ashraf, M., and Foolad, M.R. (2007). Heat tolerance in plants: An overview. *Environmental and Experimental Botany 61* (3): 199–223.

Ward, J.L., Baker, J.M., Llewellyn, A.M., Hawkins, N.D., and Beale, M.H. (2011). Metabolomic analysis of *Arabidopsis reveals* hemiterpenoid glycosides as products of a nitrate ion-regulated, carbon flux overflow. *Proceedings of the National Academy of Sciences 108* (26): 10762–10767.

Whang, K.-S., Lee, J.-C., Lee, H.-R., Han, S.-I., and Chung, S.-H. (2014). *Terriglobustenax sp.* nov., an exopolysaccharide-producing *Acidobacterium* isolated from rhizosphere soil of a medicinal plant. *International Journal of Systematic and Evolutionary Microbiology 64* (Pt_2): 431–437.

Williamson, E. (2003). Drug interactions between herbal and prescription medicines. *Drug Safety 26*: 1075–1092.

Wings, S., Müller, H., Berg, G., Lamshöft, M., and Leistner, E. (2013). A study of the bacterial community in the root system of the maytansine containing plant *Putterlickia verrucosa*. *Phytochemistry 91*: 158–164.

Winzer, T., Gazda, V., He, Z., Kaminski, F., Kern, M., Larson, T.R., and Vaistij, F.E. (2012). A *Papaver somniferum* 10-gene cluster for synthesis of the anticancer alkaloid noscapine. *Science 336* (6089): 1704–1708.

Xianzhi, Y., Shiping, G., Lingqi, Z., and Hua, S. (2003). Select of producing podophyllotoxin endophytic fungi from *podophyllin* plant. *Natural Product Research and Development 15* (5): 419–422.

Xu, Z., Xu, Q., Zheng, Z., and Huang, Y. (2012). Kribbella amoyensis sp. nov., isolated from rhizosphere soil of a pharmaceutical plant, *Typhonium giganteum* Engl. *International Journal of Systematic and Evolutionary Microbiology 62* (Pt_5): 1081–1085.

Yolcu, H., Gunes, A., Gullap, M.K., and Cakmakci, R. (2012). Effects of plant growth-promoting rhizobacteria on some morphologic characteristics, yield and quality contents of Hungarian vetch. *Turkish Journal of Field Crops 17* (2): 208–214.

Zhang, F., Yang, Y.L., He, W.L., Zhao, X., and Zhang, L.X. (2004). Effects of salinity on growth and compatible solutes of callus induced from *Populus euphratica*. *In Vitro Cellular & Developmental Biology-Plant 40* (5): 491–494.

Zhang, H.Y., Xue, Q.H., Shen, G.H., and Wang, D.S. (2013). Effects of actinomycetes agent on ginseng growth and rhizosphere soil microflora. *Ying Yong Sheng Tai Xue Bao= the Journal of Applied Ecology 24* (8): 2287–2293.

Zhang, L.-J., Liu, H.-K., Hsiao, P.-C., Kuo, L.-M.Y., Lee, I.-J., Tian-Shung, W., Chiou, W.-F., and Kuo, Y.-H. (2011). New isoflavonoid glycosides and related constituents from astragali radix (*Astragalus membranaceus*) and their inhibitory activity on nitric oxide production. *Journal of Agricultural and Food Chemistry 59* (4): 1131–1137.

Zhao, S., Ye, G., Fu, G., CHEnG, J.-X., YAnG, B.B., and PEnG, C. (2011). *Ganoderma lucidum* exerts anti-tumor effects on ovarian cancer cells and enhances their sensitivity to cisplatin. *International Journal of Oncology 38* (5): 1319–1327.

Zhou, K. and Raffoul, J.J. (2012). Potential anticancer properties of grape antioxidants. *Journal of Oncology 2012*.

Zubek, S. and Błaszkowski, J. (2009). Medicinal plants as hosts of arbuscular mycorrhizal fungi and dark septate endophytes. *Phytochemistry Reviews 8* (3): 571–580.

4

Uses of Compost in Agriculture and Bioremediation – A Review

Aparna Gunjal

Department of Microbiology, Dr. D.Y. Patil, Arts, Commerce and Science College, Pimpri, Pune, Maharashtra

4.1 Introduction

Landfill and incineration are the ways of solid waste treatment globally. Methane is the greenhouse gas that causes global warming and that needs to be minimized. Therefore, recycling wastes and organic material by returning to agricultural land is most effective. Composting is the most economical and effective way of recycling by placing organic material back in the soil to enhance soil properties (Awasthi et al. 2014). Landfill of biodegradable waste contributes to environmental degradation due to the production of methane gas (Hurrell 2015). A considerable amount of biodegradable organic waste is generated in most countries, and their management is problematic. Recycling wastes and organic material by returning to agricultural land is essential. Composting is the most cost-effective and promising method of treating wastes. Composting is a biological process that turns organic matter into a dark, rich, colored substance, termed "compost," which is used as a good soil conditioner. Microbes convert complex organic substances into simpler substances during the composting process with the production of carbon dioxide, water, and minerals and stabilized organic matter called compost. During the composting process, heat is generated that kills pathogens, weeds, etc.

Among waste-management strategies, the composting approach has become very popular in the disposal and treatment of organic wastes with many economic and environmental benefits, including the formation of a stabilized product (Varma and Kalamdhad 2014). The global compost market is studied to reach $9.2 billion by 2024, with a compound annual growth rate (CGAR) of about 6.8% from 2019 to 2024 (www.globenewswire.com). The main aim of this chapter is to highlight the various applications of compost, how it improves the soil's physico-chemical parameters and boosts plant growth and yield, and how it ultimately benefits farmers. Amendment to the soil to help plant growth is essential. It is therefore necessary to study how compost amends the soil and helps improve agriculture and horticulture.

Core Microbiome: Improving Crop Quality and Productivity, First Edition. Edited by
Javid A. Parray, Nowsheen Shameem, Elsayed Fathi Abd-Allah, and Mohammad Yaseen Mir.

4.2 Applications of Compost

4.2.1 Compost Use for Soil Amendment and as Conditioner

The role of organic amendments in agriculture is to provide minerals to the crops and improve soil properties. Many physicochemical changes in the soil are due to modifications that help minimize chemical fertilizers (Bonilla et al. 2012). Organic amendments include manure, crop and food residues, compost, and fertilizers. The compost helps maintain soil stability. This also promotes the balance within the agroecosystem. The composting process converts organic waste into a humus-like substance called compost, which can be used as a soil fertility booster (Li et al. 2013).

Compost helps improve soil properties by minimizing bulk density and enhancing the permeability and stability of the soil. As a result, it also increases the water absorption capacity of the soil (Adugna 2016) and reduces soil abrasion (Martinez-Blanco et al. 2013). Compost improves drought resistance and soil physical properties (Adugna 2016), increases the availability of nutrients and microbial populations, and enhances crop yield (Chelah et al. 2011). Compost use in soil amendment increases surface water infiltration and reduces the intensity of surface run-off from land, and also reduces transport of dissolved and suspended phosphorus and nitrate–nitrogen. The addition of compost in agricultural soil is also known to significantly increase the organic material in the soil (Bellamy et al. 2005).

Compost amendment to soil also helps delay surface degradation. Tong et al. (2018) have reported the reuse of wastes as compost for enhancing soil properties in the afforested land of Beijing plain.

4.2.2 Compost to Alleviate Soil Compaction

Soil compaction is a severe landscape management problem, as it slows turf establishment by stopping the movement of air, water, and minerals within the soil. The standard methods to alleviate soil compaction are costly and require human resources. Therefore, turf managers have been recommending the use of compost and compost mixed with bulking agents. The use of compost and compost with bulking agents helps in root penetration, and turf establishment enhances water-holding capacity and resistance to insects and pests.

4.2.3 Control of Erosion by the Addition of Compost

Soils that are rich in silt, with low clay deposits, and organic material, erode fast. Compost helps to decrease soil erodibility. The coarseness of the particles in the compost reduces flow velocity. Coarse particles are heavier and do not erode easily. Compost also helps in the establishment of rapid vegetation, and this slows surface flow velocity. Compost-amended soils have reduced erosion by several orders of magnitude over untreated soils (Anastasi et al. 2005). Mulch from wood residual reduced erosion when compared with the control. Hydro compost (consisting of three parts dairy manure compost and one part recycled paper mulch tackifier and water) helps lower soil erosion rates. It increases the seed germination percentage compared with mat fiber mulch with guar tackifier. Compost also plays a vital role in reducing surface and subsurface run-off from the land (Taiwo 2011).

Goldstein (2002) suggested that soils amended with compost have indicated delays in surface degradation and decreases in sediment run-off.

On steep embankments of roads and highways, compost is used to reduce soil erosion and establish turf, as compost forms a thick permanent growth to improve soil infrastructure. Turfgrasses are challenging to maintain and sensitive to turf diseases, pests, insects, and soil compaction. Turf managers were using chemical fertilizers, pesticides, fungicides, etc. or a combination of all three, which are rather harmful to the environment. However, turf managers now prefer to use compost for the remediation of turfgrasses.

4.2.4 Bioremediation of Organic Contaminants, Heavy Metals, and Pesticides in Soils

Soils are contaminated with harmful chemicals. Potential bioapproaches are biostimulation, i.e., fertilizers and bioaugmentation by adding organic amendments with active microorganisms in it, or the compost. Recently, compost has been used for soil bioremediation. Compost addition can comprise "super bioaugmentation" with a mixture of degrading microbes, combined with "biostimulation" (Kastner and Miltner 2016).

Compost is used in bioremediation to break down toxic pollutants and other harmful substances, minimizing the toxicity of metallic contaminants in organic residues, wastes, and by-products. After the composting process stabilizes the waste, it can be disposed to landfills or farmland (Ankidawa and Nwodo 2011). The stabilized organic matter can also form complexes and restrict metal flow and enhance absorption in plants.

The microbial activity during composting helps in the decomposition of pesticides in the soil. A report suggests that the pesticide content is present in composts following the breakdown of pesticides during composting (Wu and Ma 2000). The addition of compost is known to enhance the degradation of two herbicides viz., benthiocarb S-4-chlorobenzyl diethylthiocarbamate (benthiocarb) and 4-chloro-2-methylphenoxyacetic acid (MCPA) in soils. The soil amended with compost improved the breakdown of a fumigant (1, 3-dichloropropene) in comparison with the soil without compost. The concentration of pesticides was at safe levels for plants in soils due to the application of compost.

4.2.5 Compost in Horticulture as the Supply of Minerals to Crops and Other Plants

Compost application is beneficial in horticulture as an effective means of recycling; it also minimizes the use of chemicals and harmful fertilizers. Compost has been trialed for the grapevine (Powell et al. 2007). Based on the benefits of compost, it can enhance water absorption and minimize the need for irrigation. Compost is also an exciting peat substitute. Compost added to soil has improved the biomass of crops and has increased N, P, and K uptake (Miller et al. 2015). There is a report on a significant increase in the growth and biomass of *Anemone coronaria* L. cv. Red meron (Aydinsakir et al. 2009).

When compost nitrogen is used in the soil, organic nitrogen is converted to readily available nitrate–nitrogen for growing plants and microbes as well. Compost usually contributes 10–15 kg of nitrogen per 10 cubic meters applied per crop. This nitrogen from the compost helps plants grow. High NH_3 emission is due to NH^{4+}–N content in the compost

(Nicholson et al. 2017). Nitrogen emissions from green and food compost are relatively more minor, and they can be used to build nitrogen reserves and improve soil condition (Nicholson et al. 2017). The study's findings have revealed around a 30% reduction of N_2O emission due to manure compost in andosol soil (Mukumbuta et al. 2017).

Phosphorus is a crucial central element for plant growth. Compost usually provides 6–8 kg of plant-available phosphorus, which is around 65–90 kg super-phosphate. Compost use increases soil phosphorus supply by releasing inorganic phosphorus through the mineralization of organic phosphorus (Wang et al. 2010).

Compost normally contains 0.8–1.2% potassium. Therefore, with 1.0% (w/w) potassium content, 10 m^3 of mature compost will contain 40 kg of potassium, providing 36 kg of plant-available potassium, equivalent to 90 kg of potassium sulfate. When applied to soil, municipal waste compost increased the amount of nitrogen, phosphorus, and potassium available for plants (Soheil et al. 2012). Compost-amended soils contain nutrients (macro and micronutrients) for plant growth (Brown and Cotton 2011).

Compost is used in orchards and vineyards. There has been a report on improved growth and yield of citrus, avocado, apples, and table and wine grapes due to the application of compost during planting (Buckerfield and Webster 2003). The application of compost and compost mulches in fruit and vineyards can be advantageous when growing conditions are less stringent; soils are more impoverished, especially when less water is available. Thus, compost can be used to increase uniformity and, hence, enrich the entire orchard and vineyard.

Adding compost to vegetable cultivation has reduced the use of chemicals and harmful pesticides by 80%. It has also minimized the detrimental effects of intensive cropping on soil performance and quality of water. Compost also increased the ability to hold cations, viz., potassium due to which there was a 20% reduction in the requirement of potassium for vegetable crops.

Compost has improved the yield of lettuce, carrot, broccoli, etc. There is a report on the increase in red pepper yield due to compost application (Park et al. 2018). The soil mixed with 25% compost showed an increase in the yield of cabbage and onion (Smith et al. 1992). Increased growth on the use of compost as fertilizer has been reported on onion and lettuce (Brechin and McDonald 1994), tomatoes (Manios and Kapetanios 1992), barley (Hountin et al. 1995), and rice (Charfen 1981).

4.3 Use of Compost in Mulching

The application of compost in mulch can be used to control weeds and improve soil fertility (Ozores-Hampton et al. 2002). Mulching is mainly found in areas abounding in greenery. The effect of compost on mulching, inhibiting weeds is achieved by minimizing the light reaching the seeds of the weeds, which often germinate in the topsoil. Before compost is applied in mulching, the place must be cleared of all established weeds. Woodchip waste from young trees and compost prepared using poultry manure as mulch have been used in an organic apple orchard to control weeds in California (Smith et al. 2000). The thickness of the compost layer required is around 10 cm to be effective against germinating weeds (Hanslin et al. 2005). Mulch consists of two layers of composts: nutrient-rich and nutrient-poor, which is the top that has particles of size 20 mm. Trees and shrubs get nutrients from fine particles, and weeds can be controlled as they cannot grow on a coarse and nutrient-poor layer of the mulch. Compost is applied as mulch in orchards, and this improves the water balance and physical properties of the soil, reduces erosion, and decreases the use of inorganic fertilizers.

4.4 Perspectives

The compost-turning machines market is estimated to reach around US$ 150 million by 2028. These developments will also increase the volume of sales of waste-management equipment. The drum type compost machine will become the dominant compost machine in the market. It is expected that there will be a tremendous demand for compost-turning machines. Agriculture will benefit due to the use of compost-turning machines. Different types of waste generated in cities can be used to make value-added products, i.e., compost by solid-state fermentation (SSF) method (Farrell and Jones 2009). These value-adding products (composts) are considered to be eco-friendly, economical, and non-toxic, and they help promote sustainable agriculture without causing any pollution and not causing any harm to plants and the environment.

4.5 Conclusion

Compost is an excellent soil conditioner and can enhance the growth of crops and many plants due to the supply of various nutrients, viz., nitrogen, phosphorus, and potassium, and minerals, and this will thus be very helpful to many farmers. The use of chemical fertilizers can be minimized, thanks to the application of compost to the soil, and it is also very cost-effective. Also, compost increases microbial diversity, which is vital for the soil ecosystem. The application of compost will be beneficial for organic farming and agricultural sustainability.

References

Adugna, G. (2016). A review on impact of compost on soil properties, water use and crop productivity. *Academic Research Journal of Agricultural Science and Research* 4: 93–104.

Anastasi, A., Varese, G.C., and Marchisio, V.F. (2005). Isolation and identification of fungal communities in compost and vermicompost. *Mycologia* 97: 33–44.

Ankidawa, B.A. and Nwodo, E. (2011). Recycling biodegradable waste using composting technique. *Journal of Environment Science and Resources Management* 4: 40–49.

Awasthi, M.K., Pandey, A.K., Khan, J., Bundela, P.S., Wong, J.C., and Selvam, A. (2014). Evaluation of thermophilic fungal consortium for organic municipal solid waste composting. *Bioresource Technology* 168: 214–221.

Aydinsakir, K., Unlu, A., Yilmaz, S., and Ari, N. (2009). The effects of compost applications on yield and quality characteristics of *Anemone coronaria* L. cv. red meron. *Acta Horticulturae* 807: 359–364.

Bellamy, P.H., Loveland, P.J., Bradley, R.I., Lark, R.M., and Kirk, G.D. (2005). Carbon losses from all soils across England and Wales 1978–2003. *Nature* 437: 245–248.

Bonilla, N., Gutierrez-Barranquero, J.A., de Vicente, A., and Cazorla, F.M. (2012). Enhancing soil quality and plant health through suppressive organic amendments. *Diversity* 4: 475–491..

Brechin, J. and McDonald, G.K. (1994). Effect of form and rate of pig manure on the growth and nutrient uptake and yield of barley (cv. Gallen). *Australian Journal of Experimental Agriculture* 34: 505–510.

Brown, S. and Cotton, M. (2011). Changes in soil properties and carbon content following compost application: Results of on-farm sampling. *Compost Science and Utilization* 19: 88–97.

Buckerfield, J. and Webster, K. (2003). Composted organics for soil rehabilitation. Report to Natural Heritage Trust.

Charfen, L. (1981). The recycling of crop residues in Taiwan. *Food and Fertilizer Technology Center Extension* 177: 1–8.

Chelah, M.B., Nozulaidi, M.N., Musliania, M.I., Khanif, Y.M., and Jahan, M.S. (2011). Composting increases BRIS soil health and sustains rice production on BRIS soil. *Soil Asia* 37: 291–295.

Farrell, M. and Jones, D.L. (2009). Critical evaluation of municipal solid waste composting and potential compost markets. *Bioresource Technology* 100: 4301–4310.

Goldstein, N. (2002). Erosion control research update. *Biocycle* 43: 78–82.

Hanslin, H.M., Saebo, A., and Bergersen, O. (2005). Estimation of oxygen concentrations in the soil gas phase beneath compost mulch by means of a simple method. *Urban Forestry and Urban Greening* 4: 37–40.

Hountin, J.A., Karam, A., and Prent, L.E. (1995). Effect of peat moss-shrimp wastes compost on the growth of barley (*Hordeum Vulgare L.*) on a loamy sand soil. *Communications in Soil Science and Plant Analysis* 26: 3275–3289.

Hurrell, J.W. (2015). Climate and climate change. Climate variability: North Atlantic and Arctic Oscillation. Encyclopedia of Atmospheric Sciences. 47–60.

Kastner, M. and Miltner, A. (2016). Application of compost for effective bioremediation of organic contaminants and pollutants in soils. *Applied Microbiology and Biotechnology* 100: 3433–3449.

Li, Z., Lu, H., Ren, L., and He, L. (2013). Chemosphere experimental and modeling approaches for food waste composting: A review. *Chemosphere* 93: 1247–1257.

Manios, V.I. and Kapetanios, E. (1992). Effects of town refuse compost as soil amendment on green tomato crop. *Biological Wastes* 19: 35–62.

Martinez-Blanco, J., Lazcano, C., Christensen, T., Munoz, P., Rieradevall, J., Moller, J., Anton, A., and Boldrin, A. (2013). Compost benefits for agriculture evaluated by life cycle assessment: A review. *Agronomy for Sustainable Development* 33: 721–732.

Miller, J.J., Beasley, B.W., Drury, C.F., Larney, F.J., and Hao, X. (2015). Influence of long-term (9 yr) composted and stockpiled feedlot manure application on selected soil physical properties of a clay loam soil in Southern Alberta. *Compost Science and Utilization* 23: 1–10.

Mukumbuta, I., Shimizu, M., and Hatano, R. (2017). Mitigating global warming potential and greenhouse gas intensities by applying composted manure in cornfield: A 3-year field study in an Andosol soil. *Agriculture* 7: 1–20.

Nicholson, F., Bhogal, A., Cardenas, L., Chadwick, D., Misselbrook, T., Rollett, A., Taylor, M., Thorman, R., and Williams, J. (2017). Nitrogen losses to the environment following food-based digestate and compost applications to agricultural land. *Environmental Pollution* 228: 504–516.

Ozores-Hampton, M., Obreza, T.A., Stoffella, P.J., and Fitzpatrick, G. (2002). Immature compost suppresses weed growth under greenhouse conditions. *Compost Science and Utilization* 10: 105–113.

Park, Y., Lim, T., Park, J., and Lee, S. (2018). Long-term effects of chemical fertilizer and compost applications on yield of red pepper and soil chemical properties. *Korean Journal of Soil Science and Fertilization* 51: 111–118.

Powell, K.S., Burns, A., Norng, S., Granett, J., and McGourty, G. (2007). Influence of composted green waste on the population dynamics and dispersal of grapevine phylloxera *Daktulosphaira vitifoliae*. *Agriculture, Ecosystems and Environment* 119: 33–38.

Smith, M.W., Carroll, B.L., and Cheary, B.S. (2000). Mulch improves pecan tree growth during orchard establishment. *Horticultural Science* 35: 192–195.

Smith, S.R., Hall, J.E., and Hadley, P. (1992). Composting sewage sludge wastes in relation to their suitability for use as fertilizer materials for vegetable crop production. *Acta Horticulturae* 302: 202–215.

Soheil, R., Hossien, M.H., Gholamreza, S., Leila, H., Mozhdeh, J., and Hassan, E. (2012). Effects of composted municipal waste and its leachate on some soil chemical properties and corn plant responses. *International Journal of Agricultural Research and Reviews* 2: 801–814.

Taiwo, A.M. (2011). Composting is a sustainable waste management technique in developing countries. *Journal of Environmental Science and Technology* 4: 93–102.

Tong, J., Sun, X., Li, S., Qu, B., and Wan, L. (2018). Reutilization of green waste as compost for soil improvement in the afforested land of the Beijing plain. *Sustainability* 10: 1–17.

Varma, V.S. and Kalamdhad, A.S. (2014). Effects of leachate during vegetable waste composting using rotary drum composter. *Environmental Engineering Research* 19: 67–73.

Wang, D., Shi, Q., Wang, X., Wei, M., Hu, J., Liu, J., and Yang, F. (2010). Influence of cow manure vermicompost on the growth, metabolite contents, and antioxidant activities of Chinese cabbage (*Brassica campestris* spp. *Chinensis*). *Biology and Fertility of Soils* 46: 689–696.

Wu, L. and Ma, L.W. (2000). Relationship between compost stability and extractable organic carbon. *Journal of Environmental Quality* 31: 1323–1328.

5

Metagenomics and Microbiome Engineering

Identification of Core Microbiome and Improvement of *Rhizosphere*

Bahman Fazeli-Nasab[1,2], Nafiseh Mahdinezhad[3], and Javid A. Parray[4]

[1]Research Department of Agronomy and Plant Breeding, Agricultural Research Institute, University of Zabol, Zabol, Iran
[2]Department of Biotechnology and Plant Breeding, Faculty of Agriculture, Ferdowsi University of Mashhad, Mashhad, Iran
[3]Department of Agronomy and Plant Breeding, Faculty of Agricultural, University of Zabol, Zabol, Iran
[4]Department of Environmental Science, Higher Education Department, Government Degree Eidgah, Srinagar, India

5.1 Introduction

A variety of microorganisms, including bacteria, fungi, algae, and lichens, are found in the soil. Viruses grow and reproduce in each living cell of the soil. Therefore, the soil is an appropriate habitat for microbes, bearing a very high number of each microorganism [1, 2]. There are around 10^7 (forest soil) to 10^9 (dry soil and other soil types) bacteria in one gram of surface soil [3].

Predominant soil bacterial species include *proteobacteria*, *actinobacteria*, and *actinobacteria*. Though the soil bacterial community is affected by several factors, including pH, temperature, element composition, nutrients, and soil moisture content, the presence of the members of other bacterial phyla is less common [4, 5]. Soil bacteria are usually free-living and of particular importance due to their participation in carbon and nitrogen fixation cycles and their association with higher plants [6].

Soil fungi involve a wide variety of microorganisms, including about tens of thousands of species and 170 genera that have been identified in the soil [7]. Generally, the number of fungi found in one gram of soil is around 10^6 [8]. In fertile soil, the length of the fungal filaments may vary from 10 to 100 meters per gram [6].

The term "microbiome" refers to the habitat of a variety of bacterial and fungal species with a balanced and unique coexistence forming a part of the biological characteristics of humans, animals, and plants. The microbiome is a dynamic and completely intricate ecosystem that has been a major challenge in research. Plants have their microbiome of bacteria, fungi, and other microorganisms living within or on their surfaces. Plant microbiomes or phytobiomes have been defined as microbial communities aboveground (*phyllosphere*) and underground (*rhizosphere*) and an important determinant of plant health and productivity [9, 10]. A leaf can solely host millions of bacteria of thousands of different species and contribute to the growth and photosynthesis of plants or help them survive drought and other stressful conditions. Plant root microbiome includes microbial populations within the *rhizosphere* (the part of the soil that has direct contact with the root) and the endosphere (within the root). Microorganisms that coexist with plants are key players in the

Core Microbiome: Improving Crop Quality and Productivity, First Edition. Edited by
Javid A. Parray, Nowsheen Shameem, Elsayed Fathi Abd-Allah, and Mohammad Yaseen Mir.

natural food cycle. They help plants absorb nutrients and play a significant role in improving crops, conserving endangered species, and protecting plants against heat, drought, and pathogenic organisms [10, 11].

Plant microbiomes have a variety of beneficial effects, including increasing nutrient uptake, enhancing disease management, and promoting tolerance to abiotic stress. For example, the microbiome plays an important role in disease-inhibitory soils where the microbial community prevents the formation or persistence of the pathogen. In a phosphorus-deficient environment, *mycorrhizal* microbiomes can amplify soil potassium solution by releasing organic acids, such as oxalate [11]. Moreover, studies have shown that the enhancement of *rhizosphere* microbial diversity is directly related to increased plant organ fitness. Improving the potential of multiple reciprocal microbial species therefore results in greater adaptability and stability of plants against disorders [12]. However, other studies have shown that reducing microbial diversity does not necessarily reduce ecosystem performance, as other microbes can occupy the environment without affecting plant productivity [13]. In general, plant growth-promoting bacteria (PGPR) easily colonize the root *rhizosphere*, establishing a free and live relationship with host plants. Such interactions usually lead to increased productivity and decreased biotic and abiotic stresses through diverse mechanisms. It has been reported that ACC deaminase-containing bacterial strains can partly decrease the negative effects of ethylene-induced stress (salinity and cadmium are the most potent inducers of ethylene biosynthesis) in plants [14].

The complex challenge of linking microbiome effects to taxonomic changes has generated the functional metagenomics concept, which describes a microbiome based on its characteristics rather than taxonomy [13]. This concept of functional metagenomics is closely related to the hologenome theory of evolution, which describes an individual as the combined genome of the host and microbial communities [15]. The key point is the application of the abundance and assembly of the microbiome community in improving crop yield and understanding the efficiency of the microbiome under different conditions [11].

Microbial abundance is affected by a variety of other factors, including soil property (structure, texture, pH, nutrient content, etc.), which is a better determining factor than geographical scale [16]. The effect of seasonal factors on the microbial community of the *phyllosphere* in beans, soybeans, and canola has been shown, as well [17]. Recently, dust's ability in carrying and transporting large amounts of microorganisms from around the world has been reported, a process that may change microbial composition from region to region [1, 18–20]. In a study on the microbial community and predominant microbial species in the dust storm of the northern regions of Sistan and Balochistan, the largest community of aerobic and anaerobic bacteria (1875000 CFU/g and 156667 CFU/g, respectively) was reported in the city of Hirmand (Figure 5.1). The highest population of aerobic and anaerobic fungi was observed in Zabol city (833 CFU/g) and Zahak city (2167 CFU/g), respectively (Figure 5.2). The most common bacterial species investigated in this study is *Bacillus sp.*, followed by *Streptomyces pactum*. The most common fungal species is *Penicillium sp.*, followed by *Aspergillus*. The results have shown a high and diverse population of various microbial species, especially pathogenic fungi in the dust storm of this region [1, 2, 21].

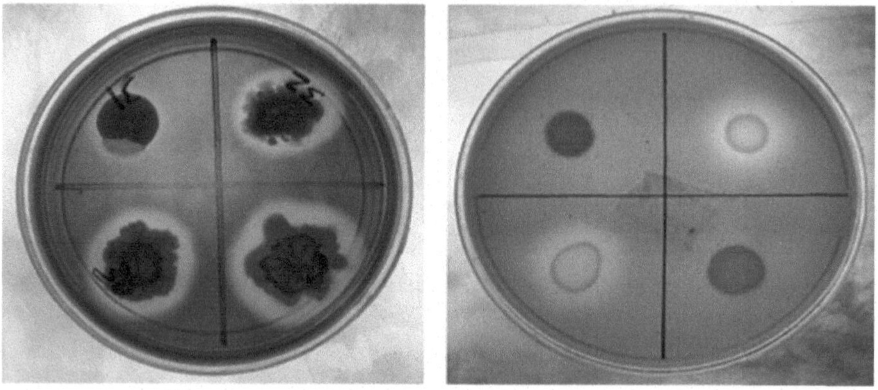

Figure 5.1 Sample of cultivated bacteria (the right side related to phosphate-solubilizing bacteria and left side related to Siderophore-producing bacteria) (Photo: Dr. E. Shirmohammadi).

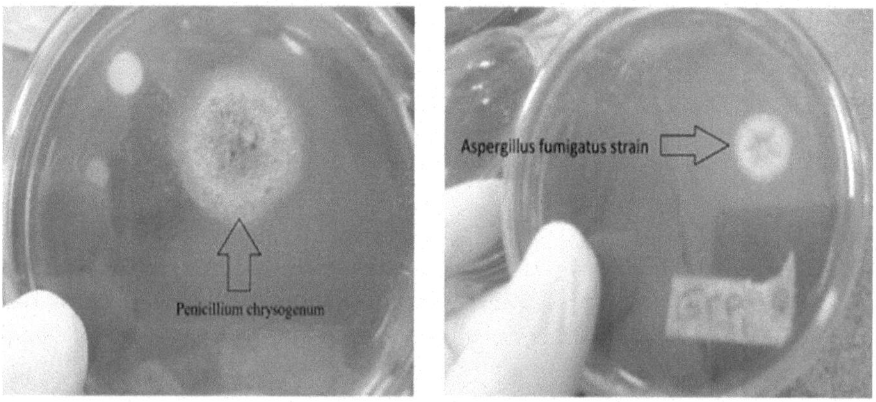

Figure 5.2 Clonic specimens of cultivated fungis (the right side related to Penicillim chrysogenum and the left side related to Aspergillu fumigatus).

Identifying the population composition of microorganisms in each region is very essential to agricultural science, as it can help reduce crop dependence on harmful chemical fertilizers. On the other hand, as the world's population grows, agriculture plays a very significant role in providing human food [22, 23].

Since the Green Revolution, the production of crops around the world has relied on using chemical fertilizers and pesticides [24]. Nowadays, the negative effects of such products on both the environment and human health are well known [25, 26]. As these chemical compounds are applied on a large scale while their oscillation efficiency is low in plants, they cause extensive environmental pollution [27]. Therefore, besides the achievement of optimal plant performance, agricultural systems require significant changes to reduce the use of these compounds. In recent decades, specific bacteria-based biofertilizers with plant growth-promoting properties have been introduced as an alternative approach to reduce the use of chemical compounds and to improve plant nutrient utilization [28].

New fertilizers have a number of benefits for their host, including increasing nutrient uptake, inhibiting pathogen attack, and promoting plant growth under stressful conditions [29]. After application, they should interact with the microbiome (the total microbes of an area at a given time) of the *rhizosphere* (the soil completely attached to the plant root), *endophytes* (microbes found within roots), and *rhizoplane* (microbes found on the root surface) to have successful colonization and provide benefits to the plant [9, 30–32]. Moreover, bacterial strains, if used in conditions similar to their natural environment, can colonize around or inside the root more effectively, affecting plant growth. Therefore, besides developing the methods of bacterial application, it is required to have a good understanding of the microbiome diversity of different plant roots in different environments. On the other hand, studies have shown that bacterial communities, especially at the species level, change from one genotype to another [33, 34]. Therefore, careful determination of the microbiome of different plant *rhizospheres* requires further assessment. To achieve this, we first need to figure out the complexity and diversity of the *rhizosphere* and endophytic indigenous microbial community of crops [35, 36].

According to conventional laboratory methods, identification of a major part of microbial populations remains unknown as less than 1% of bacterial populations in arid environments can be isolated in pure culture media. Therefore, the analysis of complex plant-associated microbiomes has technical limitations. It is now possible to accurately assess plant-associated microbial communities using high-performance sequencing methods. So, over the past few years, the microbiomes of the *rhizosphere*, *endophytes*, and *rhizoplane* have been studied in several plants including *Arabidopsis taliana* and *Hordeum vulgare* [37, 38, 39, 40]. Furthermore, the microbiota (bacterial genes) of *Glycine max rhizosphere* in Amazon fields and *Zea mays* in US fields have been investigated in several studies [41, 42].

Microorganisms make up two-thirds of Earth's biodiversity, and in many environments, 99% of microorganisms cannot be cultured by standard methods. On the other hand, understanding genetic variation, population structure, and environmental patterns of most organisms requires culture-independent methods [43]. As a result, metagenomics and next-generation DNA sequencing (NGS) methods are used to identify microbial populations in different ecosystems, independent of cultivation methods [44, 45]. As understanding the demographic structure of root-associated bacteria is a key prerequisite to determine the role of each individual of the population, metagenomics can provide imperative answers to this major uncultivable part [44].

Studying the effect of developmental stages on the *rhizosphere* microbial community structure at the vegetative and reproductive growth stages of non-rotating dryland wheat crop using the 16S rRNA gene sequencing method indicated different distribution patterns in these two stages among members of *proteobacteria*, *bacteroidetes*, *actinobacteria*, and *cyanobacteria* and certain genera of these phyla. It was concluded that plants can select a specific bacterial community to suit their developmental stage, performing specific desirable actions at that stage to maintain plant health [44].

About 96% of the sequences in the vegetative growth stage of some plants, such as wheat, were related to the members of *proteobacteria*, *bacteroidetes*, *actinobacteria*, *verrucomicrobia*, *firmicutes*, and *chloroflexi* phyla, respectively. However, during the reproductive stage, the members of *proteobacteria*, *cyanobacteria*, *bacteroidetes*, *actinobacteria*, *acidobacteria*, and *firmicutes* phyla showed the highest frequency, respectively [44], because the growth stage is a very important factor in the formation of plant microbiome [46, 47]. As the growth stage is a major factor in the formation of plant microbiome, identification of root-associated microbiomes requires information on the complexity and diversity of the *rhizosphere* and endophytic indigenous microbial community of crops. However, there is very limited information on root microbiome, requiring further studies on *rhizosphere* microbiome of commercial crops and non-model plants.

So far, two main methods to analyze microbial communities using NGS are available involving shotgun metagenomics sequencing and 16S rRNA gene sequencing. Shotgun metagenomics sequencing includes identification of the bacterial DNA sequence isolated from the whole microbial community. The reads obtained in this way are analyzed using genomics databases as the reference and each is finally assigned a specific taxon [48]. Determination of 16S rRNA gene sequence is based on the PCR, which allows the amplification of rDNA from a large number of species using degenerated primers. Differences in less conserved areas of the 16S rRNA gene make it possible to identify specific taxa based on rDNA sequence databases information [49].

In general, it is believed that 16S rRNA gene sequencing is a great and well-described method that can provide adequate knowledge on the composition of microbial communities using a comparatively small number of sequences (about 200000). However, the main limitation of this method is that the groups are assigned based on the sequence of only one single region of the bacterial genome. Besides, the selection of the primers to amplify rDNA is very vital in this method, as the results of some studies have revealed that some primers show bias for certain groups [50]. Metagenomics, on the other hand, requires higher coverage (10–30 million reads) and more complexity in data analysis. However, by collecting sequence information of a wide range of genomic regions, shotgun metagenomic sequencing allows for more accurate identification at the species level and ultimately provides a more detailed description of microbial communities.

In recent years, the application of NGS methods to analyze complex microbial communities has increased dramatically. The reasons include the continuous reduction in costs and the greater acceptance of the ability of the NGS method in a more comprehensive specification of microbial communities than traditional cultivation-based methods. Such dramatic technological developments have been accompanied by significant growth in several implementation and preparation variables, including sampling, storage, DNA extraction, primer pairs, sequencing technology, chemicals, reading length, and insert size. Considering these changes, in microbiome studies, the balance between reproducibility and comparability of the results obtained using different methods is essential to achieve sufficient accuracy.

5.2 Definitions of Genome, Metagenome, Genomics, and Metagenomics

5.2.1 Genome

In molecular biology and genetics, a genome is defined as the genetic material of an organism that can contain either DNA or RNA (in case of viruses). Genomes include genes (coding regions), intergenic regions (non-coding regions of DNA), and the genetic material of mitochondria and chloroplasts.

A genome is a complete set of the DNA of an organism, including all its genes. Compared with genetics, which refers to the study of individual genes and their role in heredity, genomics is aimed to collectively describe and determine the genes that drive the production of proteins using enzymes and messenger RNA molecules. Proteins, in turn, are responsible for the structure of the body, such as organs and tissues, controlling chemical reactions and transmitting intercellular signals.

5.2.2 Genomics

It is defined as a comprehensive study of the function and interaction of the complete gene set of an organism and its products. Specification of the complete genome sequence of organisms is one of the goals of genomics, which is useful for biological studies. *Hemophilus influenze* was the first microbial genome to be fully sequenced in 2005, with whole microbial genome sequencing expanding rapidly. Genomics methods depend on the preparation of pure culture, and as a result, these methods have limitations in studying non-cultivable organisms [51]. In general, genomics is an interdisciplinary field of science that focuses on the structure, function, evolution, mapping, and editing of genomes. Genomics also involves sequencing and analyzing genomes by using high-yield DNA and bioinformatics sequencing to collect and analyze the function and structure of all genomes. Advances in genomics have triggered a revolution in systems-based discovery and research in biology to help understand even the most complex biological systems, such as the brain.

5.2.3 Metagenome

It is defined as the whole genetic material in an environmental sample, including the genomes of many single organisms.

5.2.4 Metagenomics

In the late 1980s, it was shown that it is possible to study the environment of non-cultivable microorganisms by direct analysis and sequencing of rRNA genes. This discovery, along with modifications of DNA isolation and sequencing techniques, promoted genomics studies on microbial communities and invented the term "metagenomics" [52]. "Meta" is a Greek word meaning *beyond*. The term "metagenomics" was first used in 1998 by Handelson et al. [53].

Simply, metagenomics is a new scientific discipline that allows the genomic study of non-cultivable microorganisms obtained from the environmental samples. Metagenomics is also known as environmental genomics, eco-genomics, and community genomics. While traditional microbiology, microbial genome sequencing, and genomic rely on cultured colonies, sequencing of primary peripheral genes cloned with specific genes (often the 16S rRNA gene) creates an index of diversity in the natural sample. This revealed that the vast majority of microbial biodiversity has been destroyed by culture-based methods. Metagenomics is defined as the application of new genomics techniques to the direct study of microbial populations in their natural habitats without the need for laboratory culture (Figure 5.3). Of course, metagenomics has many advantages over genomics, including its

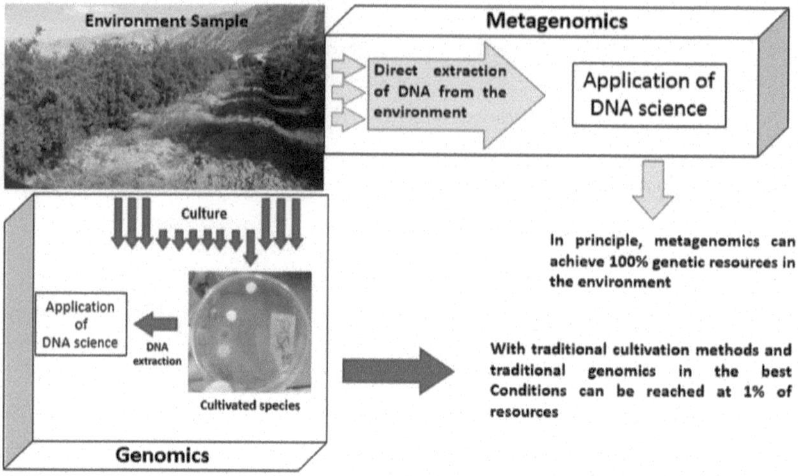

Figure 5.3 Comparison of metagenomics and genomics. Fast and inexpensive, no need for pure breeding of studied organisms, and most importantly, the analysis of about 100% of all organisms in a particular environment are the advantages of metagenomics over genomics. *Source*: Darvishi [43].

rapidity and cheapness, lack of need for pure culture, and, most importantly, analysis of about 100% of organisms in a particular environment [54]. Nowadays, metagenomics aims to assess microbial diversity in soil, deep sea, human–animal gastrointestinal tract, and to study their application in various fields, including microbiology, biotechnology, medicine, virology, and agriculture [55].

Recent advances in sequencing technologies such as shotgun metagenomics sequencing, pyro-sequencing, and computational and bioinformatics methods have led to the advancement of metagenomics in providing a brief overview of the life of non-cultivable microorganisms. Metagenomics is a new field following the merger of molecular biology and genetics. Metagenomic studies have been closely associated with gene array studies, functional genomics, gene expression studies, proteomics, metabolomics, and microscopic studies. It has provided the right insight into solving problems such as genome evolution and our inability to understand the specific position of most non-cultivable microorganisms. Metagenomics now allows microbial ecology to be studied on a much larger scale and in more detail than ever [56, 57].

Describing the microbial ecology of Earth remains one of the most important challenges in biology, with profound implications for human health, agriculture, chemistry, energy, and other fields. While estimating the number of ancient bacterial species to have reached millions, we have so far recorded only a very small fraction of the world's microbial diversity. However, our understanding of microbial communities has improved dramatically through the application of high-throughput DNA sequencing technologies [58–60].

rRNA sequencing, especially small subunits (SSUs), has been widely used for more than 30 years to investigate the structure of the microbial community, despite the limitations imposed by DNA sequencing technologies. This was the only method available for many years that, despite the difficulty, cloned each desired gene, linking the results together. Metagenomics allows the sequencing of millions of reads in a single run, albeit at a lower cost as compared with other sequencing technologies. Therefore, the application of NGS methods in metagenomics significantly increased the amount of data that can be generated from non-cultivable samples, providing a basic approach for metagenomics, microbial ecology, and soil microbiome projects [61].

5.3 Basis of Metagenomics

Metagenomics is a method in which microbial diversity in a given environment is studied without culturing microorganisms [62] (Figure 5.4). Conversely, in genomics, culturing is performed for a specific species or organism [63].

Metagenomics is the study of genetic material obtained directly from environmental samples. This broad discipline may be called environmental genomics, economist, or community genomics. Traditional microbiology and sequencing the microbial genome are based on the culture of colonies and sequencing clone-specific genes (usually the 16S rRNA gene) to obtain a diverse appearance in the natural sample [63, 64].

Different studies have shown that the vast majority of microbial biodiversity is lost by cultivation-based methods. Recent studies have used PCR-directed sequencing or shotgun sequencing to obtain largely unbiased samples of all genes from all members of a sampled community. Because of its ability to reveal the unknown diversity of microscopic life, metagenomics provides a powerful tool to inspect the microbial world, which has the potential to change our understanding of the entire living world. With lowering DNA sequencing cost, metagenomics allows the study of microbial ecology at a much higher and more detailed level than before [29, 64].

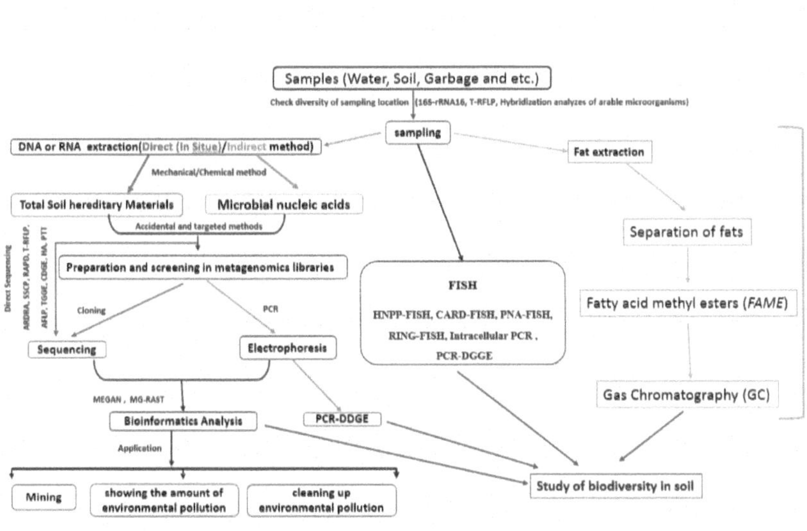

Figure 5.4 Metagenomics analysis in a microbial environment based on nucleic acids.

This technology is a multi-step process that relies on four major steps: isolation of the genetic material of all living microscopic organisms in the study environment, transfer of genetic material to vectors, construction of a genomics library, and analysis of genetic material in the metagenomics library (Figures 5.5–5.8). Recent advances in sequencing technologies, such as shotguns, pyroscanning, computational methods, and bioinformatics, have led to the advancement of metagenomics. The exploitation extent of information provided by metagenomics technology is such that it can be used to enhance the interaction between human society and the living environment [62].

5.4 Steps Involved in the Metagenomics Study

5.4.1 Sampling

The most important factor in the assessment of a particular population is the target that needs to be considered. Therefore, to assess diversity in the initial steps of culture-based research projects, changeability and the opportunity to perform different technologies with the highest efficiency are required. Accordingly, by using culture-based methods or molecular probing, a preliminary study was carried out to identify the sampling area and the correct sampling method in metagenomics [65]. Soils change from one place to another at a micrometer distance due to the physical heterogeneity of minerals and biomaterials. But usually, surface areas have more cellular diversity than the underlying layers. Moreover, depending on the type of population being studied, sampling time in metagenomics analysis should be selected in such a way that the greatest microbial

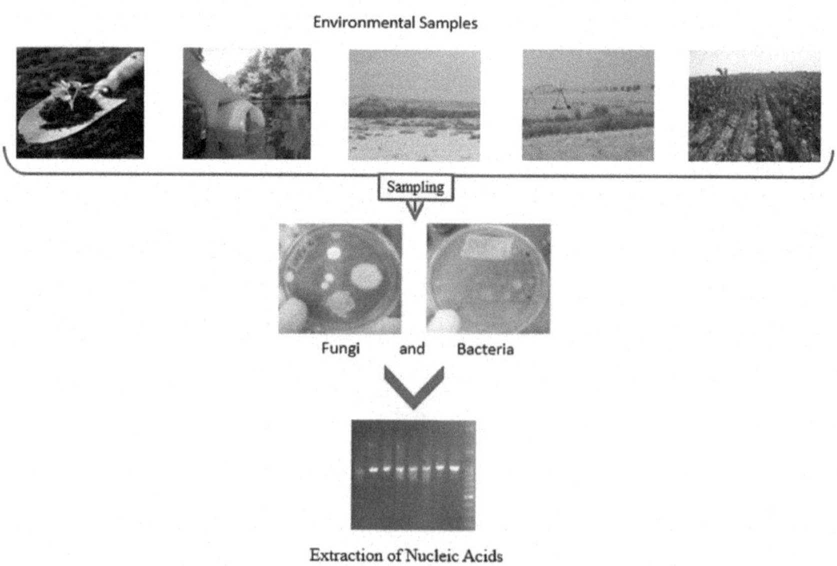

Figure 5.5 Steps involved in metagenomics (sampling and extraction of nucleic acids from environmental samples).

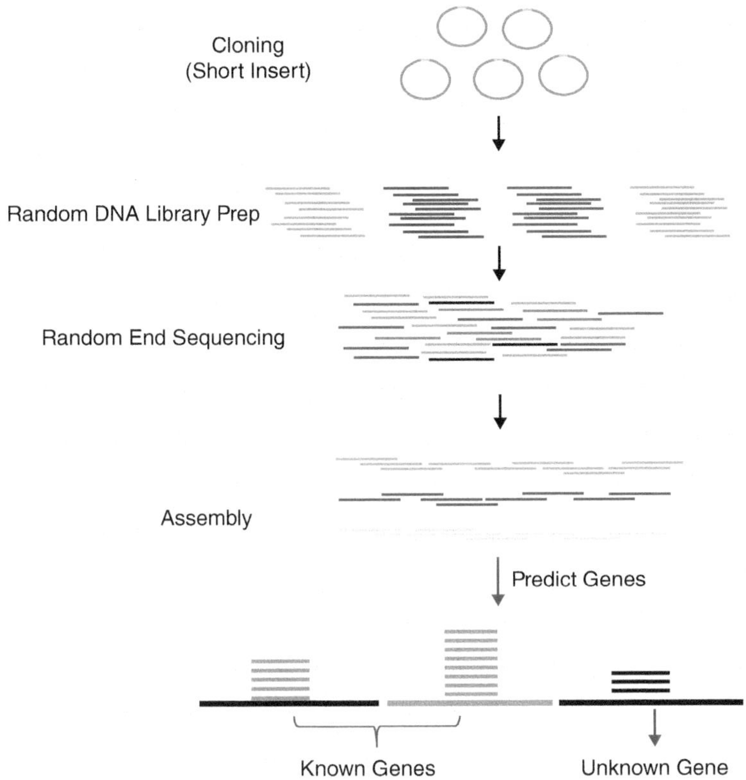

Figure 5.6 Steps involved in metagenomics (making a metagenomics library in random ways).

diversity is obtained [66]. Also, when population diversity or microbial community function is being considered, soil sampling and its volume are very important and should be carefully determined depending on the objectives of the study (Figure 5.4). In metagenomics, samples from any environment, such as water, soil, and animal digestive system, can be used to extract nucleic acids (Figures 5.4 and 5.5). Three categories of environments are available, which include highly diverse environments, such as soil and seawater, natural environments for target genes/enriched biocatalysts, as well as extreme environments [67].

Microbial populations are usually a combination of archaea, bacteria, viruses, and protozoa that require particular methods to extract their nucleic acids due to the diversity of their structure and cell wall. Although a variety of commercial kits are available to isolate nucleic acids from the environmental samples [68], no protocol for extracting a single DNA (or RNA) is available due to the physicochemical diversity of the environmental samples. Therefore, it is imperative to select the right method that specifies a particular genetic sequence or physiological function. General criteria should often be considered to retain diversity, avoid DNA sharing, and prevent contamination with certain substances, such as polyphenols [69].

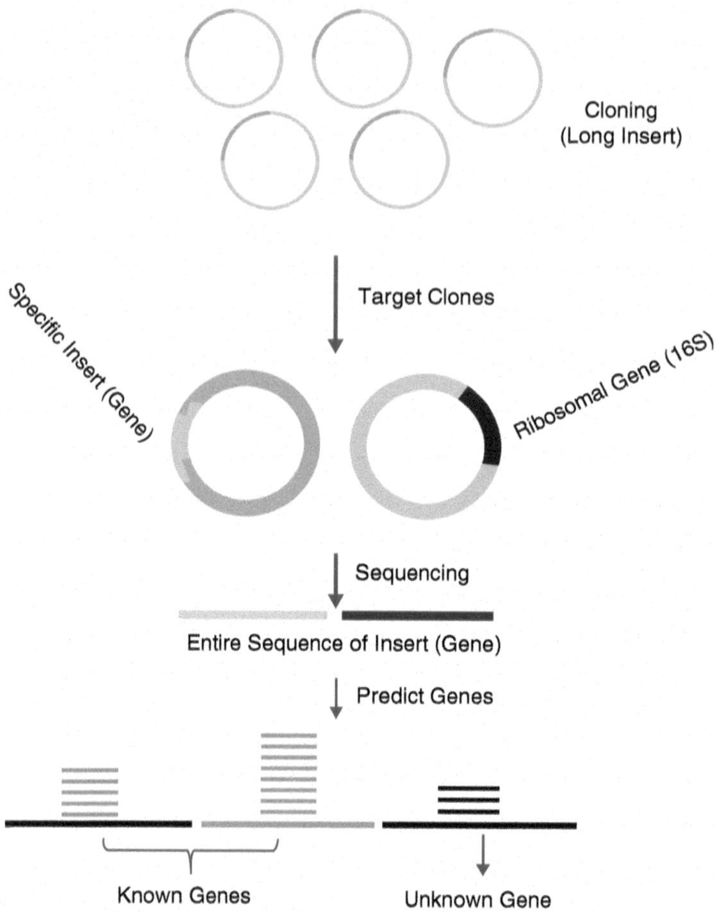

Figure 5.7 Steps involved in metagenomics (making a metagenomics library in targeted ways).

5.4.2 DNA Extraction and Purification

Depending on the goals specified as follows, the methods of sampling from the environment, DNA extraction and purification, library construction, and screening are different. Generally, to fully cover the microbial population, it is necessary to isolate and clone a large amount of DNA from a small sample. Therefore, it is assumed that all species of an environment have the same frequency. However, computational studies on microbial diversity have shown that the number of most microorganisms among other populations is small [70].

The quality and accuracy of the information obtained from the metagenomics study of each population depend only on the quality of the DNA extraction process. Meanwhile, the general question is what methods should be used to extract DNA directly from different soil types so that the quality of DNA assembly is the same for all members of other complex samples? For example, the cells of different species differ in their ability to break down under different conditions, and even some members of the same species have different

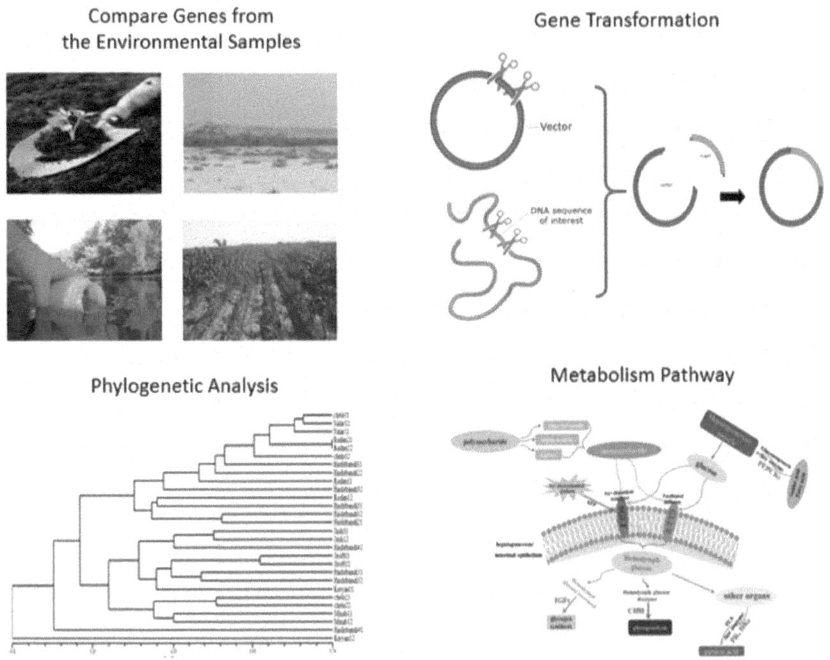

Figure 5.8 Steps involved in metagenomics. *Source*: Metagenomics Library Analysis [52]

denaturation ability in different physiological states. In addition, the conditions necessary for the breakdown of resistant cells in a population may severely denature other group members. Another aspect that is being considered at present is the development of methods that differentiate between the DNA of living and non-living cells, and this can have an outstanding effect on designing the metabolic capacity of a microbial population. One of these approaches is the use of environmental RNA analysis. The first step is to choose the cell to break down a method that depends on the type of organism's biological structure, the size of the required fragments, and the properties of the environmental matrix. However, guidelines that are compatible with most environmental organisms usually lead to the greatest possible environmental genetic variation [66].

Of course, achieving the right environmental samples genome for metagenomic studies poses a number of challenges, as it is difficult to isolate the environmental genome from complex environmental matrixes, such as sludge, wastewater, sediment, or soil. Depending on whether the purpose of the study is sequencing, PCR amplification, or metagenomics library construction, the required fragment and its extraction method will vary. In general, there are two theories for extracting nucleic acids from environmental samples [66] (Figure 5.4): (1) direct method in which the cells are lysed in the soil sample and then nucleic acids are extracted and (2) indirect method in which, the cells are first isolated from the soil, enriched, and then lysed to obtain nucleic acids [63].

The type of DNA extraction method used in metagenomics studies has a very effective role in validating and ensuring the results of studies on the richness and microbial diversity in the sample. The main strategy to obtain metagenomic DNA is to prepare the cells and directly lyse them, which involves both mechanical and chemical lysis. Studies have shown that mechanical methods provide greater diversity as compared with chemical treatments [71]. However, chemical lysis can extract DNA with a higher molecular weight, performing better in some taxa due to some of their unique biochemical characteristics, such as cell wall composition [67].

The indirect method in which the microbial cells in the sample matrix are isolated before DNA extraction minimizes the problems caused by mineral and particularly organic pollutants [72]. However, the quantity of DNA extracted by this method is diminished. Moreover, the extraction efficiency is either reduced or negative, depending on the strength of cell binding to the environmental sample matrix [73]. Hence, most microbial studies are carried out based on direct DNA extraction methods. A major advantage of the direct DNA extraction method is its higher yield as compared with the indirect method. Alternatively, because of extracting the whole study sample, the possibility of deleting a part of the sample is ruled out as it is possible through indirect methods, resulting in research deviation and subsequent complications in microbial studies [72].

In the direct extraction method, although the isolation time is reduced and genome variety is increased, a higher percentage of other materials, which are often inhibitors, are provided. For example, in soil samples with chemical and physical heterogeneity, an isolated genome is often associated with different organic compounds, which would be an inhibitor to downstream processes, such as DNA extraction and construction of the metagenomics library. These contaminants may include humic acid, phenolic polymers, polysaccharides, and nucleases. Therefore, for successful subsequent manipulations, these materials must be removed during the purification process. However, the physiological state of microbes in the direct extraction method is important in determining the quality of the extracted genome. Indirect extraction, on the other hand, overcomes these limitations, by yielding only a minimum quantity of other substances in the genome sample [66].

To achieve greater microbial populations, enrichment methods are sometimes used to increase the likelihood of nucleic acids isolation. Although enriched cultures create limited biodiversity, they are very effective in rapidly isolating large DNA fragments and cloning new genes with high biotechnological value [74].

Genome enrichment approaches can be used to target the energetic components of the microbial population [75, 76]. The SIP technique involves the use of a labeled stable-isotope substrate in which the separation is based on the slope of the density gradient, separating DNA and RNA based on their weight. For example, using the CH_3OH marker, forest soil metagenomic DNA was examined, identifying *methylotroph proteobacteria* and a new gene, mxaF (novel methanol dehydrogenase), which belongs to the *acidobacteria* taxon [77].

5.4.3 Preparation and Screening in Metagenomics Libraries

Metagenomic libraries are powerful tools to discover the diversity of non-cultivable microbes along with genetic and phylogenetic studies and to understand the practical relationship between the environment and these microbes [43, 67]. Metagenomics library construction is carried out in two random and targeted methods (Figure 5.7). The classic metagenomics library construction method involves inserting sequences of less than 10 kg into a vector. However, large vectors such as cosmid, phagemid, or BACs, which can carry DNA fragments of approximately 40–200 kb, replaced older vectors [78].

To generate genomic libraries from metagenomic samples, several factors must be considered; in shotgun sequencing studies, small fragments are often used, providing useful information in some cases, though where sequencing is performed for larger fragments, the need for creating genomic libraries is more felt. The segment inserted in the vectors should be large enough to cover the entire gene or cluster of genes needed to achieve great performance. The presence of a promoter and a ribosome junction are also essential in constructing a vector. For the expression process, *cis* sequences should be identified by *trans* factors provided by the host. A variety of vectors, including vectors based on lambda phage, cosmids, fosmids, and BACs, have been used in metagenomics studies [79].

Information obtained from metagenomic libraries can be effective in the development of practical aspects of microorganisms in industry, agriculture, medicine, and the environment. The exploitation extent of the information provided by metagenomics technology is such that it can be used to create a better interaction between human society and the living environment [64].

Screening in metagenomic libraries is carried out based on either sequence or function. The sequence-based screening method is used when screening is performed based on the sequence of a specific gene; this will be done using PCR-based techniques and probes in which the sequence of primers and probes will be designed based on the conserved gene sequences. A variety of selection methods, including growth-based selection or detectable phenotypes, are used in metagenomic libraries where performance-based screening is carried out based on the expression of an inducible gene [63, 80].

5.4.4 Sequencing Methods

In a general outline of the successive stages of sequencing, the genome is first extracted by direct or indirect methods. In the next step, the DNA is purified and fragmented to be sequenced directly or to be cloned by inserting into linear vectors and introducing into an appropriate host cell (Figure 5.4). Accordingly, by cloning DNA fragments of the environmental samples, metagenomics libraries will be constructed. Libraries are then screened to find out new sequences or specific functions. Thus, using gene expression methods and by identifying proteins and small molecules with biological activities, the expected gene functions are obtained. On the other hand, new sequences will be identified using metagenomic annotation, hybridization, and PCR with specific primers [67, 81].

One of the major objectives in the metagenomic study is to sequence environmental sample genes directly or following library construction. But metagenomic sequencing is different from traditional sequencing for many reasons. In other words, in metagenomics, special attention is paid to the sampling method and the diversity of the sampling area should be well illustrated by previous studies including 16S rRNA gene sequencing, t-RFLP fingerprinting molecular analysis, and hybridization analysis of cultivable microorganisms. The extraction of suitable nucleic acids from the sample is another challenging step in the metagenomics library construction process. On the other hand, the resulting sequences are compared with those of other microbial populations in databases to identify new genes and pathways. Meanwhile, data storage and computer analysis are essential steps in the metagenomic process, which should be continued throughout the process. Overall, the genetic process is the answer to questions, such as, "What is there?" and "What does it do?" [82].

In metagenomics studies, third-generation sequencing methods are used, and a single strand of DNA is directly sequenced during synthesis. This can eliminate errors during replication. As a result, these techniques are much simpler and cheaper than previous methods [45, 63].

5.4.5 Bioinformatics Analysis

A large database of information generated by the above systems must be processed to make them meaningful to users (Figure 5.4, Figure 5.8); therefore, one of the challenges is to develop data analysis tools. Data analysis is performed using two computational methods, including alignment and assembly. If a reference genome for the sample sequence is available (another genome at an appropriate evolutionary distance), sequence alignment analysis is mainly performed [83, 84]. For DNA samples without a reference, sequence assembly is required to perform further analysis (Figure 5.7).

Bioinformatic tools, on the other hand, should be used for phylogenetic analysis. Metagenomics analysis tools include MEGAN and the MG-RAST server (http://metagenomics.nmpdr.org) [85].

5.5 Metagenomics Applications

Though metagenomics is a nascent science, it is considered as one of the leading and applicable sciences in various fields, including basic sciences, medicine, agriculture, and industry due to its high capabilities. Metagenomics methods can be used to identify uncultivable microorganisms for mining, to show the extent of environmental pollution, to identify the positive and negative interactions of microorganisms with plants, and to use them to increase crop yields [86–89].

5.6 Metagenomics: a Way to Study Soil Biodiversity

Soil as a rich source of genetic diversity can greatly improve our knowledge of microbial ecology, enzymes, and bioactive compounds beneficial for human society. There are different types of microorganisms in soils that coexist or are independent of plant roots. However, only a small portion of soil microorganisms can be cultured. Microorganisms in natural environments may involve genes that encode biodegradation or biological pathways. In recent years, various molecular and culture-independent methods, such as metagenomics, have been developed to study many soil microbes [59, 90, 91].

Studying soil microorganisms using metagenomics methods can provide useful information on soil biodiversity and the performance of these microorganisms in the ecosystem. On the other hand, it can be provided as a soil health indicator to the protectors of biosphere reserves. Nowadays, scientists and researchers, with the help of modern science, have been able to carry out many large-scale projects that seemed impossible until a few years back. This section examines soil metagenomics strategies, new studies, and new perspectives [59, 90, 92].

Microorganisms play a key role in soil function. Different bacterial types modulate energy circulation in soils, playing an essential role in soil biodiversity. To specify soil microbial composition, most studies in the past have used 16srRNA gene sequencing or specific gene-based techniques that assess the composition of the soil microbial community [79, 93]. Although these techniques are very useful, they do not provide much information on the physiology spectrum of society. With the rapid development of molecular techniques, metagenomics is used to identify and specify the composition and function of microbial communities [94, 95]. By using metagenomics, more details on microbial population transmission, food decomposition, or soil contaminants may be obtained. Besides, the status of soil health can be determined by using this approach and the microbial profile as biological indicators that specify soil health status properly.

Nowadays, screening methods are designed to select specific genes from metagenomic libraries to specify biocatalysts as well as new bioactive molecules. Moreover, various vectors, artificial bacterial chromosomes, and biological data processing tools and databases have been developed to entirely investigate genes and biodiversity [43].

Shortly, metagenomics will become one of the most dynamic and applicable sciences, especially in the field of biological sciences, because of the use of comprehensive research methods [43]. Numerous metagenomic projects are in process, some of which are available on the server at www.ebi.ac.uk/metagenomics/projects.

To overcome the problems and limitations of cultivation methods, various DNA-based molecular and phylogenetic methods have been used to identify microbial species and their communities. These methods have dramatically increased our knowledge of microbial diversity and ecology. In general, the existing methods, which are based on 16S rRNA gene sequencing in prokaryotes, 18S rRNA gene sequencing in eukaryotes, and some other important functional genes analysis, provide extensive information on the classification and identification of the species in an environment [96–99]. However, this data often gives us little information on the genome and the practical role of various microorganisms within a community. The sequencing of ribosomal RNAs (rRNAs) and their encoding genes as evolutionary clocks have begun a new age in microbial science to explain uncultured microorganisms in the environment. More than 13,000 new prokaryotes have been identified using 16srRNA gene sequencing alone. Also, the 23S rRNA gene sequence has been suggested as an additional sequence due to its length and specific deletions and additions, which can make it a better molecular marker for phylogenetic identifications [100]. Using different vectors, such as cosmids, fosmids, or BACs; high molecular weight DNA samples; and genomic libraries, microbial ecologists have been able to receive helpful information on non-culturable microorganisms [74].

5.7 Methods Used in Metagenomics

Nowadays, methods based on direct sequence analysis of the small start unit (SSU) of a ribosomal gene (rRNA) have rapidly replaced laboratory culture methods, as in these methods, a suitable solution is provided to compare the composition and structure of microbial communities. Such methods, including ARDRA, SSCP, RAPD, T-RFLP, AFLP, TGGE, CDGE, HA, PTT, and ultimately DNA sequencing, are widely applied in nucleic acid analysis. Besides, other chemical analysis methods, such as biological markers, are highly effective; for example, we can use the analysis of phospholipid fatty acid (PLFA) to determine the characteristics of microbial communities (Figure 5.4). On the other hand, for the microscopic specification of microbial cells, the FISH technique is used. Accordingly, some changes have been made to enhance signal intensity, some of which include HNPP-FISH, CARD-FISH, PNA-FISH, RING-FISH, and intra-tissue PCR. One of the best methods for targeting genes in metagenomics is PCR-DGGE (Figure 5.4), which, in addition to its application in environmental sample studies, is a well-known method of investigating human and animal gut microbes [101, 61]. Denaturing gradient gel electrophoresis (DGGE) is the best offer in this type of studies as it is reliable, repeatable, fast, and somewhat inexpensive. It is also possible to use this method to analyze multiple samples at the same time, making it possible to track changes in the microbial populations [102].

5.7.1 Denaturing Gradient Gel Electrophoresis

One of the best ways to target a gene in metagenomics is through PCR-DGGE, in which the PCR product of the DNA that is extracted from microbial communities is achieved using specific 16S rRNA primers in major microbial groups, such as bacteria and archaea, and the use of 18S rRNA primers in eukaryotic groups, resulting in the multiplication of segments of the same size and length. DGGE is a technique used for isolating nucleic acid fragments that are of the same size but of different sequence content. Using the DGGE method, detection of microbial diseases and identification of microorganisms on a large scale can be achieved more simply and rapidly [103].

The DGGE method was developed by Stuart Fischer and Leonard Lerman at the University of New York in Albania [104]. DGGE works with a small amount of metagenomic DNA (or RNA) samples and is electrophoresed on acrylamide gel containing DNA denaturing agents (urea and formaldehyde). Researchers have shown that fragments in which mutations have occurred could be identified by comparing the position of DNA melting along the gel gradient [105].

In 1979, Gerard Muyzer was the first who used the small ribosome unit to encode genes in DGGE. Since then, this model has become a widespread and efficient method in microbial ecology [104]. The basis of such approaches is the achievement of DNA, which is extracted from microbial communities using specific 16S rRNA primers in major microbial groups, such as bacteria and archaea, and the use of 18S rRNA primers in eukaryotic groups, resulting in the multiplication of segments of the same size and length. Therefore, the aforementioned PCR product cannot be separated by agarose gel electrophoresis to form the expected polymorphism. However, sequence changes (guanine–cytosine [GC] content) between different microbial rRNAs show different properties in changing the state of DNA molecules, which is the basis of polymorphisms in microbial groups in DGGE [73, 106, 107].

Studies in recent decades have shown that DGGE for functional genes (such as genes that are effective in sulfur reduction, nitrogen fixation, and ammonium oxidation) can also provide relevant information on the microbial function and their genetic relationships with plants or animals at the same time. Furthermore, it has been possible to assess intricate microbial environments such as soil, wastewater, and industrial effluents, such as palm oil plant effluent using DGGE [107, 108].

DGGE is a special type of electrophoresis in which a constant temperature (about 60°C) and different concentration of DNA denaturing chemicals are used and PCR products are usually loaded on vertical gels containing 8% polyacrylamide (w/v), with a thickness range of 1.0–1.5 mm [102]. The acrylamide gel denaturing gradient provided in the presence of urea and formamide is gradually increasing from zero at the top of the gel to 100% at its bottom. Following electrophoresis, PCR products are separated by the gel as individual bands and stained with either ethidium bromide or silver nitrate stain. When the sequences of target genes reach their melting threshold, the two strands of DNA are separated. Adenine/thymine and cytosine/guanine base pairs are chemically differentiated by melting. The hydrogen bond between base pairs is broken by the heat of the device and increasing denaturing gradient provided by denaturing chemicals (urea and formaldehyde) [102].

In general, GC-rich DNA fragments are more stable and remain two-stranded until a higher concentration of denaturing chemicals are used along with the gel. While two-stranded DNA fragments can migrate better along with the acrylamide gel, denatured DNA molecules move at a slower rate or stop migration in the gel. Any variation in DNA sequence results in melting at different temperatures and subsequent migration of different fragments to different positions in the gel. This gives DGGE the power to differentiate and track mutations in sequences without prior knowledge. Hence, in addition to investigating genetic diversity, this method is also employed to spot mutations in closely related living organisms [107]. It should be mentioned that on knowing the position of a particular sequence of one or more microbes in acrylamide gel, it is possible to notice their presence or absence in microbial samples, taking advantage of the acquired patterns in various studies with different microbiological and environmental purposes.

5.8 Comparison of DGGE and PLFA Performance in Soil Microbial Diversity Assessment

The most popular way to determine the microbial diversity of an environment in metagenomic studies depends on the sequence 16S rRNA gene. However, this approach can lead to biased estimates due to variations in the number of 16S rRNA gene copies among different organisms, even closely related ones, and also due to PCR errors [83]. Besides, 16S rRNA gene sequences of the genome of most microbes in an environmental sample lack related classified sequences in reference databases [80]. Therefore, it is difficult to accurately identify the amount of soil microbial diversity based on the sequence of ribosomal genes.

Generally, efficient methods for soil management, agricultural production, and environmental quality assessment require techniques and methods that can identify and assess soil microbial communities. These techniques are divided into biochemical and molecular techniques [109, 110]. Each of these techniques has its advantages and disadvantages. For example, microbial culture-based methods cannot identify non-culturable microorganisms (not all microorganisms are grown in the culture medium) [111]. Moreover, the extraction of total DNA or RNA from microbial communities and subsequent assessment of the microbial community using techniques, such as DGGE, TGGT, or T-RTLP, has some disadvantages due to low repeatability and efficiency of the PCR method [112, 113]. Besides, these molecular techniques are mainly qualitative and molecular techniques and PCR-based techniques require a lot of time and money [112]. A quick way to assess the biomass and structure of the soil microbial community is to use PLFA. Molecular techniques do not provide as much information on diversity, nutritional–physiological status, and stress as PLFA patterns do [109, 112, 114].

The PLFA uses the following four steps (Figure 5.4):
Step 1: fat extraction
Step 2: fat isolation
Step 3: methylation and formation of FAME
Step 4: gas chromatography (GC)

PLFA analysis provides a fingerprint of the microbial community structure. PLFA analysis estimates the biomass and nutritional–physiological status of each group of the microbial community structure [112, 114]. This method has been widely used in soil microbial community studies [115, 116].

The resolution of the PCR method is more limited than that of the PLFA because less statistically relevant knowledge is obtained from the fingerprint patterns of the PCR methods than those of the PLFA. For example, in a typical DGGE analysis, 20–50 quantifiable bands may differ in intensity by one or two orders of magnitude (because of detection and imaging limitations), whereas in a characteristic PLFA profile, over 70 continuous PLFA peaks can be noticed in concentrations ranging more than at least three orders of magnitude. Therefore, quantifying the changes in the population density of an organism needs other supplementary analysis (such as quantitative PCR on specific groups or species) to confirm [117].

Therefore, it seems that PLFA is the best way to elucidate and estimate the changes in the structure of the microbial community. Certain PLFAs can also be used as biomarkers for particular microbial groups. In this method, a fatty acid is extracted directly from the bacterial culture medium or the environmental samples and following methylation, it is identified by a GC device. There are typically several PLFA that are used as bio-indicators of specific microbial groups. Fatty acid analysis has been used successfully to characterize the microbial community of agricultural soils, areas contaminated with heavy metals or circular compounds, acidic and basic soils, and other diverse and dissimilar environments [118, 119].

Metagenomics and microbiome engineering provide key answers to environmental questions. They also enable scientists to find the potential functions associated with specific microorganisms in multi-species societies [69, 120]. However, it has been reported that the type of DNA extraction method (Illumina Miseq 300PE and Hiseq4000, 100 bp PE) and the type of software selected for taxonomic classification can play a vital role, even showing greater resolution than the type of technology chosen in metagenomics sequencing, and the type of region selected for amplification and sequencing of the target gene [121]. Therefore, meticulous care in the optimization of DNA extraction methods, types and locus of conserved genes, types of fatty acids, and types of taxonomic classification software should be taken when metagenomics and microbiome engineering are used.

5.9 Factors Affecting Soil Microbiome

As our understanding of the importance of soil as an important biosphere component increases, much research is being done to assess the quality of soil resources. Soil not only plays a significant role in food production and clothing, but it also affects the quality of the environment. Therefore, quality assessment and soil quality definitions are of particular importance. On the other hand, soil microbial organisms and activities are considered as the most dynamic soil characteristics and indicators of soil health [122].

Soil microbial organisms make up the living part of the soil, which usually makes up less than 1% of the soil's volume, while their numbers are very high. Despite occupying a small part of the soil, their performance is imperative for soil and ecosystems. Mobility and immobility of minerals and heavy metals within the soil are also the result of microbial activity. The number and activity of soil microbial organisms depend on plant growth, soil type, soil treatment type, plowing, and the climate of each area. Soil microflora is a combination of different groups of organisms, including bacteria, fungi, algae, and actinomycetes [123].

Various factors affect soil microbial communities so that metagenomic studies on wheat root-related microbes have revealed that pH and nitrogen content are associated with bacterial diversity under a variety of field conditions [124]. It was shown that increasing soil uranium levels decreases the population of bacteria, fungi, and *actinomycetes* significantly [122]. Soil salinization also diminishes the growth, reproduction, and activity of soil organisms, mainly soil microflora, in different ways [125]. However, the presence of plants in saline environments may help lessen the destructive effects of low to moderate salinity on soil microbial activity by increasing the continuous root secretions and production of young roots [126].

Though pesticides have provided many pest control services to humans by controlling weeds, plant diseases, and harmful insects, high concentrations of some pesticides can be harmful to humans, plants, microorganisms, and other organisms. On the other hand, organisms present in the shallower layers of Earth are most affected [127]. The entry of such toxins into the environment can have destructive effects on a specific group of organisms, an effect that may not have been aimed at these chemical consumers. Therefore, the conditions and methods of using these toxins should be adjusted in such a way that no damage or at least minimum damage is posed to the environment, so that the characteristics of the soil biosocial community, which are the most important indicators of health, sustainability, and fertility of the soil and ecosystem, are not endangered [128].

The toxicity of heavy metals negatively affects the biological quality of soil, plant growth, and human health. Lead is found mainly in soils contaminated with mineral wastes or resulting from agricultural and industrial activities around the world. One of the most important solutions to this problem is phytoremediation, which has been considered as a new, cost-effective, and very promising method to clean up contaminated soils [129], though the efficiency of this method can be low due to the low solubility and release of heavy metals, especially lead, in the soil, limiting their absorption by plants. Therefore, phytoremediation has been proposed using chelating agents, such as ethylenediaminetetraacetic acid (EDTA), to artificially increase the solubility of heavy metals in the soil, and, consequently, to increase their availability for plants [130, 131].

Ethylenediaminetetraacetic acid is one of the strongest metal chelators with a strong effect on metal chemical groups and their subsequent mobility, solubility, and bioavailability in the soil solution phase [132]. Studies have shown that EDTA can release lead from various soils, especially iron and manganese oxides and soil organic matter [133]. When EDTA is added to the soil, the formation of the lead–EDTA complex changes the balance of sedimentation and adsorption toward increased lead solubility in the soil due to its high stability constant ($pK_a = 17.8$). Wu et al. [134] and Nascimento [135] reported an EDTA-induced 600- and 217-fold increase in soil lead concentration, respectively.

Adding EDTA as a chelating agent to soil contaminated with heavy metals may increase heavy metals availability and subsequent absorption by plants, but it can have negative effects on the soil's biological properties, as well so that the negative effect of EDTA on the activity of soil dehydrogenase and urease enzymes has been confirmed. On the other hand, addition of citric acid has a positive effect on the activity of dehydrogenase, urease, and phosphomonoesterase enzymes in alkaline soils [136].

Metals in soil-soluble form have a more toxic effect on soil microorganisms than when they are in the form of complexes with organic compounds, or iron and manganese oxides [137]. Also, EDTA and ion-EDTA complexes show low biological, optical, and chemical degradability and high solubility and toxicity in the soil environment; they may remain in the soil for several months [138].

It has been reported that EDTA does not act specifically to form a complex with heavy metals and is exposed to multiple reactions with other cations present in the soil, thereby adversely affecting the chemical properties of the soil (such as changes in the consumption of macro- and micro-nutritional elements); therefore, it is of concern that EDTA may not be a suitable option for phytoremediation of lead-contaminated soil [139]. Therefore, nowadays, attention to low molecular weight organic acids (LMWOAs) has been increased as these organic acids have high degradability, less toxicity, and high chelating power, thereby showing less adverse effects on the environment. Among organic acids, while forming decomposable complexes with heavy metals [132], citric acid significantly increases the solubility and adsorption of metals by plants [140], thereby indirectly improving soil microbial activity and *rhizosphere* physical and chemical properties [134].

In general, soil microorganisms are the first living organisms to experience the negative effects of pollutants, and thereby their diversity and population can be used as an indicator to assess environmental pollution [141]. Soil microorganisms are the main sources of soil enzymes, and despite the relatively small quantity of soil enzymes, they play a key role in the cycle of the main soil nutrients. On the other hand, the addition of EDTA and NTA to the soil increases microbial respiration and DOC availability of heavy metals and reduces the microbial biomass carbon and the mineralization of nitrogen. Nevertheless, little is known about the effect of these chelating agents on the activity of soil enzymes. Soil enzyme activity may be a good indicator of soil quality as these properties are more dynamic and often more sensitive than the physicochemical soil properties, reflecting the environmental conditions [129].

In general, a reduction in microbial biomass specifies the relative mobility and mineralization of nutrients in the microbial buffet [142]. The carbon–nitrogen ratio of the microbial biomass indicates the soil relative ratio of fungi compared with bacteria, determining how the changes in the soil or pollutants may affect the abundance of these organisms.

Depending on the composition of the soil microbial population, this ratio usually varies between 5 and 15; a high ratio indicates a higher ratio of fungi compared with bacteria and vice versa [143]. qCO_2 or metabolic quotient is expressed as the ratio of CO_2 obtained from respiration per unit of microbial biomass carbon per unit of time [143]. Increased microbial community stress due to the entry of pollutants into the soil raises this biological coefficient. Therefore, this coefficient is a good indicator to assess the effect of environmental stresses, including pesticides on microbial biomass [143]. This physiological parameter has been suggested to assess the effect of contamination on microbial energy requirements and probably carbon loss through microbial biomass as well [142].

Soil microbial communities play a crucial role in the breakdown of organic matter, nutrients cycle, homeostasis, plant coexistence, soil fertility, and plant health [109]. It is therefore logical that the ecological balance of the structure of soil microbial communities is used as an indicator of the soil ecosystem health. Indicators used to assess soil ecosystem status should be able to reflect changes in the management of climate and stress conditions in the ecosystem. Until recently, changes in the physico-chemical properties, including pH, EC, the state of available soil elements, and soil organic carbon have been used to evaluate the response to stress conditions [144]. However, these characteristics are not sensitive enough to predict changes in the soil ecosystem in the early stages. On the other hand, the microbial characteristics of the soil respond very quickly to environmental disturbances and changes. Normally, the stress response is investigated using a microbial scale in terms of the number of microbes, the intensity of microbial respiration, and microbial activity [145]. However, these are not very sensitive indicators as there are many activities and very complex interactions within microbial communities [144].

Estimation of soil biologic indicators helps define soil quality. Enzymatic activities can be considered as indicators of soil health and fertility [122]. Enzymes are protein molecules that, when attached to a precursor, convert it into one or more products [146]. Soil enzymes are similar to enzymes in other systems in that the rate of reaction depends considerably on pH, ionic strength, temperature, and the presence or absence of inhibitory factors [147]. The most important and valuable application of soil enzymes is the assessment of the materials introduced into the soil and the relative soil health condition.

Several studies have been carried out to identify soil enzymatic changes mediated by acid rain, heavy metals, pesticides, and other chemicals used in industry and agriculture. Heavy metals have lasting inhibitory effects on soil microbial activity [122, 148]. Decreased activity of microbial organisms is a clear result of heavy metals availability, which, in turn, reduces enzymatic activities, especially those of extracellular enzymes [149].

Phosphatase enzyme is a soil enzyme that has been considered as one of the appropriate indicators to measure soil quality in various studies. In the soil, phosphorus is mainly found as an organic compound. In surface soil, the share of organic phosphorus is between 15% and 80% of total soil phosphorus content. The very low soil phosphorus concentration and the need for its recycling changes soil organic matter phosphorus into mineral phosphorus, in which the role of microbes and the secretion of phosphatase enzyme by them cannot be overlooked. Microbial phosphatase is produced mostly by organotrophy of the soil microflora [150]. As mentioned in various sources, due to the chemical similarity of uranium and phosphorus, this element can affect the soil phosphorus cycle and, ultimately, microorganisms [151].

Microorganisms are the basis of life on Earth and offer a high degree of biodiversity. This huge genetic and biological diversity can be used to increase the recognition of new genes, metabolic pathways, and their products. Identifying the genetic and biological diversity of microorganisms is an important part of scientific research. Many microorganisms cannot be grown in the laboratory, which limits the understanding of the physiology, genetics, and ecology of the microbial community. The best solution to this problem is to use a comprehensive knowledge base (and tool) called *metagenomics*, which involves cloning culturing-independent methods as well as microbiome engineering [103].

5.10 Plant Growth-Affecting Bacteria

In commercial agriculture, the unreasonable and unbalanced use of fertilizers and chemical pesticides, which destroys soil and its living organisms, has reduced soil production potential, soil fertility, and plant quality. In recent years, sustainable organic farming has been given more importance to maintain natural life, biodiversity, soil and water resources, and the environment. Microbiome technology allows for the simultaneous reduction of these environmental impacts and the sustainable increase in the quantity and quality of agricultural products.

Microbial engineering helps plants by providing nutrients, performing biological control, producing plant-like hormonal substances, reducing plant ethylene levels, making the plant resistant to various environmental stresses, including water and nutrient deficiencies, and reducing the toxicity of heavy elements. The use of bacteria (*azotobacter* and *azospirillum*) and *mycorrhiza* fungus as biofertilizers has been proposed to increase nitrogen and phosphorus uptake and thereby improve the growth of a number of crops [152, 153]. These bacteria may accumulate in the *rhizosphere*, on the root surface, or even within the intracellular space of plant roots.

Different types of *rhizobacteria* with a negative effect on plant growth and yield have been identified as deleterious *rhizobacteria*. Many of these bacteria belong to the genus *Desulfovibrio*, *Agrobacterium*, *Erwinia*, *Enterobacter*, and *Chromobacter*. The harmful effects of this group of bacteria are due to the accumulation of metabolites, such as aliphatic acids, phenolic acids, sulfuric acids, and mainly hydrocyanic acid. In contrast, those that increase plant growth and major crop product yield are called PGPR [9, 154]. These beneficial bacteria are also called yield-increasing bacteria (YIB). Using various mechanisms, they can affect plant growth fertility in a wide range of crops, including cereals, ornamental plants, vegetables, and some trees. Plant treatment with PGPR bacteria increases germination percentage, seedling strength, plant vigor, buds and roots growth, plant biomass, and seed weight. It also accelerates the flowering process and increases the yield of cereals, herbaceous plants, and fruit trees [154–156]. *Pseudomonas*, *azotobacter*, and *azospirillum* are the most important PGPRs [157].

These bacteria are able to exert various effects on the plant through various mechanisms, as for example, some directly stimulate plant growth. They do this by generating and secreting PGPRs, such as auxins, gibberellins, and cytokines, or by providing required plant nutritional elements, including phosphorus or nitrogen [158, 154].

Some other PGPRs affect plant growth indirectly or only through indirect mechanisms. Indirect effects are mainly biological control of the inhibition of the growth of pathogenic bacteria, fungi, and nematodes. These are also referred to as "biocontrol" factors [159]. Control mechanisms include competition for colonization and occupancy of suitable spaces on plant root system or competition for nutrient use, and antibiosis in relationships, secretion of hydrogen cyanide and other volatile compounds of this type, secretion of lubricating enzymes, such as chitinases and glucanases, and ultimately increasing plant systemic resistance [160–162].

5.11 How Bacteria Move Toward Plants

The chemicals secreted by the plant are involved in the movement of bacteria toward the plant. Root, bud, seed, and root hair secretions are absorbents that include sugars, amino acids, organic acids, and phenolic substances. Most bacteria make a symbiotic relationship by moving toward the roots of plants. Rhizobia form nodules and fix nitrogen for plants. Agrobacteria form gall in plants. *Azospirilla* makes a symbiotic relationship with plant root. They are present in the root *rhizosphere* and by producing plant growth stimulants, such as auxin, cytokinin, and gibberellin phytohormones, they enhance root germination, plant growth, and increase water and mineral uptake levels. *Acetobacter* and *Herbaspirillum* enter the roots through intercellular spaces at the tip of the roots and then are transmitted to the plant [163, 164]. The movement of rhizobium toward some legume family plants and the movement of *Azospirillium*, *Acetobacter*, *Azotobacter*, and *Azoarcus* toward plants of *Gramineae* family such as wheat, barley, corn, sorghum, and sugarcane have been shown. Although some bacteria settle in the *rhizosphere* of certain plants, various factors can slow or accelerate the movement of the bacterium toward the root and change the placement of the bacterium within the root.

It has been reported that mutants of *Rhizobium meliloti* that have lost their movement toward water, acids, and sugars show no movement toward the roots [165]. Bacterial movement and chemotaxis toward the roots create competition between different microbial strains, and the strains that have better chemotaxis and motility form more nodules [166, 167]. Compounds, such as isoflavones, *genistein*, and *daidzein*, in *Fabaceae* family plants, attract bacteria toward the roots [168]. Loss of rhizobium plasmid does not stop bacterial chemotaxis toward naringin and apigenin showing that, in addition to plasmid genes, other genes are involved in bacterial uptake by the root. Rhizobium O antigen (lipopolysaccharide) stimulates the production of genistein in plants [169]. In addition to studying chemotaxis in nodule-inducing bacteria, chemotaxis has been studied in other bacteria, such as the movement of *Pseudomonas aeruginosa* toward tomato or the movement of fluorescent *Pseudomonas* toward soybeans in the soil [170].

5.12 Soil Improvement to Increase Microbiome

Salinity, soil pH, drought, nutrient deficiencies, and flooding conditions will each cause changes in soil microbial communities, resulting in decreased biological indicators, such as respiration and microbial biomass with increasing salinity [109, 171]. Drought stress and nutrient deficiencies increase saturated/unsaturated fatty acids ratio, the *trans*/*cis* fatty acids ratio, and cyclopropyl fatty acid/monoenoic acid ratio [172]. Soil pH affects the composition of soil microbial communities by affecting fatty acids [119]. Under flooding conditions, some PLFA is significantly reduced as the indicator of saprophytic fungi, while some others show an increase as the indicators of gram-positive bacteria. However, it does not have a significant effect on stress indicators (cyclopropyl and the *trans*/*cis* fatty acids ratio) [173]. In addition, some environmental conditions, such as prolonged saturation (flood and oxygen-free conditions), are more harmful to fungi [174]. Therefore, agricultural management is the most important factor in improving soil microbes.

Various agricultural management practices, such as rotation, fallow, burning residues of previous crops to cultivate subsequent crops, and adding animal manure to the soil, can affect the PLFA of the membranes of soil microbes. Burning changes the prevalence of microbial populations from bacteria to fungi and actinomycetes [175, 176]. Some PLFA as the indicators of saprophytic fungi is less in cultivated lands than in fallow lands as plowing interferes with the growth of fungi. Moreover, the amount of bacterial fatty acids was also higher in fallow lands [175, 177]. Soil phospholipid fatty acid shows a highly positive correlation ($r = 0.82$) with the amount of organic carbon [109]. Also, the fungi/bacteria ratio shows a highly positive correlation with the amount of organic carbon ($r = 0.94$) [119].

Although there are a variety of mechanisms that can affect the overall diversity of bacteria and fungi in a *rhizosphere* microbiome, carbon rhizodeposition of root exudates is considered to be the largest contributor to microbial abundance and profiling. Through carbon deposition in root exudates, plants are able to manage the type and abundance of the total microbiome in the soil. In addition to concentrations, plants can also manipulate certain types of carbon in the *rhizosphere*, effectively extracting these exudates (such as malate, citrate, and oxalate) by using root exudates as the selective adsorbent chemical and affect the composition of the microbiome community. This mechanism is the product of convergence between host and reciprocal microbiota. In addition, other conditions, such as soil type, seasonality, host growth stage, and perturbation disorder (long-term water deficit, management practices, etc.), can have a profound effect on the microbiome community [11].

In general, the main factor controlling the microbial population in the soil is organic carbon because of the general availability of organic carbon limits microbial production. Organic carbon has been reported to be not only a nutrient for the growth of microorganisms but also plays a major role in regulating soil environmental conditions, including salinity [178]. The results of some studies have revealed that organic matter can enhance salinity tolerance in microbes as microbes synthesize and produce osmotic pressure regulating compounds (organic compounds such as glutamine, proline, glycine, and betaine) to combat salinity stress and to produce these compounds, they need a lot of energy, which they get from organic matter [109, 179, 180]. These microorganisms, which have evolved in saline environments, are able to overcome the destructive effects of high salt concentrations as intracellular compounds, enzymes, ribosomes, membranes, etc. are altered in different ways so that they can withstand harsh conditions [181]. With increasing organic carbon, PLFA related to bacteria, such as *actinomycetes*, and saprophytic fungi are also increased, showing a high positive correlation with all or most of the microbial groups.

Diverse microbial communities exist in all ecosystems, and the features of ecosystems are specified by the microbiomes they have. The distinctive characteristic of each ecosystem is the result of a complex network of interactions in microbial communities and between microorganisms and hosts. Furthermore, it is the response to metabolites produced by microbial communities. Disruption of microbiomes causes adverse phenotypes in hosts, resulting in disease pathogenesis, disturbing the balance of related ecosystems. Therefore, disruption of microbiomes can cause dramatic changes, including reduced soil fertility in the environment. Microbiome engineering can modify microbial compounds to improve soil quality and take advantage of it [10, 182, 183]. Microbiome engineering can be used to amend microbial structure and restore environmental balance. As a result, microbiome engineering has been used to ameliorate human health and agricultural productivity. The current importance and applications of microbiome engineering, especially in plants and soil, have been investigated [184].

Soil microbiome is mainly influenced by agricultural management practices. By using soil microbiome engineering, very diverse and balanced soil microorganisms have been made to promote soil health and plant fertility [184]. One of the activities of soil microbiome engineering in agricultural management is the implementation of organic farming. This method proposes to improve and strengthen soil microbial diversity, biological cycles, and biological activity. By decreasing the use of synthetic fertilizers and herbicides and long-term use of natural fertilizers, it improves soil richness and decreases soil uniformity and dispersion, changing the soil microbial composition [185, 186]. Another way to improve soil microbiome is land-use change, i.e. to change vegetation or practice crop rotation. A meaningful change in the composition and function of soil microbiome can take place following a change in vegetation. Other agricultural applications, such as tillage, fallow, and rotation, play a role in making structural and functional changes in soil microbiome, changing the soil yield. Since many agricultural management methods in question have been successfully implemented for decades, recent efforts to realize their effects on soil microbiota composition has made optimizing these methods for optimal crop yield by soil microbiome engineering possible [187, 188, 189, 190].

Bacteria with the desired functions can be selected and combined to create artificial microbiomes for plant microbial engineering. For example, five root-related bacteria have been shown to protect nicotina palati lesions against sudden wilting. Also, rhizobacteria have been widely used to improve the growth and function of numerous plants either combined or individually. The most important benefit of artificial microbiomes is that their microbial composition can be controlled. However, the improvement of artificial microbiomes is limited by our knowledge of the microbiome population, and only microbes that can be grown in artificial microbiomes can be used [191, 192].

One of the methods of plant disease management is mixing soil with disease-suppressing microbes. Recently, this method has been used to control various plant diseases, such as sugar beet infection, common scab of potato, and tobacco black root rot. It is also used to reduce drought stress and to increase biomass in dry conditions [193, 194].

In a study, two bacterial strains, including *Enterobacter* sp. (16i) and *Bacillus sp.* (12D6), have delayed the onset of drought stress in wheat (*Triticum estivum* subsp. *Estivumum* TAM 111) and corn seedlings (B73). Plant roots inoculated with PGPR significantly changed drought tolerance-associated morphological traits and increased water retention capacity.

It was concluded that understanding the environmental implications of the interactions between plants and microbes in a water-deficient environment could be used directly to reduce drought stress in grain products [11].

Each microbiome can potentially be used as an inoculator to improve host conditions, affecting host phenotype. As microbiomes undergo more drastic ecological changes in the environment, selecting the type of microbiome is important so that common microbiomes are not usually used for inoculation. For example, the intestinal microbiome is not used to inoculate *rhizospheres*. Accordingly, selection can target the host phenotype for indirect differentiation between microbiomes that affect the host phenotype in different ways. In essence, the host is used as an explorer to evaluate the microbiome's performance that affects the host's fitness. Using microbiome engineering, new and improved microbiome functions can be selected without knowing the composition of the microbiome or their synergetic interaction. Synergy can also be due to host–microbial interactions. For example, in the symbiosis between root nodules and nitrogen-fixing bacteria, nitrogen fixation occurs only because of incompatible pairs of plant and bacterial genotypes. Besides, none of the partners show nitrogen fixation in the absence of another [10].

Genetically, microbiome engineering is the synthetic selection happening directly on the host (as the host phenotype is measured directly). Microbiomes, on the other hand, are indirectly elected by their effects on the host phenotype. Such an indirect choice can be more efficient and cost-effective than direct selection. Indirect selection is especially useful when the selected attribute is indirectly difficult or costly. Moreover, host-based selection can be designed to simplify analysis. For example, the host genotype can be kept constant so that artificial selection can only focus on microbiomes [10].

Microbiome engineering can be an option for designing microbiome utilization tools and to achieve beneficial results in agriculture and the food industry. This technology suggests that we do not create the microbiome, but manipulate the existing microbiome, moving beyond the present technologies that use antibiotics and selective probiotics. Combined biology plays a very significant role in guiding the new but predictable probiotic products performance, which, in addition to plants and soil, manipulates the microbiome and its activities in a predictable way [10].

Microbiome engineering can alter microbiomes through both ecological and evolutionary processes. Environmental processes involve modifications in the diversity and uniformity of society, the abundance of relative species, and the structure of host–microbial and microbial–microbial interaction networks. Evolutionary processes involve the extinction of a microbial species in the habitat, changes in allele frequency, mutations, and horizontal gene transfer, which rebuild the structure of the microbial genome. While vertical gene transfer can help stabilize a microbial community over time, rapid circulation in the microbial community can limit the productivity of host–microbiome engineering. In particular, accidental loss of microbial genotypes or new genotypes can destroy the microbial properties that are formed by artificial selection [10].

By making use of microbiome knowledge, microbiome engineering has proved to be successful in treating disease and improving agricultural productivity. Moreover, microbiome engineering has made it possible to generate certain types of biological products with unique properties through the metabolic activities of each of the microbiome bacteria. Advances in microbiome engineering could potentially benefit the emergence of synthetic biology in recent decades. With synthetic biology, intelligent microbes with new properties can be engineered. However, microbial engineering is still in its initial way, showing great potential for unlocking [184].

Although phages have been used in agriculture and medicine for antibacterial purposes, they can be designed to change host characteristics and acquire new functions. Thus, phage manipulation, such as using engineered phages to kill pathogenic microbes, could be a potential way to modify the bacterial community and affect microbial engineering [195–198].

5.13 Conclusion

Soil as a component of natural ecosystems, similar to other habitats, has not been considered for its biodiversity. On the other hand, if soil is considered a biological reservoir, and national and international programs for protection, management, and rational operation are prepared and implemented, knowing soil biodiversity can prevent its intentional or unintended destruction by humans and other factors. Organic carbon is also a major factor in controlling soil microbial communities as the availability of organic carbon generally affects microbial production. Therefore, to maintain soil microbial population, it is recommended to check for soil carbon and organic materials, and supplementing soil if the soil is deficient in organic materials.

Besides controlling organic carbon, other soil microbiome engineering functions in agricultural management include organic farming and vegetation change or crop rotation. These methods aim to promote and reinforce soil microbial diversity, biological cycles, and biological activity. Reducing the use of chemical fertilizers and herbicides and the long-term use of organic fertilizers enhances soil richness and reduces soil uniformity and dispersion, changing the soil microbial structure.

Therefore, optimal care in the optimization of DNA extraction methods, types and locii of conserved genes, types of fatty acids, and types of taxonomic classification software should be considered when metagenomics and microbiome engineering are used. Moreover, various methods can be used to investigate soil biodiversity, but DGGE and PLFA methods have proved to be the most important and most efficient among the ones reported.

Abbreviations

AFLPs: amplified fragment length polymorphisms
ARDRA: amplified ribosomal DNA restriction analysis
BACs: bacterial artificial chromosomes
C/N: carbon to nitrogen
DGGE: denaturing gradient gel electrophoresis
DOC: dissolved organic carbon
DRB: deleterious *rhizobacteria*
FAME: fatty acid methyl ester

FISH: fluorescence *in-situ* hybridization
HCN: hydrogen cyanide
LMWOAs: low molecular weight organic acids
NGS: next-generation sequencing
PCR: polymerase chain reaction
PFGE: pulsed field gel electrophoresis
PGPR: plant growth-promoting *rhizobacteria*
PLFA: phospholipid fatty acids
RAPDs: random amplified polymorphic DNAs
rRNA: ribosomal RNA gene
SIP: stable-isotope probing
SSCP: single-strand conformation polymorphism
SSUs: small subunits
TGGE: temperature gradient gel electrophoresis
T-RFLP: terminal restriction fragment length polymorphism
YIB: yield-increasing bacteria

References

1 Abbasi-Moghadam, J., Shahriari, A., and Fazeli-Nasab, B. Investigation of bacteria and fungi populations associated with airborne dust during "wind of 120 days" blowing in the urban areas of Sistan plain. 15th Iranian Soil Science Congress, Isfahan University of Technology, Isfahan, Iran, Congress COI: SSCI15, Article COI: SSCI15_687; 2017.

2 Ali-Soufi, M., Shahriari, A., Shirmohammadi, E., and Fazeli-Nasab, B. (2019). Investigation of Dust Microbial Community and Identification of its Dominance Species in Northern Regions of Sistan and Baluchestan Province. *Journal of Water and Soil Science (Science and Technology of Agriculture and Natural Resources)* 23 (1): 309–320.

3 Whitman, W.B., Coleman, D.C., and Wiebe, W.J. (1998). Prokaryotes: The unseen majority. *Proceedings of the National Academy of Sciences* 95 (12): 6578–6583.

4 Guieu, C., M-d, L.-P., Ridame, C., and Thomas, C. (2002). Chemical characterization of the Saharan dust end-member: Some biogeochemical implications for the western Mediterranean Sea. *Journal of Geophysical Research: Atmospheres* 107: D15.

5 Janssen, P.H. (2006). Identifying the dominant soil bacterial taxa in libraries of 16S rRNA and 16S rRNA genes. *Applied and Environmental Microbiology* 72 (3): 1719–1728.

6 Fallah Nosratabad, A.R. (2013). Evaluation of relationships between soil properties and total bacteria and fungi in soils of Guilan. *Journal of Soil Management and Sustainable* 2 (2): 49–68.

7 Shahoei, S.S. (2006). The nature and characteristics of soils (Translation). Kurdistan University Press, 900.

8 Paul, E.A. (2014). *Soil Microbiology, Ecology and Biochemistry*. Academic press.

9 Berendsen, R.L., Pieterse, C.M., and Bakker, P.A. (2012). The rhizosphere microbiome and plant health. *Trends in Plant Science* 17 (8): 478–486. https://doi.org/10.1016/j.tplants.2012.04.001

10 Mueller, U.G. and Sachs, J.L. (2015). Engineering microbiomes to improve plant and animal health. *Trends in Microbiology* 23 (10): 606–617. https://doi.org/10.1016/j.tim.2015.07.009

11 Jochum, J.M.D. (2019). Enhanced drought tolerance through plant growth promoting rhizobacteria and microbiome engineering applications, Doctoral dissertation, Texas A & M University. Available electronically from https://hdl.handle.net/1969.1/184937.

12 Lau, J.A. and Lennon, J.T. (2012). Rapid responses of soil microorganisms improve plant fitness in novel environments. *Proceedings of the National Academy of Sciences* 109 (35): 14058–14062.

13 Nannipieri, P., Ascher, J., Ceccherini, M., Landi, L., Pietramellara, G., and Renella, G. (2017). Microbial diversity and soil functions. *European Journal of Soil Science* 68 (1): 12–26. https://doi.org/10.1111/ejss.4_12398

14 Fazeli-Nasab, B. and Sayyed, R. (2019). *Plant Growth-Promoting Rhizobacteria and Salinity Stress: A Journey into the Soil. Plant Growth Promoting Rhizobacteria for Sustainable Stress Management*, 21–34. Springer. https://doi.org/10.1007/978-981-13-6536-2_2

15 Zilber-Rosenberg, I. and Rosenberg, E. (2008). Role of microorganisms in the evolution of animals and plants: The hologenome theory of evolution. *FEMS Microbiology Reviews* 32 (5): 723–735. https://doi.org/10.1111/j.1574-6976.2008.00123.x

16 Lareen, A., Burton, F., and Schäfer, P. (2016). Plant root-microbe communication in shaping root microbiomes. *Plant Molecular Biology* 90 (6): 575–587. https://doi.org/10.1007/s11103-015-0417-8

17 Copeland, J.K., Yuan, L., Layeghifard, M., Wang, P.W., and Guttman, D.S. (2015). Seasonal community succession of the phyllosphere microbiome. *Molecular Plant-microbe Interactions* 28 (3): 274–285. https://doi.org/10.1094/MPMI-10-14-0331-FI

18 Ali-Soufi, M., Shahriari, A., Shirmohammadi, E., and Fazeli-Nasab, B. Investigation of biological properties and microorganism identification in susceptible areas to wind erosion in Hamoun wetlands. Conference on the Hamoon International Wetland Reclamation Solutions and Approaches, University of Zabol, Zabol, Iran, Article COI: HAMOUN01_022; Conference COI: HAMOUN01; 2016.

19 Ali-Soufi, M., Shahriari, A., Shirmohammadi, E., and Fazeli-Nasab, B. Identification and isolation of associated microorganisms with airborne dust loaded over Sistan plain. 15th Iranian Soil Science Congress, Isfahan University of Technology, Isfahan, Iran, Congress COI: SSCI15, Article COI: SSCI15_895; 2017.

20 Ali-Soufi, M., Shahriari, A., Shirmohammadi, E., and Fazeli-Nasab, B. (2017). Investigation of biological properties and microorganism identification in susceptible areas to wind erosion in Hamoun wetlands. Congress on restoration policies and approaches of Hamoun international wetland Zabol. 231–240.

21 Shahriari, A., Ali-Soufi, M., and Shirmohammadi, E. Impact of airborne dust abundance on number of hospital admissions for respiratory diseases in Sistan region. The First of International Conference on Dust, 2e4 March; 2016; Shahid Chamran University, Ahvaz, Iran; 2016.

22 Dawwam, G., Elbeltagy, A., Emara, H., Abbas, I., and Hassan, M. (2013). Beneficial effect of plant growth promoting bacteria isolated from the roots of potato plant. *Annals of Agricultural Sciences* 58 (2): 195–201. https://doi.org/10.1016/j.aoas.2013.07.007

23 MdVB, F., Seldin, L., De Araujo, F.F., and Mariano, R. (2010). *Plant Growth Promoting Rhizobacteria: Fundamentals and Applications. Plant Growth and Health Promoting Bacteria*, 21–43. Springer. https://doi.org/10.1007/978-3-642-13612-2_2

24 Raja, N. (2013). Biopesticides and biofertilizers: Ecofriendly sources for sustainable agriculture. *Journal of biofertilizers & biopesticides* 4 (1): 17–27.

25 Conway, G.R. and Pretty, J.N. (2013). *Unwelcome Harvest: Agriculture and Pollution*. Routledge, Taylor & Francis.

26 McGuinness, H. (1993). *Living Soils: Sustainable Alternatives to Chemical Fertilizers for Developing Countries*. Consumer Policy Institute, Consumers Union.

27 Altieri, M.A. and Nicholls, C.I. (2005). Agroecology and the search for a truly sustainable agriculture [online]: United Nations Environmental Programme, Environmental Training Network for Latin America and the Caribbean.

28 Bhardwaj, D., Ansari, M.W., Sahoo, R.K., and Tuteja, N. (2014). Biofertilizers function as key player in sustainable agriculture by improving soil fertility, plant tolerance and crop productivity. *Microbial Cell Factories* 13 (1): 66. https://doi.org/10.1186/1475-2859-13-66

29 Culligan, E.P. (2014). Functional metagenomic analysis of the human gut microbiome to identify novel salt tolerance genes. University College Cork.

30 Lakshmanan, V., Selvaraj, G., and Bais, H.P. (2014). Functional soil microbiome: Belowground solutions to an aboveground problem. *Plant Physiology* 166 (2): 689–700. https://doi.org/10.1104/pp.114.245811

31 Moe, L.A. (2013). Amino acids in the rhizosphere: From plants to microbes. *American Journal of Botany* 100 (9): 1692–1705. https://doi.org/10.3732/ajb.1300033

32 Turner, T.R., James, E.K., and Poole, P.S. (2013). The plant microbiome. *Genome Biology* 14 (6): 209. https://doi.org/10.1186/gb-2013-14-6-209

33 Aleklett, K., Hart, M., and Shade, A. (2014). The microbial ecology of flowers: An emerging frontier in phyllosphere research. *Botany* 92 (4): 253–266. https://doi.org/10.1139/cjb-2013-0166

34 Brusetti, L., Francia, P., Bertolini, C., Pagliuca, A., Borin, S., Sorlini, C., Abruzzese, A., Sacchi, G., Viti, C., and Giovannetti, L. (2005). Bacterial communities associated with the rhizosphere of transgenic Bt 176 maize (*Zea mays*) and its non-transgenic counterpart. *Plant and Soil* 266 (1-2): 11–21. https://doi.org/10.1007/s11104-005-5399-x

35 De-la-peña, C. and Loyola-Vargas, V.M. (2014). Biotic interactions in the rhizosphere: A diverse cooperative enterprise for plant productivity. *Plant Physiology* 166 (2): 701–719. https://doi.org/10.1104/pp.114.241810

36 Tkacz, A. and Poole, P. (2015). Role of root microbiota in plant productivity. *Journal of Experimental Botany* 66 (8): 2167–2175. https://doi.org/10.1093/jxb/erv157

37 Bulgarelli, D., Garrido-Oter, R., Münch, P.C., Weiman, A., Dröge, J., Pan, Y., McHardy, A.C., and Schulze-Lefert, P. (2015). Structure and function of the bacterial root microbiota in wild and domesticated barley. *Cell Host & Microbe* 17 (3): 392–403. https://doi.org/10.1016/j.chom.2015.01.011

38 Bulgarelli, D., Rott, M., Schlaeppi, K., Evl, V.T., Ahmadinejad, N., Assenza, F., Rauf, P., Huettel, B., Reinhardt, R., and Schmelzer, E. (2012). Revealing structure and assembly cues for Arabidopsis root-inhabiting bacterial microbiota. *Nature* 488 (7409): 91–95. https://doi.org/10.1038/nature11336

39 Lundberg, D.S., Lebeis, S.L., Paredes, S.H., Yourstone, S., Gehring, J., Malfatti, S., Tremblay, J., Engelbrektson, A., Kunin, V., and Del Rio, T.G. (2012). Defining the core *Arabidopsis thaliana* root microbiome. *Nature* 488 (7409): 86–90. https://doi.org/10.1038/nature11237

40 Schlaeppi, K., Dombrowski, N., Oter, R.G., Evl, V.T., and Schulze-Lefert, P. (2014). Quantitative divergence of the bacterial root microbiota in *Arabidopsis thaliana* relatives.

Proceedings of the National Academy of Sciences 111 (2): 585–592. https://doi.org/10.1073/pnas.1321597111

41 Aira, M., Gómez-Brandón, M., Lazcano, C., Bååth, E., and Domínguez, J. (2010). Plant genotype strongly modifies the structure and growth of maize *rhizosphere* microbial communities. *Soil Biology and Biochemistry* 42 (12): 2276–2281. https://doi.org/10.1016/j.soilbio.2010.08.029

42 Mendes, L.W., Kuramae, E.E., Navarrete, A.A., Van Veen, J.A., and Tsai, S.M. (2014). Taxonomical and functional microbial community selection in soybean rhizosphere. *The ISME Journal* 8 (8): 1577–1587. https://doi.org/10.1038/ismej.2014.17

43 Darvishi, F. (2017). Screening of New Microorganisms and Their Useful Genes: From Traditional Methods to Metagenomics. *Iranian Journal of Biology (Biological Science Promotion)* 1 (1): 43–51.

44 Maleki-Tabrizi, N., Shahnejat-Bushehri, A.A., and Hosseini Salekdeh, G. (2018). Effect of developmental stages on microbial community during in wheat rhizosphere. *Modern Genetics Journal* 13 (2): 181–188.

45 Mardis, E.R. (2008). Next-generation DNA sequencing methods. *Annual Review of genomics and human Genetics* 9: 387–402.

46 Dombrowski, N., Schlaeppi, K., Agler, M.T., Hacquard, S., Kemen, E., Garrido-Oter, R., Wunder, J., Coupland, G., and Schulze-Lefert, P. (2017). Root microbiota dynamics of perennial *Arabis alpina* are dependent on soil residence time but independent of flowering time. *The ISME Journal* 11 (1): 43–55. https://doi.org/10.1038/ismej.2016.109

47 Edwards, J., Johnson, C., Santos-Medellín, C., Lurie, E., Podishetty, N.K., Bhatnagar, S., Eisen, J.A., and Sundaresan, V. (2015). Structure, variation, and assembly of the root-associated microbiomes of rice. *Proceedings of the National Academy of Sciences* 112 (8): E911–E920. https://doi.org/10.1073/pnas.1414592112

48 Truong, D.T., Franzosa, E.A., Tickle, T.L., Scholz, M., Weingart, G., Pasolli, E., Tett, A., Huttenhower, C., and Segata, N. (2016). Erratum: MetaPhlAn2 for enhanced metagenomic taxonomic profiling. *Nature Methods* 13 (1): 101-101. https://doi.org/10.1038/nmeth0116-101b

49 Albanese, D., Fontana, P., De Filippo, C., Cavalieri, D., and Donati, C. (2015). MICCA: A complete and accurate software for taxonomic profiling of metagenomic data. *Scientific Reports* 5: 9743. https://doi.org/10.1038/srep09743

50 Rintala, A., Pietilä, S., Munukka, E., Eerola, E., Pursiheimo, J.-P., Laiho, A., Pekkala, S., and Huovinen, P. (2017). Gut microbiota analysis results are highly dependent on the 16S rRNA gene target region, whereas the impact of DNA extraction is minor. *Journal of Biomolecular Techniques: JBT* 28 (1): 19. https://doi.org/10.7171/jbt.17-2801-003

51 Otero, J.M. and Nielsen, J. (2010). Industrial systems biology. *Biotechnology and Bioengineering* 105 (3): 439–460. https://doi.org/10.1002/bit.22592

52 Ghazanfar, S., Azim, A., Ghazanfar, M.A., Anjum, M., and Begum, I. (2010). Metagenomics and its application in soil microbial community studies: Biotechnological prospects. *Journal of Animal & Plant Sciences* 6 (2): 611–622.

53 Handelsman, J., Rondon, M.R., Brady, S.F., Clardy, J., and Goodman, R.M. (1998). Molecular biological access to the chemistry of unknown soil microbes: A new frontier for natural products. *Chemistry & Biology* 5 (10): R245–R249.

54 Wooley, J.C., Godzik, A., and Friedberg, I. (2010). A primer on metagenomics. *PLoS Computational Biology* 6 (2): e1000667. https://doi.org/10.1371/journal.pcbi.1000667

55 Chistoserdova, L. (2010). Recent progress and new challenges in metagenomics for biotechnology. *Biotechnology Letters* 32 (10): 1351–1359. https://doi.org/10.1007/s10529-010-0306-9

56 Allen, E.E. and Banfield, J.F. (2005). Community genomics in microbial ecology and evolution. *Nature Reviews Microbiology* 3 (6): 489–498. https://doi.org/10.1038/nrmicro1157

57 Pontes, D.S., Lima-Bittencourt, C.I., Chartone-Souza, E., and Nascimento, A.M.A. (2007). Molecular approaches: Advantages and artifacts in assessing bacterial diversity. *Journal of Industrial Microbiology & Biotechnology* 34 (7): 463–473. https://doi.org/10.1007/s10295-007-0219-3

58 Ercolini, D. (2013). High-throughput sequencing and metagenomics: Moving forward in the culture-independent analysis of food microbial ecology. *Applied and Environmental Microbiology* 79 (10): 3148–3155. doi: 10.1128/AEM.00256-13

59 Jacobsen, C.S. and Hjelmsø, M.H. (2014). Agricultural soils, pesticides and microbial diversity. *Current Opinion in Biotechnology* 27: 15–20. https://doi.org/10.1016/j.copbio.2013.09.003

60 Schloss, P.D. (2009). A high-throughput DNA sequence aligner for microbial ecology studies. *PLoS One* 4 (12). doi: 10.1371/journal.pone.0008230. PMCID: PMC2788221; PMID: 20011594

61 Possemiers, S., Verthé, K., Uyttendaele, S., and Verstraete, W. (2004). PCR-DGGE-based quantification of stability of the microbial community in a simulator of the human intestinal microbial ecosystem. *FEMS Microbiology Ecology* 49 (3): 495–507. https://doi.org/10.1016/j.femsec.2004.05.002

62 Segata, N., Izard, J., Waldron, L., Gevers, D., Miropolsky, L., Garrett, W.S., and Huttenhower, C. (2011). Metagenomic biomarker discovery and explanation. *Genome Biology* 12 (6): R60. https://doi.org/10.1186/gb-2011-12-6-r60

63 Schmeisser, C., Steele, H., and Streit, W.R. (2007). Metagenomics, biotechnology with non-culturable microbes. *Applied Microbiology and Biotechnology* 75 (5): 955–962. https://doi.org/10.1007/s00253-007-0945-5

64 Shelswell, K.J. (2004). Metagenomics: The science of biological diversity. Sci. Creat. Quart. Available at: http://www.scq.ubc.ca/metagenomics-thescience-of-biological-diversity (accessed 4 November 2011).

65 Uhlík, O., Jecná, K., Leigh, M.B., Macková, M., and Macek, T. (2009). DNA-based stable isotope probing: A link between community structure and function. *Science of the Total Environment* 407 (12): 3611–3619. https://doi.org/10.1016/j.scitotenv.2008.05.012

66 Delmont, T.O., Robe, P., Clark, I., Simonet, P., and Vogel, T.M. (2011). Metagenomic comparison of direct and indirect soil DNA extraction approaches. *Journal of Microbiological Methods* 86 (3): 397–400. https://doi.org/10.1016/j.mimet.2011.06.013

67 Fooladvand, Z. and Fazeli-Nasab, B. Meta-genomics method to the study of biodiversity in wetlands. Conference on the Hamoon International Wetland Reclamation Solutions and Approaches. Iran (Zabol); 2016. p. 274–281.

68 Kauffmann, I.M., Schmitt, J., and Schmid, R.D. (2004). DNA isolation from soil samples for cloning in different hosts. *Applied Microbiology and Biotechnology* 64 (5): 665–670. https://doi.org/10.1007/s00253-003-1528-8

69 Simon, C. and Daniel, R. (2011). Metagenomic analyses: Past and future trends. *Applied and Environmental Microbiology* 77 (4): 1153–1161. https://doi.org/10.1128/AEM.02345-10

70 Eyers, L., George, I., Schuler, L., Stenuit, B., Agathos, S.N., and El Fantroussi, S. (2004). Environmental genomics: Exploring the unmined richness of microbes to degrade xenobiotics. *Applied Microbiology and Biotechnology* 66 (2): 123–130. https://doi.org/10.1007/s00253-004-1703-6

71 Niemi, R.M., Heiskanen, I., Wallenius, K., and Lindström, K. (2001). Extraction and purification of DNA in rhizosphere soil samples for PCR-DGGE analysis of bacterial consortia. *Journal of Microbiological Methods* 45 (3): 155–165. https://doi.org/10.1016/S0167-7012(01)00253-6

72 Bey, B.S., Fichot, E.B., Dayama, G., Decho, A.W., and Norman, R.S. (2010). Extraction of high molecular weight DNA from microbial mats. *Biotechniques* 49 (3): 631–640.

73 Tabatabaei, M., Rahim, R.A., Abdullah, N., Wright, A.-D.G., Shirai, Y., Sakai, K., Sulaiman, A., and Hassan, M.A. (2010). Importance of the methanogenic archaea populations in anaerobic wastewater treatments. *Process Biochemistry* 45 (8): 1214–1225. https://doi.org/10.1016/j.procbio.2010.05.017

74 Singh, J., Behal, A., Singla, N., Joshi, A., Birbian, N., Singh, S., Bali, V., and Batra, N. (2009). Metagenomics: Concept, methodology, ecological inference and recent advances. *Biotechnology Journal: Healthcare Nutrition Technology* 4 (4): 480–494. https://doi.org/10.1002/biot.200800201

75 Borneman, J. (1999). Culture-independent identification of microorganisms that respond to specified stimuli. *Applied and Environmental Microbiology* 65 (8): 3398–3400. PubMed: 10427025

76 Urbach, E., Vergin, K.L., and Giovannoni, S.J. (1999). Immunochemical detection and isolation of DNA from metabolically active bacteria. *Applied and Environmental Microbiology* 65 (3): 1207–1213. PubMed: 10049885

77 Radajewski, S., Webster, G., Reay, D.S., Morris, S.A., Ineson, P., Nedwell, D.B., Prosser, J.I., and Murrell, J.C. (2002). Identification of active methylotroph populations in an acidic forest soil by stable-isotope probing. *Microbiology* 148 (8): 2331–2342.

78 Turnbaugh, P.J. and Gordon, J.I. (2008). An invitation to the marriage of metagenomics and metabolomics. *Cell* 134 (5): 708–713. https://doi.org/10.1016/j.cell.2008.08.025

79 Blatny, J.M., Brautaset, T., Winther-Larsen, H.C., Karunakaran, P., and Valla, S. Improved broad-host-range RK2 vectors useful for high and low regulated gene expression levels in gram-negative bacteria. *Plasmid*, 1997; 38(1): 35–51. doi: 10.1006/plas.1997.1294

80 Mande, S.S., Mohammed, M.H., and Ghosh, T.S. (2012). Classification of metagenomic sequences: Methods and challenges. *Briefings in Bioinformatics* 13 (6): 669–681. https://doi.org/10.1093/bib/bbs054

81 Malik, S., Beer, M., Megharaj, M., and Naidu, R. (2008). The use of molecular techniques to characterize the microbial communities in contaminated soil and water. *Environment International* 34 (2): 265–276. https://doi.org/10.1016/j.envint.2007.09.001

82 Brennerova, M.V., Josefiova, J., Brenner, V., Pieper, D.H., and Junca, H. (2009). Metagenomics reveals diversity and abundance of meta-cleavage pathways in microbial communities from soil highly contaminated with jet fuel under air-sparging bioremediation. *Environmental Microbiology* 11 (9): 2216–2227. https://doi.org/10.1111/j.1462-2920.2009.01943.x

83 Liu, B., Gibbons, T., Ghodsi, M., and Pop, M. (2010). MetaPhyler: Taxonomic profiling for metagenomic sequences. 2010 IEEE international conference on bioinformatics and biomedicine (BIBM). IEEE, INSPEC Accession Number: 11794950 2010. 95–100, doi: 10.1109/BIBM.2010.5706544

84 Porazinska, D.L., Giblin-davis, R.M., Faller, L., Farmerie, W., Kanzaki, N., Morris, K., Powers, T.O., Tucker, A.E., Sung, W., and Thomas, W.K. (2009). Evaluating high-throughput sequencing as a method for metagenomic analysis of nematode diversity. *Molecular Ecology Resources* 9 (6): 1439–1450. https://doi.org/10.1111/j.1755-0998.2009.02611.x

85 Huson, D.H., Auch, A.F., Qi, J., and Schuster, S.C. (2007). MEGAN analysis of metagenomic data. *Genome Research* 17 (3): 377–386. https://doi.org/10.1101/gr.5969107

86 Baldrian, P., Head, I.M., Prosser, J.I., Schloter, M., Smalla, K., and Tebbe, C.C. (2011). Ecology and metagenomics of soil microorganisms. *FEMS Microbiology Ecology* 78 (1): 1–2. https://doi.org/10.1111/j.1574-6941.2011.01184.x

87 Caro-Quintero, A. and Konstantinidis, K.T. (2012). Bacterial species may exist, metagenomics reveal. *Environmental Microbiology* 14 (2): 347–355. https://doi.org/10.1111/j.1462-2920.2011.02668.x

88 Steele, H.L. and Streit, W.R. (2005). Metagenomics: Advances in ecology and biotechnology. *FEMS Microbiology Letters* 247 (2): 105–111. https://doi.org/10.1016/j.femsle.2005.05.011

89 Valenzuela, L., Chi, A., Beard, S., Orell, A., Guiliani, N., Shabanowitz, J., Hunt, D.F., and Jerez, C.A. (2006). Genomics, metagenomics and proteomics in biomining microorganisms. *Biotechnology Advances* 24 (2): 197–211. https://doi.org/10.1016/j.biotechadv.2005.09.004

90 Andújar, C., Arribas, P., Ruzicka, F., Crampton-Platt, A., Timmermans, M.J., and Vogler, A.P. (2015). Phylogenetic community ecology of soil biodiversity using mitochondrial metagenomics. *Molecular Ecology* 24 (14): 3603–3617. https://doi.org/10.1111/mec.13195

91 E Liu, G. (2011). Recent applications of DNA sequencing technologies in food, nutrition and agriculture. *Recent Patents on Food, Nutrition & Agriculture* 3 (3): 187–195.

92 Jung, J., Philippot, L., and Park, W. (2016). Metagenomic and functional analyses of the consequences of reduction of bacterial diversity on soil functions and bioremediation in diesel-contaminated microcosms. *Scientific Reports* 6 (1): 1–10. https://doi.org/10.1038/srep23012

93 Biddle, J.F., Fitz-Gibbon, S., Schuster, S.C., Brenchley, J.E., and House, C.H. (2008). Metagenomic signatures of the Peru Margin subseafloor biosphere show a genetically distinct environment. *Proceedings of the National Academy of Sciences* 105 (30): 10583–10588. https://doi.org/10.1073/pnas.0709942105

94 Boël, G., Pichereau, V., Mijakovic, I., Mazé, A., Poncet, S., Gillet, S., Giard, J.-C., Hartke, A., Auffray, Y., and Deutscher, J. (2004). Is 2-phosphoglycerate-dependent automodification of bacterial enolases implicated in their export? *Journal of Molecular Biology* 337 (2): 485–496. https://doi.org/10.1016/j.jmb.2003.12.082

95 Brikun, I.A., Reeves, A.R., Cernota, W.H., Luu, M.B., and Weber, J.M. (2004). The erythromycin biosynthetic gene cluster of *Aeromicrobium erythreum*. *Journal of Industrial Microbiology and Biotechnology* 31 (7): 335–344. https://doi.org/10.1007/s10295-004-0154-5

96 Darvishi, F., Golbang, N., Hojati, Z., and Motevali-Bashi, M. (2006). Isolation of the Streptomycin antibiotic production regulatory gene (strR) by PCR. *Iranian Journal of Biology* 19 (3): 264–271.

97 Hojati, Z., Motovali-Bashi, M., Golbang, N., and Darvishi, F. (2007). Detection of cloned strR, an antibiotic regulatory gene, using RFLP and Nested PCR. *Pakistan Journal of Biological Sciences* 10 (18): 3079–3084.

98 Mafakher, L., Mirbagheri, M., Darvishi, F., Nahvi, I., Zarkesh-Esfahani, H., and Emtiazi, G. (2010). Isolation of lipase and citric acid producing yeasts from agro-industrial wastewater. *New Biotechnology* 27 (4): 337–340. https://doi.org/10.1016/j.nbt.2010.04.006

99 Mirbagheri, M., Nahvi, I., Emtiazi, G., Mafakher, L., and Darvishi, F. (2012). Taxonomic characterization and potential biotechnological applications of *Yarrowia lipolytica* isolated from meat and meat products. *Jundishapur Journal of Microbiology (JJM)* 5 (1): 346–351.

100 Handelsman, J. (2004). Metagenomics: Application of genomics to uncultured microorganisms. *Microbiology and Molecular Biology Reviews* 68 (4): 669–685. doi: 10.1128/MMBR.68.4.669-685.2004

101 Kim, D.H., Brunt, J., and Austin, B. (2007). Microbial diversity of intestinal contents and mucus in rainbow trout (*Oncorhynchus mykiss*). *Journal of Applied Microbiology* 102 (6): 1654–1664. https://doi.org/10.1111/j.1365-2672.2006.03185.x

102 Muyzer, G. and Smalla, K. (1998). Application of denaturing gradient gel electrophoresis (DGGE) and temperature gradient gel electrophoresis (TGGE) in microbial ecology. *Antonie Van Leeuwenhoek* 73 (1): 127–141. https://doi.org/10.1023/A:1000669317571

103 Tabatabaei, M. and Pourmazaheri, H. (2012). Metagenomics and its application in identification of genetic diversity of microbial ecosystems. *Modern Genetics Journal* 7 (4): 313–324.

104 Fischer, S.G. and Lerman, L.S. (1979). Length-independent separation of DNA restriction fragments in two-dimensional gel electrophoresis. *Cell* 16 (1): 191–200. https://doi.org/10.1016/0092-8674(79)90200-9

105 Myers, R.M., Maniatis, T., and Lerman, L.S. (1987). *[31] Detection and Localization of Single Base Changes by Denaturing Gradient Gel Electrophoresis. Methods in Enzymology*, 501–527. Elsevier. https://doi.org/10.1016/0076-6879(87)55033-9

106 Hughes, J.B., Hellmann, J.J., Ricketts, T.H., and Bohannan, B.J. (2001). Counting the uncountable: Statistical approaches to estimating microbial diversity. *Applied and Environmental Microbiology* 67 (10): 4399–4406. doi: 10.1128/AEM.67.10.4399-4406.2001.

107 Tabatabaei, M., Zakaria, M.R., Rahim, R.A., Wright, A.-D.G., Shirai, Y., Abdullah, N., Sakai, K., Ikeno, S., Mori, M., and Kazunori, N. (2009). PCR-based DGGE and FISH analysis of methanogens in an anaerobic closed digester tank for treating palm oil mill effluent. *Electronic Journal of Biotechnology* 12 (3): 12–13. doi: 10.2225/vol12-issue3-fulltext-1.

108 Tabatabaei, M., Zakaria, M.R., Rahim, R.A., Abdullah, N., Wright, A.D.G., Shirai, Y., Shamsara, M., Sakai, K., and Hassan, M.A. (2010). Comparative study of methods for extraction and purification of environmental DNA from high-strength wastewater sludge. *African Journal of Biotechnology* 9 (31): 4926–4937.

109 Barin, M., Aliasgharzad, N., and Rasoli-Sadaghiani, M.H. (2015). Using of lipid biomarkers for assessing soil microbial community structure. *Journal of Agricultural Biotechnology* 7 (2): 11–36.

110 Kirk, J.L., Beaudette, L.A., Hart, M., Moutoglis, P., Klironomos, J.N., Lee, H., and Trevors, J.T. (2004). Methods of studying soil microbial diversity. *Journal of Microbiological Methods* 58 (2): 169–188. https://doi.org/10.1016/j.mimet.2004.04.006

111 Poormazaheri, H., Salehi-Jouzani, G.H., Khayam Nekouei, S.M., Tabatabaei, M., Maali Amiri, R., Soheilivand, S., Karimi, E., Ghanavati, H., and Mirdamadian, S.H. (2013). Evaluation of some native bacteria isolated from composting process. *Journal of Agricultural Biotechnology* 5 (1): 1–11.

112 Cardon, Z.G. and Gage, D.J. (2006). Resource exchange in the *rhizosphere*: Molecular tools and the microbial perspective. *Annu Rev Ecol Evol Syst* 37: 459–488.

113 Gelsomino, A., Keijzer-Wolters, A.C., Cacco, G., and Van Elsas, J.D. (1999). Assessment of bacterial community structure in soil by polymerase chain reaction and denaturing gradient gel electrophoresis. *Journal of Microbiological Methods* 38 (1-2): 1–15. https://doi.org/10.1016/S0167-7012(99)00054-8

114 Zelles, L. (1999). Fatty acid patterns of phospholipids and lipopolysaccharides in the characterisation of microbial communities in soil: A review. *Biology and Fertility of Soils* 29 (2): 111–129. https://doi.org/10.1007/s003740050533

115 Griffiths, B.S., Heckmann, L.H., Caul, S., Thompson, J., Scrimgeour, C., and Krogh, P.H. (2007). Varietal effects of eight paired lines of transgenic Bt maize and near-isogenic non-Bt maize on soil microbial and nematode community structure. *Plant Biotechnology Journal* 5 (1): 60–68. https://doi.org/10.1111/j.1467-7652.2006.00215.x

116 Velasco, A.G.-V., Probanza, A., Fjg, M., Solano, B.R., and Lucas, J.A. (2010). Characterization of the rhizosphere microbial community from different *Arabidopsis thaliana* genotypes using phospholipid fatty acids (PLFA) analysis. *Plant and Soil* 329 (1-2): 315–325. https://doi.org/10.1007/s11104-009-0160-5

117 Ramsey, P.W., Rillig, M.C., Feris, K.P., Holben, W.E., and Gannon, J.E. (2006). Choice of methods for soil microbial community analysis: PLFA maximizes power compared to CLPP and PCR-based approaches. *Pedobiologia* 50 (3): 275–280. https://doi.org/10.1016/j.pedobi.2006.03.003

118 Aliasgharzad, N., L-m, M., and Olsson, P.A. (2010). Acidification of a sandy grassland favours bacteria and disfavours fungal saprotrophs as estimated by fatty acid profiling. *Soil Biology and Biochemistry* 42 (7): 1058–1064. https://doi.org/10.1016/j.soilbio.2010.02.025

119 Å, F., Bååth, E., and Tunlio, A. (1993). Shifts in the structure of soil microbial communities in limed forests as revealed by phospholipid fatty acid analysis. *Soil Biology and Biochemistry* 25 (6): 723–730. https://doi.org/10.1016/0038-0717(93)90113-P

120 Zhang, X., Li, H., Li, C.-J., Ma, T., Li, G., and Liu, Y.-H. (2013). Metagenomic approach for the isolation of a thermostable β-galactosidase with high tolerance of galactose and glucose from soil samples of Turpan Basin. *Bmc Microbiology* 13 (1): 237.

121 Vahidi, M., Hosseini Salekdeh, G., Afraz, F., and Behmanesh, M. (2019). Study of rice straw-adherent sheep rumen microbiome diversity: A comparative 16S rDNA and whole metagenome analysis. *Modern Genetics Journal* 13 (4): 459–477.

122 Bagherifam, S., Lakziyan, A., Ahmadi, S.J., Fotovvat, A., and Rahimi, M. (2010). The effects of different uranium concentrations on soil microbial populations and enzymatic activities. *Journal of Nuclear Science and Technology* 31 (1): 52-57.

123 Lamas, C. (2005). Factors affecting the uranium availability in soils. fall agricultural research.

124 Rascovan, N., Carbonetto, B., Perrig, D., Díaz, M., Canciani, W., Abalo, M., Alloati, J., González-Anta, G., and Vazquez, M.P. (2016). Integrated analysis of root microbiomes of soybean and wheat from agricultural fields. *Sci Rep* 6: 28084. https://doi.org/10.1038/srep28084

125 Killham, K. (1994). *Soil Ecology*. Cambridge University Press.

126 Boyerahmadi, M., Raiesi, F., and Mohammadi, J. (2010). Influence of Different Salinity Levels on Some Microbial Indices in the Presence and Absence of Plant's Living Roots. *Journal of Water and Soil Science* 14 (51): 103–115.

127 Mansourzadeh, M. and Raiesi, F. (2012). The effect of eradican (EPTC) on microbial biomass C and N, and urease and arylsulphatase activities in a calcareous soil under field conditions. *Journal of Science and Technology of Agriculture and Natural Resources* 16 (59 (B)): 153–167. Record Number: 20123256142

128 Mansourzadeh, M. and Raiesi, F. (2013). The ecological effects of different loading rates of metalaxyl on microbial biomass in unplanted and planted soils under field conditions. *Agroecology* 5 (3): 220–230. https://doi.org/10.22067/jag.v5i3.28991

129 Usman, A.R., Almaroai, Y.A., Ahmad, M., Vithanage, M., and Ok, Y.S. (2013). Toxicity of synthetic chelators and metal availability in poultry manure amended Cd, Pb and As contaminated agricultural soil. *Journal of Hazardous Materials* 262: 1022–1030. https://doi.org/10.1016/j.jhazmat.2013.04.032

130 Huang, J.W., Chen, J., Berti, W.R., and Cunningham, S.D. (1997). Phytoremediation of lead-contaminated soils: Role of synthetic chelates in lead phytoextraction. *Environmental Science & Technology* 31 (3): 800–805. https://doi.org/10.1021/es9604828

131 Tandy, S., Schulin, R., and Nowack, B. (2006). Uptake of metals during chelant-assisted phytoextraction with EDDS related to the solubilized metal concentration. *Environmental Science & Technology* 40 (8): 2753–2758. https://doi.org/10.1021/es052141c.

132 Shahid, M., Pinelli, E., and Dumat, C. (2012). Review of Pb availability and toxicity to plants in relation with metal speciation; role of synthetic and natural organic ligands. *Journal of Hazardous Materials* 219: 1–12. https://doi.org/10.1016/j.jhazmat.2012.01.060

133 Udovic, M. and Pb, L.D. (2009). Zn and Cd mobility, availability and fractionation in aged soil remediated by EDTA leaching. *Chemosphere* 74 (10): 1367–1373. https://doi.org/10.1016/j.chemosphere.2008.11.013

134 Wu, L., Luo, Y., Christie, P., and Wong, M.H. (2003). Effects of EDTA and low molecular weight organic acids on soil solution properties of a heavy metal polluted soil. *Chemosphere* 50 (6): 819–822. https://doi.org/10.1016/S0045-6535(02)00225-4

135 Nascimento, C. (2006). Organic acids effects on desorption of heavy metals from a contaminated soil. *Scientia Agricola* 63 (3): 276–280. https://doi.org/10.1590/S0103-90162006000300010

136 Hosseini, S.S., Lakzian, A., and Halajnia, A. (2017). The Effect of EDTA and Citric acid on Soil Enzymes Activity, Substrate Induced Respiration and Pb Availability in a Contaminated Soil. *Journal of Water and Soil* 30 (6): 2032–2045.

137 Zhang, F., Zhu, Z., Dong, Z., Cui, Z., Wang, H., Hu, W., Zhao, P., Wang, P., Wei, S., and Li, R. (2011). Magnetically recoverable facile nanomaterials: Synthesis, characterization and application in remediation of heavy metals. *Microchemical Journal* 98 (2): 328–333. https://doi.org/10.1016/j.microc.2011.03.005

138 Shahid, M., Zia-Ur-Rehman, M., Sabir, M., and Ahmad, H.R. (2015). *Phytoremediation of Pb-Contaminated Soils Using Synthetic Chelates: Soil Remediation and Plants*. Elsevier Inc.. http://dx.doi.org/10.1016/B978-0-12-799937-1.00014-0

139 Zhang, H., Zhao, F.-J., Sun, B., Davison, W., and Mcgrath, S.P. (2001). A new method to measure effective soil solution concentration predicts copper availability to plants. *Environmental Science & Technology* 35 (12): 2602–2607. https://doi.org/10.1021/es000268q

140 Freitas, E.V., Nascimento, C.W., Souza, A., and Silva, F.B. (2013). Citric acid-assisted phytoextraction of lead: A field experiment. *Chemosphere* 92 (2): 213–217. https://doi.org/10.1016/j.chemosphere.2013.01.103

141 Ultra, J.V.U., Yano, A., Iwasaki, K., Tanaka, S., Kang, Y., and Sakurai, K. (2005). Influence of chelating agent addition on copper distribution and microbial activity in soil and copper uptake by brown mustard (*Brassica juncea*). *Soil Science & Plant Nutrition* 51 (2): 193–202. https://doi.org/10.1111/j.1747-0765.2005.tb00023.x

142 Landi, L., Renella, G., Moreno, J., Falchini, L., and Nannipieri, P. (2000). Influence of cadmium on the metabolic quotient, L-: D-glutamic acid respiration ratio and enzyme activity: Microbial biomass ratio under laboratory conditions. *Biology and Fertility of Soils* 32 (1): 8–16. https://doi.org/10.1007/s003740000205

143 Suman, A., Lal, M., Singh, A., and Gaur, A. (2006). Microbial biomass turnover in Indian subtropical soils under different sugarcane intercropping systems. *Agronomy Journal* 98 (3): 698–704. https://doi.org/10.2134/agronj2005.0173

144 Kaur, A., Chaudhary, A., Kaur, A., Choudhary, R., and Kaushik, R. (2005). Phospholipid fatty acid – a bioindicator of environment monitoring and assessment in soil ecosystem. *Current Science* 89 (7): 1103–1112.

145 Shahbazi, F., Aliasgharzad, N., Ebrahimzad, S., and Najafi, N. (2013). Geostatistical analysis for predicting soil biological maps under different scenarios of land use. *European Journal of Soil Biology* 55 ((March–April)): 20–27. https://doi.org/10.1016/j.ejsobi.2012.10.009.

146 Tabatabai, M. and Bremner, J. (1969). Use of p-nitrophenyl phosphate for assay of soil phosphatase activity. *Soil Biology and Biochemistry* 1 (4): 301–307. https://doi.org/10.1016/0038-0717(69)90012-1

147 Stevenson, F.J. and Cole, M.A. (1999). *Cycles of Soils: Carbon, Nitrogen, Phosphorus, Sulfur, Micronutrients*. John Wiley & Sons.

148 Deng, S. and Tabatabai, M. (1995). Cellulase activity of soils: Effect of trace elements. *Soil Biology and Biochemistry* 27 (7): 977–979. https://doi.org/10.1016/0038-0717(95)00005-Y

149 Bandick, A.K. and Dick, R.P. (1999). Field management effects on soil enzyme activities. *Soil Biology and Biochemistry* 31 (11): 1471–1479. https://doi.org/10.1016/S0038-0717(99)00051-6

150 Maguire, R., Sims, J., and Coale, F. (2000). Phosphorus solubility in biosolids-amended farm soils in the Mid-Atlantic region of the USA. *Journal of Environmental Quality* 29 (4): 1225–1233. https://doi.org/10.2134/jeq2000.00472425002900040028x

151 Sheppard, S., Evenden, W., and Anderson, A. (1992). Multiple assays of uranium toxicity in soil. *Environmental Toxicology and Water Quality* 7 (3): 275–294. https://doi.org/10.1002/tox.2530070307

152 Ahemad, M. and Ms, K. (2012). Effects of pesticides on plant growth promoting traits of Mesorhizobium strain MRC4. *Journal of the Saudi Society of Agricultural Sciences* 11 (1): 63–71. https://doi.org/10.1016/j.jssas.2011.10.001.

153 Shaharoona, B., Arshad, M., Zahir, Z.A., and Khalid, A. (2006). Performance of Pseudomonas spp. containing ACC-deaminase for improving growth and yield of maize (*Zea mays* L.) in the presence of nitrogenous fertilizer. *Soil Biology and Biochemistry* 38 (9): 2971–2975. https://doi.org/10.1016/j.soilbio.2006.03.024

154 Harman, G.E. and Uphoff, N. (2019). Symbiotic root-endophytic soil microbes improve crop productivity and provide environmental benefits. *Scientifica* 2019. Article ID 9106395 https://doi.org/10.1155/2019/9106395

155 Gray, E. and Smith, D. (2005). Intracellular and extracellular PGPR: Commonalities and distinctions in the plant–bacterium signaling processes. *Soil Biology and Biochemistry* 37 (3): 395–412. https://doi.org/10.1016/j.soilbio.2004.08.030

156 Podile, A.R. and Kishore, G.K. (2007). *Plant growth-promoting Rhizobacteria. Plant-associated Bacteria*, 195–230. Springer. https://doi.org/10.1007/978-1-4020-4538-7_6

157 Banerjee, M.R., Yesmin, L., and Vessey, J.K. (2006). *Plant-growth-promoting Rhizobacteria as Biofertilizers and Biopesticides. Handbook of Microbial Biofertilizers*, 137–181. New York: Food Products Press. ISBN; 1560222700

158 Chauhan, H., Bagyaraj, D., Selvakumar, G., and Sundaram, S. (2015). Novel plant growth promoting rhizobacteria—Prospects and potential. *Applied Soil Ecology* 95: 38–53. https://doi.org/10.1016/j.apsoil.2015.05.011

159 Tank, N. and Saraf, M. (2010). Salinity-resistant plant growth promoting rhizobacteria ameliorates sodium chloride stress on tomato plants. *Journal of Plant Interactions* 5 (1): 51–58. https://doi.org/10.1080/17429140903125848

160 Ahemad, M. and Khan, M.S. (2012). Ecological Assessment of Biotoxicity of Pesticides towards Plant Growth Promoting Activities of Pea (*Pisum Sativum*)-Specific *Rhizobium* Sp. Strain mrp1. *Emirates Journal of Food and Agriculture* 24 (4): 334–343.

161 Ahemad, M. and Khan, M.S. (2012). Effect of fungicides on plant growth promoting activities of phosphate solubilizing *Pseudomonas putida* isolated from mustard (*Brassica compestris*) rhizosphere. *Chemosphere* 86 (9): 945–950. https://doi.org/10.1016/j.chemosphere.2011.11.013

162 Ahemad, M. and Khan, M.S. (2012). Evaluation of plant-growth-promoting activities of rhizobacterium *Pseudomonas putida* under herbicide stress. *Annals of Microbiology* 62 (4): 1531–1540. https://doi.org/10.1007/s13213-011-0407-2

163 Hauwaerts, D., Alexandre, G., Das, S.K., Vanderleyden, J., and Zhulin, I.B. (2002). A major chemotaxis gene cluster in *Azospirillum brasilense* and relationships between chemotaxis operons in α-proteobacteria. *FEMS Microbiology Letters* 208 (1): 61–67. https://doi.org/10.1111/j.1574-6968.2002.tb11061.x

164 Steenhoudt, O. and Vanderleyden, J. (2000). *Azospirillum*, a free-living nitrogen-fixing bacterium closely associated with grasses: Genetic, biochemical and ecological aspects. *FEMS Microbiology Reviews* 24 (4): 487–506. https://doi.org/10.1111/j.1574-6976.2000.tb00552.x

165 Hawes, M.C. and Smith, L.Y. (1989). Requirement for chemotaxis in pathogenicity of *Agrobacterium tumefaciens* on roots of soil-grown pea plants. *Journal of Bacteriology* 171 (10): 5668–5671. https://doi.org/10.1128/jb.171.10.5668-5671.1989

166 De Weert, S., Vermeiren, H., Mulders, I.H., Kuiper, I., Hendrickx, N., Bloemberg, G.V., Vanderleyden, J., De Mot, R., and Lugtenberg, B.J. (2002). Flagella-driven chemotaxis towards exudate components is an important trait for tomato root colonization by *Pseudomonas fluorescens*. *Molecular Plant-microbe Interactions* 15 (11): 1173–1180. https://doi.org/10.1094/MPMI.2002.15.11.1173

167 Scharf, B.E., Hynes, M.F., and Alexandre, G.M. (2016). Chemotaxis signaling systems in model beneficial plant–bacteria associations. *Plant Molecular Biology* 90 (6): 549–559. https://doi.org/10.1007/s11103-016-0432-4

168 Barbour, W., Hattermann, D., and Stacey, G. (1991). Chemotaxis of *Bradyrhizobium japonicum* to soybean exudates. *Applied and Environmental Microbiology* 57 (9): 2635–2639.

169 Song, S.-C. and Lin, L.-P. (1999). The transition of *Rhizobium fredii* lipopolysaccharides induced by soybean root exudation. *Botanical Bulletin of Academia Sinica* 40: 73–78.

170 Zhang, X., Hughes, J.G., Subuyuj, G.A., Ditty, J.L., and Parales, R.E. (2019). Chemotaxis of *Pseudomonas putida* F1 to alcohols is mediated by the carboxylic acid receptor McfP. *Applied and Environmental Microbiology* 85 (22). https://doi.org/10.1128/AEM.01625-19

171 Wichern, J., Wichern, F., and Joergensen, R.G. (2006). Impact of salinity on soil microbial communities and the decomposition of maize in acidic soils. *Geoderma* 137 (1-2): 100–108. https://doi.org/10.1016/j.geoderma.2006.08.001

172 Moore-Kucera, J. and Dick, R.P. (2008). PLFA profiling of microbial community structure and seasonal shifts in soils of a Douglas-fir chronosequence. *Microbial Ecology* 55 (3): 500–511. https://doi.org/10.1007/s00248-007-9295-1

173 Bossio, D.A. and Scow, K. (1998). Impacts of carbon and flooding on soil microbial communities: Phospholipid fatty acid profiles and substrate utilization patterns. *Microbial Ecology* 35 (3-4): 265–278. https://doi.org/10.1007/s002489900082

174 Langer, U. and Rinklebe, J. (2009). Lipid biomarkers for assessment of microbial communities in floodplain soils of the Elbe River (Germany). *Wetlands* 29 (1): 353–362. https://doi.org/10.1672/08-114.1

175 Hedlund, K. (2002). Soil microbial community structure in relation to vegetation management on former agricultural land. *Soil Biology and Biochemistry* 34 (9): 1299–1307. https://doi.org/10.1016/S0038-0717(02)00073-1

176 Steenwerth, K.L., Jackson, L.E., Calderón, F.J., Stromberg, M.R., and Scow, K.M. (2002). Soil microbial community composition and land use history in cultivated and grassland ecosystems of coastal California. *Soil Biology and Biochemistry* 34 (11): 1599–1611. https://doi.org/10.1016/S0038-0717(02)00144-X

177 Zornoza, R., Guerrero, C., Mataix-Solera, J., Scow, K., Arcenegui, V., and Mataix-Beneyto, J. (2009). Changes in soil microbial community structure following the abandonment of agricultural terraces in mountainous areas of Eastern Spain. *Applied Soil Ecology* 42 (3): 315–323. https://doi.org/10.1016/j.apsoil.2009.05.011

178 Kang, S. (2005). *Spatial Structure of Soil Microbial Communities and Their Controlling Factors*, 249. University of Virginia.

179 Hoyle, F., Murphy, D., and Brookes, P. (2008). Microbial response to the addition of glucose in low-fertility soils. *Biology and Fertility of Soils* 44 (4): 571–579. https://doi.org/10.1007/s00374-007-0237-3

180 Oren, A. (1999). Bioenergetic aspects of halophilism. *Microbiology and Molecular Biology Reviews* 63 (2): 334–348. doi: 10.1128/MMBR.63.2.334-348.1999.

181 Zahran, H. (1997). Diversity, adaptation and activity of the bacterial flora in saline environments. *Biology and Fertility of Soils* 25 (3): 211–223. https://doi.org/10.1007/s003740050306

182 Navarrete, A.A., Tsai, S.M., Mendes, L.W., Faust, K., De Hollander, M., Cassman, N.A., Raes, J., Van Veen, J.A., and Kuramae, E.E. (2015). Soil microbiome responses to the short-term effects of Amazonian deforestation. *Molecular Ecology* 24 (10): 2433–2448. https://doi.org/10.1111/mec.13172

183 Waldor, M.K., Tyson, G., Borenstein, E., Ochman, H., Moeller, A., Finlay, B.B., Kong, H.H., Gordon, J.I., Nelson, K.E., and Dabbagh, K. (2015). Where next for microbiome research? *PLoS Biology* 13 (1): e1002050. https://doi.org/10.1371/journal.pbio.1002050

184 Foo, J.L., Ling, H., Lee, Y.S., and Chang, M.W. (2017). Microbiome engineering: Current applications and its future. *Biotechnology Journal* 12 (3): 1600099. https://doi.org/10.1002/biot.201600099

185 Chaparro, J.M., Sheflin, A.M., Manter, D.K., and Vivanco, J.M. (2012). Manipulating the soil microbiome to increase soil health and plant fertility. *Biology and Fertility of Soils* 48 (5): 489–499. https://doi.org/10.1007/s00374-012-0691-4

186 Hartmann, M., Frey, B., Mayer, J., Mäder, P., and Widmer, F. (2015). Distinct soil microbial diversity under long-term organic and conventional farming. *The ISME Journal* 9 (5): 1177–1194. https://doi.org/10.1038/ismej.2014.210

187 Hartmann, M., Niklaus, P.A., Zimmermann, S., Schmutz, S., Kremer, J., Abarenkov, K., Lüscher, P., Widmer, F., and Frey, B. (2014). Resistance and resilience of the forest soil microbiome to logging-associated compaction. *The ISME Journal* 8 (1): 226–244. https://doi.org/10.1038/ismej.2013.141

188 Souza, R.C., Mendes, I.C., Reis-Junior, F.B., Carvalho, F.M., Nogueira, M.A., Vasconcelos, A.T.R., Vicente, V.A., and Hungria, M. (2016). Shifts in taxonomic and functional microbial diversity with agriculture: How fragile is the Brazilian Cerrado? *Bmc Microbiology* 16 (1): 42. https://doi.org/10.1186/s12866-016-0657-z

189 Xiong, W., Li, Z., Liu, H., Xue, C., Zhang, R., Wu, H., Li, R., and Shen, Q. (2015). The effect of long-term continuous cropping of black pepper on soil bacterial communities as determined by 454 pyrosequencing. *PLoS One* 10: 8. https://doi.org/10.1371/journal.pone.0136946

190 Xiong, W., Zhao, Q., Zhao, J., Xun, W., Li, R., Zhang, R., Wu, H., and Shen, Q. (2015). Different continuous cropping spans significantly affect microbial community membership and structure in a vanilla-grown soil as revealed by deep pyrosequencing. *Microbial Ecology* 70 (1): 209–218. https://doi.org/10.1007/s00248-014-0516-0

191 Glick, B.R. (2012). Plant growth-promoting bacteria: Mechanisms and applications. *Scientifica* 2012. Article ID 963401 https://doi.org/10.6064/2012/963401

192 Santhanam, R., Weinhold, A., Goldberg, J., Oh, Y., and Baldwin, I.T. (2015). Native root-associated bacteria rescue a plant from a sudden-wilt disease that emerged during continuous cropping. *Proceedings of the National Academy of Sciences* 112 (36): E5013–E5020.

193 Gopal, M., Gupta, A., and Thomas, G.V. (2013). Bespoke microbiome therapy to manage plant diseases. *Frontiers in Microbiology* 4: 355. https://doi.org/10.3389/fmicb.2013.00355

194 Zolla, G., Badri, D.V., Bakker, M.G., Manter, D.K., and Vivanco, J.M. (2013). Soil microbiomes vary in their ability to confer drought tolerance to Arabidopsis. *Applied Soil Ecology* 68: 1–9. https://doi.org/10.1016/j.apsoil.2013.03.007

195 Ando, H., Lemire, S., Pires, D.P., and Lu, T.K. (2015). Engineering modular viral scaffolds for targeted bacterial population editing. *Cell Systems* 1 (3): 187–196. https://doi.org/10.1016/j.cels.2015.08.013

196 Frampton, R.A., Pitman, A.R., and Fineran, P.C. (2012). Advances in bacteriophage-mediated control of plant pathogens. *International Journal of Microbiology* 2012. doi: 10.1155/2012/326452.

197 Lu, T.K. and Collins, J.J. (2007). Dispersing biofilms with engineered enzymatic bacteriophage. *Proceedings of the National Academy of Sciences* 104 (27): 11197–11202.

198 Summers, W.C. (2001). Bacteriophage therapy. *Annual Reviews in Microbiology* 55 (1): 437–451.

6

Core Microbiome

Plant Growth and Development

Thirunarayanan P[1], *Uday Kumar Thera, Tulasi Korra*[1], *and Manoj Kumar V*[2]

[1]*Department of Mycology and Plant Pathology, IAS, Banaras Hindu University, Varanasi, Uttar Pradesh*

[2]*Department of Plant Pathology, ANGRAU, Bapatla, Andhra Pradesh*

6.1 Introduction

The plant microbiome, consisting of bacteria, fungi, archaea, protists, and viruses, have a dynamic interactive effect on the physiology of hosts. Our area of focus has been on bacteria and fungi. The advantages to plants comprise 20–40% plant photosynthate that is secreted into the rhizosphere; lesser organisms survive under the phyllosphere, but due to root colonization, greater communities of organisms survive in the endosphere. Transmission is either vertical or horizontal through air or soil. The main functions of the core microbiome are nutrient mobilization, exhibition of antagonism toward the pathogen, plant stress alleviation, plant growth promotion, co-metabolism, and degradation of toxins. However, the core plant microbiome comprises genes present in a given habitat associated with a given plant. Likewise, other than beneficial microbes, specific plant pathogenic bacteria caused by infection over humans, such as *Pseudomonas aeruginosa*, *Salmonella* spp., and *Stenotrophomonas maltophilia*, can be examined in plant tissue and root-associated microbiome. Beneficial microbes mainly depend on heterogeneous groups associated with roots to protect from disease and enhance growth [1]. Microbial communities make essential nutrients like nitrogen, phosphorus, and potassium important for plant growth in agricultural production. The current overuse of large-scale chemical fertilizers has been causing an adverse environmental impact as reported by Adesemoye and co-workers [2].

Further, biofertilizers are inexpensive while ensuring better plant growth by utilizing bacteria inoculants. Furthermore, plant growth-promoting rhizobacteria (PGPR) inoculants can be easily obtained, and their strategic benefit factor is that their use can reduce the use of chemicals and increase sustainable management. The successful rates of bacterial inoculants are influenced by various factors, such as temperature, pH, and soil health. Soil health is an essential factor that affects root colonization efficiency in PGPR [3]. The focus of this chapter is to uncover knowledge related to the enhancement of plant growth and development, host–microbe interaction, and alleviation of biotic and abiotic stresses for beneficial microbes.

Core Microbiome: Improving Crop Quality and Productivity, First Edition. Edited by Javid A. Parray, Nowsheen Shameem, Elsayed Fathi Abd-Allah, and Mohammad Yaseen Mir.

6.2 Types of Microbiome

6.2.1 Root Microbiome

Many microorganisms thrive on the outermost layer of the root zone known as the rhizosphere, where a good association with bulk soils is not possible [4]. The rhizosphere, therefore, is a habitat for varied microorganisms [5]. Most soils are carbon-starved with inadequate plant nutrients because microflora secrete up to 40% of their photosynthetic products into the rhizosphere other than bulk soil. Plants select specified microbes for balanced beneficial ratio from the core microbiome. Although it is not similar in all soils, it might differ for varied soils with different microbial communities. Disease-suppressive soils have both beneficial and harmful microbes [6]. The monocropping system was found to suppress disease to some extent. Multiple-cropping systems are effective in building up disease suppressiveness because the changing cropping pattern is likely to reduce host-specific pathogens. It helps build beneficial plant–microbe interaction in the rhizosphere.

Nevertheless, reports on disease-suppressive soils, involving *Fusarium* wilt of broad host range, *Gaeumannomyces graminis* var *tritici* (take-all disease of wheat), seedling stage (dumpling-off), potato tubers (*Streptomyces scabies*) were presented by Mendes and co-workers [5]. Suppressive soils are antagonistic to reducing plant diseases mainly root diseases. Bacteria can colonize to form bacteroids, which fight against virulent pathogenic fungi, suggesting microflora immunity [7]. For example, to manage the take-all disease of wheat, farmers commonly used antibiotics, such as *Pseudomonads* and *Actinobacteria*. Soil microbiome keeps changing from time to time because plants can select rhizosphere microorganisms that are antagonistic to pathogens. Groups of bacteria such as *Gamma-Proteobacteria, Betaproteobacteria*, and *Firmicutes* play an essential role in disease-suppressive soils [8].

6.2.2 Leaf Microbiome

The term *phyllosphere* comprises the "Aerial part of a plant or a part of a plant above the ground, the leaf surface, which is considered a habitat for microorganisms" or the place where various microorganisms (bacteria, yeast, and fungi) colonize. The total leaf area corresponds to the top and bottom surfaces, about twice the surface area of land [9]. Leaf microbiome is the environment in which colonization of and interaction with epiphytes and plants occur. First, it was recently [10] found that *Sphingomonads* and *Methylobacteria* are the dominant taxa when working with the corn leaf microbiome in 300 corn lines and that the leaf microbiota ratio of the phyllosphere is adversely dependent on biotic factors. Second, in leaf microbiota, the four types of classes are *Firmicutes, Proteobacteria, Actinobacteria*, and *Bacteriodes* [11]. When studying leaf microbiota, researchers looked for the presence of methylotrophic bacteria [12–14]. Similarly, *Archaea* is associated with endophytic plants,such as *Euryarchaeota, Thaumarchaeota*, and *Crenarchaeota* [15]. Bacterial characterization of leaf microbiome under *Firmicutes* has been reported [11]. Fungal nutrition may be via intracellular or intercellular tissue, causing diseases in plant systems [16,17]. Although several authors have mentioned that leaf microbiota plays a reflective role in plant ecosystem, a microbiome indicate the presence of many a diverse species with various physiological and biochemical processes, involving the decomposition of dried leaves [18,19]. Third, the fungal leaf microbiome predominantly house underclass Ascomycetes (Capnodiales and Plesporales) and Basidiomycetes (Tremellales, Ustiaginales). Genera *Hannaella, Dioszegia, Cladosporium*, and *Moesziomyces* were frequently found. *Hannaella* is a high rootstock variety of tomato (Ganbarune), and in ungrafted tomatoes, the genus *Deinococcus* appeared [20]. In Portugal, 68 families with 149 genera in leaf microbiome of olive tree were reported, and in the Mediterranean region [21], 13 fungal taxa in twigs and leaves were reported [22].

The pathogen-triggered immunity signaling network regulates the diversity of intracellular leaf microbiome, and Arabidopsis plant mutant changed the leaf microbiome by triggering signaling pathways *cad1* mutant [23]. Fourth, *Actinobacteria* were isolated from crops and medicinal and terrestrial plants [24]. For example, mangrove plants such as *Actinomaduraglauciflava, Actinoplanemisouriensis, Nocardia* spp., and *Streptomyces* spp. were separated from cereal crops [14].

6.3 Hormone Interactions with Microbiota

Plant hormones help with the growth, protection, and symbiosis processes in our ecosystem. The core microbiome's hormonal contribution to the host plant is decreased in the phyllosphere compared with the rhizosphere [25]. Nevertheless, information from plant microbiome research has proved useful in studies on the upper parts of crop plants [26]. Auxin and cytokinin (CK) interaction improved in the plant immune system [27]. Indole acetic acid (IAA) production increased the growth of microbes that boost crop yield productivity [28]. However, the first analysis of yeast from leaf microbiome involving the predatory *Drosera* spp. showed enhanced IAA production.

Their findings strongly indicate that yeast has TRP independent signal pathways for IAA biosynthesis and that IAA-producing yeast is used as a biofertilizer inoculant for plant growth stimulation [29]. In addition, in IAA production, the microorganisms involved are *Bradyrhizobium japonicum, Mycobacterium, Enterobacter asburiae, Acinetobacter* spp., and *Serratia marcescens* [30]. Additionally, cytokinins help nitrogen transport to the upper parts of the plant to improve the biomass content [31]. The virulent microorganism releases a low range of cytokinin and reduces the salicylic acid and changes the plant's innate immunity [32] to attack cytokines in cell growth and immune mediation. Ethylene, which is at the core of these processes, is an easily diffusible phytohormone in living plant tissue, and has proved to be effective at a low concentration. The first pathogen, *R solanacearum*, produces ethylene for the premature ripening process in banana crop [33]. Ethylene regulates many plant growth methods [34], but it also involves bacterial-induced protection and shares a symbiotic platform, affecting bacteriome assembly [35]. Furthermore, several studies have shown that 1-aminocyclopropane-1-carboxylate (ACC) acts as a direct precursor for ethylene production and affects by controlling the developmental stages of crop and defense responses [36,37]. Plant microbiome indicated an increase in ethylene production by using engineered *Escherichia coli* bacteria to produce ACC oxidase, and auxin and cytokinin to upregulate the coding regions of ACC [38,39].

6.4 Enhancement of Plant Growth

Rhizosphere microorganisms are an alternate economical and environmentally friendly choice farmers can make to minimize crop damage [40]. Yet, farmers use chemical fertilizers, which, despite being helpful for a short period, have an adverse impact on the environment in the long run. Therefore, plant growth-promoting bacteria (PGPB) are an obvious choice. For example, *Azotobacter, Arthobacter, Azospirillum, Burkholderia, Bacillus, Enterobacteria, Pseudomonas*, and *Klebsiella* are the most repeatedly used microbial genera, which have shown an ability to increase crop growth and yield. Hence, the plant–microbiome interaction facilitates either a symbiotic or a non-symbiotic association. Microbiomes colonize around the root zone during plant growth, where plant roots release exudates, or secondary metabolites, into the soil. Additionally, the crop assesses the microbial-known pool and changes according to the safeguarding mechanism of development for plant growth in relation to the microbes. But the bacteria can act as free-living organisms that show a symbiotic interaction. Microbes help develop the outer layer of the root system during the nodulation process, fixing nitrogen to ammonia [41]. Previous studies on *Rhizobium* spp. confirmed that the PGPR bacteria positively correlate with nitrogen fixation in legume plants in direct and indirect ways: in the former, entry into the plant root system is gained through the modulation of hormone levels, multiplication in plant cells, nitrogen fixation in soluble form, and production of siderophore, whereas in the latter, systemic resistance is induced and antagonistic substances are produced to defend against pathogens. This is used as a biocontrol agent because it has a rapidly growing *in-vitro* condition. It quickly recognizes root exudates during root colonization and releases bioactive molecules into the soil. For example, *Agrobacterium tumefaciens* is a gram-negative rod-shaped phytopathogen that causes the crown gall disease in the root, which mainly affects woody, herbaceous plants. *A. radiobacter* (Agrocin K 84) acts as an antibiotic compound and prevents crown gall formation. Soil microbiome moderates plant health [42]. For example, the microbiome can secrete plant hormones, such as the IAA precursor in the auxin pathway, which enhances root growth. Microbes that enhance antifungal activity and assist in the alleviation of stress are listed in Table 6.1.

6.5 Nitrogen Fixation and the Possible Role of Core Microbiota

Nitrogen is the primary (macro) nutrient for plant growth and productivity – despite 78% of nitrogen being available in the atmosphere, it is unavailable in absorbable form to crops. Nitrogen-fixing organisms are classified into two: symbiotic (S) and non-symbiotic (NS) groups. Symbiotic nitrogen-fixing microorganisms, such as *Bradyrhizobium, Rhizobium, Mesorhizobium, Sinorhizobium*, and *Azorhizobium* fall under Rhizobiaceae [66]. In contrast, NS microorganisms, such as *Azospirillum, Azotobacter, Acetobacter*, and blue-green algae, fix nitrogen in non-leguminous crops [67]. In the root structure of legume plants, a nitrogen-fixing bacterium forms nodules at different stages of development within the root zone. First, the release of exudates into the soil is recognized by microbes to form a rhizosphere zone, which is divided into four sub-zones or layers: (i) active cell division takes place in meristematic tissue; (ii) formation of rhizobia and differentiation with infected

Table 6.1 List of beneficial microbes showing antimicrobial activity and alleviation of biotic and abiotic stresses.

Antimicrobial activity				
No.	**Beneficial microbes**	**Isolated from different host microbiome**	**Effect**	**References**
1	*Phoma* spp.	Finger millet	Shows antifungal activity toward *F. graminearum*	[43]
2	*Enterobacter* spp.	Finger millet	Improves antimicrobial effect toward *Fusarium* spp.	[44]
3	*Paenibacillus polymyxa*	Teosinate–wild maize	Suppresses the activity of *F. graminearum*	[43]
4	*Rhizophagus irregularis* (previously known as *Glomus intraradices*)	Wild rice	Improves resistance to *Magnaporthe oryzae*	[45]
5	*Actinobacteria* and *Proteobacteria*	Kodo millet	Produces antimicrobial compounds against biotic stresses	[46]
6	*Bacillus amyloliquefaciens*FZB42	Maize	Releases maize root exudates; produces antimicrobial compounds	[47]
7	*Bacillus magaterium*	Maize	Stays antagonistic to *F. verticillioides*	[48]
9	*Pseudomonas piscium*	Wheat	Stays antagonistic to *F. graminearum*; inhibits growth and virulence	[49]
10	*Streptomyces* spp.DEF09	Wheat	Stays antagonistic to *F. graminearum*; decreases disease severity	[50]
11	*Bacillus subtilis* ZD01	Potato	Stays antagonistic to *Alternaria solani*	[51]
12	*Pseudomonas piscium*	Wheat	Stays antagonistic to *F. graminearum*; inhibits growth and virulence	[49]
13	*Streptomyces* spp.DEF09	Wheat	Stays antagonistic to *F. graminearum*; decreases disease severity	[50]
14	*Clonostachys rosea* BAFC3874	Soybean	Shows antagonist mechanism in *Sclerotinia sclerotiorum*	[52]
15	*C.rosea* IK 726	Barley	Stays antagonistic to *F. graminearum*	[53]

(Continued)

Table 6.1 (Continued)

No.	Beneficial microbes	Isolated from different host microbiome	Effect	References
Antimicrobial activity				
16	*Bacillus amyloliquefaciens*ZDS-1	Wheat	Degrades *Fusarium* spp.; produces mycotoxin zearalenone	[54]
17	*Bacillus* and *Serratia*	Ginger	Reduces incidence of bacterial wilt of ginger	[51]
18	*C. rosea* IK 726	Barley	Reduces disease incidence in *Bipolaris sorokiniana*	[55]
19	*C. rosea* IK 726	Wheat	Enhances resistance to *Xanthomonas translucens*	[56]
Abiotic stresses				
1	*Trichoderma longibrachiatum*	Wheat	Promotes seedling tolerance in salt stress	[57]
2	*Acidiphilium, Mortierella,* and *Glomus*	Common wild rice	Adapts to low temperatures	[58]
3	Comamonadaceae and Pseudomonadaceae	Maize	Improves tolerance to cold	[59]
4	*Paenibacillus polymyxa* and *Bacillus circulans*	Maize	Improves tolerance to copper stress	[60]
5	*Paraburkholder iatropica, Herbaspirillum frisingense, P. flourescens,* and *Kosakoniaradic incitans*	Sugarcane	Reduces aluminum stress	[61]
Growth enhancement				
1	*B. amyloliquefaciens*LL2012 and *Bradyrhizobium japonicum*	Soybean	Enhances number of nodulations	[62]
2	*Bacillus siamensis*	Bean	Enhances growth	[63]
3	*Herbaspirillum seropedicae* ZAE94	Maize	Increases yield and biological nitrogen fixation (BNF)	[64]
4	*Delftia acidovorans* IAC-BECa-023	Sugarcane	Promotes growth	[65]
5	*Funneliformis mosseae*	Wheat	Enhances biomass and yield	[56]

cells; (iii) central nitrogen fixation occurs during which atmospheric nitrogen (N_2) is converted into ammonical form (NH_3) by enzyme nitrogenase; (iv) in the zone of senescence, leghemoglobin nodule formation is structurally identical to the hemoglobin in blood. The function of leghemoglobin is to balance the plant body of oxygen tension in the nodule. Many microbes, such as *Actinobacter, Azotobacter, Azosprillum, Pseudomonas, Klebsiella,* and *Rhizobium* spp., are involved in nitrogen fixation. For example, *Azosprillum* is used as a biofertilizer in many crops, such as rice, wheat, maize, and sorghum, to help with nutrients and minerals uptake. Some microbial genera can solubilize organic phosphates or inorganic phosphates. In earlier decades, BNF was studied to acquire knowledge on mineral nutrition, using model plants to also assess genes, hormones, and different signaling pathways. Farmers who used BNF byproducts have observed improved growth in developed and developing countries [68]. As regards ecological concerns, their most significant applications comprise the deduction of fossils and restoration and reforestation of barren areas to resume productive agriculture [69]. Minimal dosage of BNF byproducts is recommended for fields to avoid the release of toxic substances as eutrophication occurs in lakes and rivers, leading to serious pollution problems. Biological nitrogen fixation, which involves the conversion of atmospheric nitrogen to ammonia by the enzyme nitrogenase, was simultaneously discovered by Hermann Hellriegel, a German agronomist, and Martinus Bejierinck, a Dutch microbiologist.

6.5.1 Non-symbiotic Nitrogen Fixation

Free-living microorganisms assist with nitrogen fixation. They are aquatic or terrestrial in their habitat [70]. They can be with or without oxygen or blue-green algae. Free-living nitrogen fixers can be aerobic as *Azotobacter* and *Beijerenckia*, anaerobic as *Clostridium*, photosynthetic nitrogen fixers as *Chorobium* and *Rhodopseudomonas*, and chemosynthetic fixers as *Desulfovibro* and *Thiobacillus* (sulfur bacteria). Examples of blue-green algae may be unicellular (*Synechococcus*), filamentous non-heterocystous (*Oscillatoria*), and filamentous heterocystous (*Nostoc, Anabaena*) ones [71].

6.5.2 Symbiotic Fixation in Diazotrophs

Symbiotic fixation occurs in heterotrophs and photoautotrophs. Anton De Bary coined the term "symbiosis," which can be categorized into two categories, i.e., (i) nodulation in (a) leguminous and(b) non-leguminous plants and (ii) non-nodulation. Nodule formation in non-leguminous plants was found in plants other than legumes and nodulated roots [72], such as *Pavettazinumerm anniana*, which is a leaf nodule in some gymnosperms, *Myrica gale*, and *Casuarina equisetifolia*; non-nodulation examples include *Azolla-Anabaena azollae*, lichens – blue-green algae (Cyanobacteria), *Anthoceros–Nostoc*, Cycas (gymnosperms)–*Nostoc*, and *Anabaena* [73]. In the nodule-forming category, *Rhizobium* spp. will infect the roots of legume plants. *Rhizobium* is a free-living bacteria; the roots of legumes secrete exudates like growth factors and can be helpful for the rapid multiplication of bacteria near the rhizosphere zone. For example, *Pisum sativum* (pea) releases exudates, such as homoserine and carbohydrate-containing proteins (CCC). The modulation process involves the following steps: (i) quorum-sensing and sending chemical signals between root and bacteria; (ii) changing the shape of the root into a root curl; (iii) forming infection threads; (iv) penetrating the roots with the help of the infection threads; (v) forming the nodule tissue, after which, in the cells, rhizobium bacteria convert into bacteroids to form nitrogenase enzyme [74]. Microbes need carbon for their growth. Rhizobia provide the fixed nitrogen to the plant, true to the function of a nodule or the bacterial zone covered by cortical cells. The content of nitrogen fixation is portioned to the size of the nodule. Effective nodules are pinkish due to the formation of leghemoglobin, due to the heme protein, which contains a heme moiety attached to a peptide chain in the globin. The amount of leghemoglobin in the nodules (MW-16,000-17,000 Da) has a direct relation to atmospheric nitrogen [75,76], which is different from that in the nodulation process that occurs in the roots (Figure 6.1).

6.6 Biochemistry of Nitrogen Fixation

Nitrogen fixation is an androgenetic reaction process that consumes 160 kcal energy. Nitrogen fixers utilize atmospheric nitrogen to form ammonia [77]. During the synthesis of nitrogen, the triple bonds are broken down by the enzyme nitrogenase. When free nitrogen is released, it combines with hydrogen atom mediated by the hydrogenase enzyme to release 13 kcal energy. This reaction is called an exergonic response. BNF requires approximately 16 mol of ATP per molecule of N atom and a net input of 147 kcal energy [78].

6.7 Pathway of Nitrogen Fixation in Root Nodules

Glucose-6-phosphate acts as an electron donor. With the help of NADP combined with a hydrogen atom, it is converted to phosphoglucose acid [79]. In the following reaction:

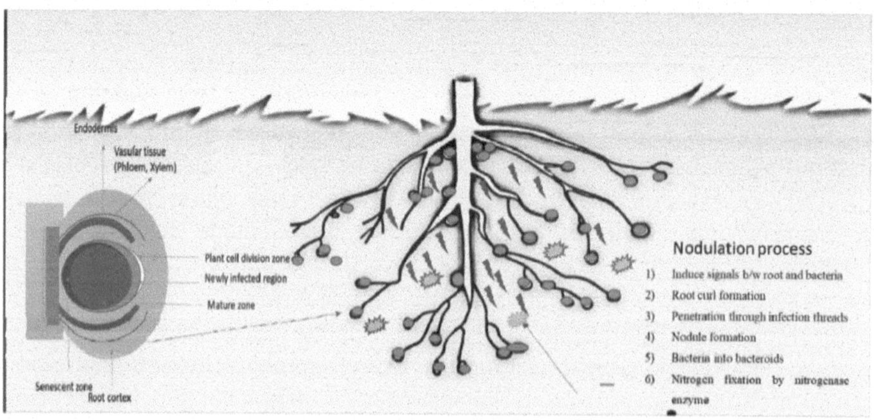

Figure 6.1 Underground parts of roots showing the nodulation process and internal nodule parts (zones).

$$\text{Glucose-6-phosphate} + \text{NADP} + H_2O \xrightarrow[0]{\textit{Hydrogenase}} \text{6 Phosphogluconic acid} + \text{NADPH} + H^+$$

NAPDH donates electrons to ferredoxin. The reduced ferredoxin acts as an electron carrier. Thus, oxidation takes place by the release of electrons by ferredoxin. During the reduction process, Fe proteins are combined with ATP molecules as a result of Mg^{+2} ions. Furthermore, the second subunit is activated and reduced. In this way, it donates electrons to convert free N_2 into ammonia [80].

Dinitrogen → **Diimide** → **Hydrazine** → **Hydroxylamide** → **Ammonia**

$$N_2 \xrightarrow[2H+,2e^-]{4\,ATP} HN{=}HN \xrightarrow[2H+,2e^-]{4\,ATP} H_2N{-}{-}NH_2 \xrightarrow[2H+,2e^-]{4\,ATP} HO_2HN{-}NH_2OH \xrightarrow[2H+,2e^-]{4\,ATP} 2NH_3$$

Overall reaction:

$$N_2{+}16ATP{+}8H^+{+}8e^- \longrightarrow 2NH_3{+}H_2{+}16ADP{+}16Pi$$

The complete reduction process is as follows:

6.8 Plant Genes for Nodule Development

Sym 28. 29, 35 is required for nodulation in *P. sativum* (pea); ortholog of *LIN* of Birds foot trefoil [81,82] and *sym* 33, 40 [83] is used to deactivate the nodulation. Following prolonged infection periods, *LIN*, the barrel clover gene alters the nodule [84]. While *ENOD93* genes are helpful in the early nodulation development process in soybean, miRNAs play a vital regulatory role and control gene expression [85,86]. In soybean–*B. japonicum* interaction, the gene expression of miRNAs increases the nodulation number, root length, and nodule primordia [87]. The early gene *NIN* regulatory protein is required for nodule development in *Lotus japonicus*. The *nin* transposon tagged mutant gene ceases early development and recognition [88]. Physiological changes occur during the establishment of the nodule. Several phytohormones play an important role in the new organs known as nodules. Gibberellin is a phytohormone involved in both steps, at an early stage of a microbial infection and later at nodulation formation [89]. Similarly, the regulation by *Agrobacterium*, *Rhizobium*, and *Azosprillum* showed positive effects for phytohormones biosynthesis. They produce IAA via the indole 3-acetamide pathway (IAP) and the indole-3 pyruvic acid pathway (IPP), and phytohormones help the stimulation of root growth.

6.9 Role of Biofertilizers

Biofertilizers comprise living or latent cells of microorganisms that are helpful for the conservation and mobilization of crop nutrients in the soil. In other words, biofertilizers are substances that contain beneficial microorganisms that promote crop growth and uptake of significant nutrients when applied to soil in absorbable form to seed or plant surfaces in the rhizosphere [90]. With regard to fungal biofertilizers, the role of a few microorganisms, such as *Pseudomonas fluourescens, Bradyrhizobium, Rhizobium* spp., *Chryseobacterium, Phyllobacterium, Streptomyces, Enterobacter asburiae*, and *Serratia mercescens*, are significant in plant growth promotion, with indirect BNF release into the siderophore and the central part of the siderophore, by converting Fe^{+2} to Fe^{3+} ions [91]. Moreover, some microorganisms, such as *Mesorhizobium mediterraneum, Bradyrhizobium* spp., *Enterobacter* spp., and *Arthobacter* spp., assist in the solubilization of phosphate, while some assist in the mobilization of phosphate, e.g. *Glomus*. In nitrogen fixation, the organisms involved are *Azocarus, Azorhizobium, Beijerinckia, Burkholderia, Frankia*, and *Gluconacetobacter*. For example, *Frateuria aurantia* acts as a potassium solubilizer; *Acidiothiobaciilus* spp. acts as silicate dissolver, and *Pseudomonas* spp. acts as a biofertilizer [92,93]. Different biofertilizers and their trade names are listed in Table 6.2.

6.10 Conclusion

Some scientists have tailored the use of the core microbiome, which has been found to have a positive effect on plants while having an eligible adverse impact on the environment. While this topic has uncovered available information on plant growth and development, it has also left the door open to some possible future research. To better understand plant–microbe interaction, we require molecular and phenotypic characterization microbiome strains to enhance nitrogen-use efficiency for different

Table 6.2 Trade names of different biofertilizers.

Microbiome	Trade name
Pseudomonas fluroscens	Bioshield, TNAUfp-7, Sudomonas, and Biomonas
Rhizobium sp.	Effect grow Rhizobium, Nodular, Rhizofer, Rhizoum, Green Rhizo, and Green Lift-R
Gluconacetobacter	Nitromax, NITRO-FIX, GluconN_2, and Green Bio Phosphorus
Burkholderia	OMNIBACT

environmental studies, such as the in-vitro plant–microbe interaction study. With the new functional genomics approach to legumes, we can easily gain access to an analysis of the symbiotic interactions within the microbiome. While map-based cloning is used to identify genes in root nodules and DNA chips or microarrays are used to provide information on gene functions and pathways, diverse new techniques involving core microbiome interactions are yet to be explored.

References

1 Glick, B.R. (2012). Plant growth-promoting bacteria: Mechanisms and applications. Scientifica, 2012.

2 Adesemoye, A.O., Torbert, H.A., and Kloepper, J.W. (2009). Plant growth-promoting rhizobacteria allow reduced application rates of chemical fertilizers. *Microbial Ecology* 58 (4): 921–929.

3 Carvalhais, L.C., Dennis, P.G., Fan, B., Fedoseyenko, D., Kierul, K., Becker, A., and Borriss, R. (2013). Linking plant nutritional status to plant–microbe interactions. *PLoSOne* 8 (7): e68555.

4 Jones, D.L., Hodge, A., and Kuzyakov, Y. (2004). Plant and mycorrhizal regulation of rhizodeposition. *New Phytologist* 163: 459–480.

5 Mendes, R. et al. (2011). Deciphering the rhizosphere microbiome for disease-suppressive bacteria. *Science* 332: 1097–1100.

6 Raaijmakers, J.M. et al. (2009). The rhizosphere: A playground and battlefield for soilborne pathogens and beneficial microorganisms. *Plant Soil* 321: 341–36.

7 Carri6n, V.J., Perez-Jaramillo, J., Cordovez, V., Tracanna, V., de Hollander, M., Ruiz-Buck, D., Mendes, L.W., van Ijcken, W.F.J., Gomez-Exposito, R., Elsayed, S.S., Mohanraju, P., Arifah, A., van der Oost, J., Paulson, J.N., Mendes, R., Van Wezel, G.P., Medema, M.H., and Raaijmakers, J.M. (2019). Pathogen-induced activation of disease-suppressive functions in the endophytic root microbiome. *Science* 336: 606–612. https://doi.org/10.1126/science.aaw9285

8 Hazen, T.C. et al. (2010). Deep-sea oil plume enriches Indigenous oil-degrading bacteria. *Science* 330: 204–208.

9 Vorholt, J.A. (2012). Microbial life in the phyllosphere. *Nature Reviews* 10: 828–840.

10 Wallace, J.G., Kremling, K.A., Kovar, L.L., and Buckler, E.S. (2018). Quantitative genetics of the maize leaf microbiome. *Phytobiomes Journal* 2 (4): 208–224.

11 Durand, A., Maillard, F., Alvarez-Lopez, V., Guinchard, S., Bertheau, C., Valot, B. et al. (2018). Bacterial diversity associated with poplar trees grown on a Hg-contaminated site: Community characterization and isolation of Hg-resistant plant growth-promoting bacteria. *Science of the Total Environment* 622: 1165–1177. DOI: 10.1016/j.scitotenv.2017.12.069

12 Kwak, M.J., Jeong, H., Madhaiyan, M., Lee, Y., Sa, T.M., Oh, T.K., and Kim, J.F. (2014). Genome information of *Methylobacterium oryzae*, a plant-probiotic methylotroph methylotroph in the phyllosphere. *PLoS O.N.E.* 9 (9): e106704. https://doi.org/10.1371/journal.pone.0106704

13 Krishnamoorthy, R., Kwon, S.W., Kumutha, K., Senthilkumar, M., Ahmed, S., Sa, T., and Anandham, R. (2018). Diversity of culturable methylotrophic bacteria in different genotypes of groundnut and their potential for plant growth promotion. *3 Biotech* 8: 275.

14 Sivakumar, N., Sathishkumar, R., Selvakumar, G., Shyamkumar, R., and Arjunekumar, K. (2020). Phyllospheric microbiomes: diversity, ecological significance, and biotechnological applications. *Plant Microbiomes for Sustainable Agriculture* 25: 113–172. 10.1007/978-3-030-38453-1_5

15 Muller, H., Berg, C., Landa, B.B., Auerbach, A., Moissl-Eichinger, C., and Berg, G. (2015). Plant genotype-specific archaeal and bacterial endophytes but similar Bacillus antagonists colonize Mediterranean olive trees. *Frontiers in Microbiology* 6: 138.

16 Yadav, A.N., Mishra, S., Singh, S., and Gupta, A. (2019b). *Recent Advancement in White Biotechnology through Fungi Volume1: Diversity and Enzymes Perspectives.* Cham: Springer International Publishing.

17 Yadav, A.N., Singh, S., Mishra, S., and Gupta, A. (2019d). *Recent Advancement in White Biotechnology through Fungi.Volume 3: Perspective for Sustainable Environments.* Cham: Springer International Publishing.

18 Guerreiro, M.A., Brachmann, A., Begerow, D., and Peršoh, D. (2018). Transient leaf endophytes are the most active fungi in 1-year-old beech leaf litter. *Fungal Diversity*89: 237–251. 10.1007/s13225-017-0390-4

19 Yao, H., Sun, X., He, C., Maitra, P., Li, X.C., and Guo, L.D. (2019). Phyllosphere epiphytic and endophytic fungal community and network structures differ in a tropical mangrove ecosystem. *Microbiome* 7 (1): 1–15.

20 Toju, H., Okayasu, K., and Notaguchi, M. (2019). Leaf-associated microbiomes of grafted tomato plants. *Scientific Reports* 9: 1787. https://doi.org/10.1038/s41598-018-38344-2

21 Martins, F., Pereira, J.A., Bota, P., Bento, A., and Baptista, P. (2016). Fungal endophyte communities in above- and belowground olive tree organs and the effect of season and geographic location on their structures. *Fungal Ecology* 20: 193–201. 10.1016/j.funeco.2016.01.005

22 Abdelfattah, A., Li DestriNicosia, M.G., Cacciola, S.O., Droby, S., and Schena, L. (2015). Metabarcoding analysis of fungal diversity in the phyllosphere and carposphere of olive (*Olea europaea*). *PLoSOne* 10 (7): e0131069.

23 Chen, T., Nomura, K., Wang, X. et al. (2020). A plant genetic network for preventing dysbiosis in the phyllosphere. *Nature* 580: 653–657.

24 Nalini, M.S. and Prakash, H.S. (2017). Diversity and bioprospecting of actinomycete endophytes from the medicinal plants. *Letters in Applied Microbiology* 64 (4): 261–270.

25 Abd El-Daim, I.A., Bejai, S., and Meijer, J. (2014). Improved heat stress tolerance of wheat seedlings by bacterial seed treatment. *Plant and Soil* 379: 337–350.

26 Mahmood, A., Turgay, O.C., Farooq, M., and Hayat, R. (2016). Seed biopriming with plant growth-promoting rhizobacteria: A review. *FEMS Microbiology Ecology* 92.

27 Stone, B., Weingarten, E., and Jackson, C. (2018). The role of the phyllosphere microbiome in plant health and function. *Annual Plant Reviews Online* 5: 533–5561.

28 Romero, F.M., Marina, M., and Pieckenstain, F.L. (2016). Novel components of leaf bacterial communities of field-grown tomato plants and their potential for plant

growth promotion and biocontrol of tomato diseases. *Research in Microbiology* 167: 222–233.

29 Sun, P.F., Fang, W.T., Shin, L.Y., Wei, J.Y., Fu, S.F. et al. (2014). Indole-3-Acetic Acid-Producing Yeasts in the phyllosphere of the Carnivorous Plant *Drosera indica* L. *PLOSONE* 9 (12): e114196. https://doi.org/10.1371/journal.pone.0114196

30 Mahanty, T., Bhattacharjee, S., Goswami, M., Bhattacharyya, P., Das, B., Ghosh, A., and Tribedi, P. (2017). Biofertilizers: A potential approach for sustainable agriculture development. *Environmental Science and Pollution Research* 24 (4): 3315–3335.

31 Holland, M. (2011). Nitrogen: Give and take from phyllosphere microbes. In: *Ecological Aspects of Nitrogen Metabolism in Plants*, 1e (ed. J.C.J.C. Polacco and C.D. Todd), 217–230. Wiley.

32 Argueso, C.T., Ferreira, F.J., Epple, P. et al. (2012). Two-component elements mediate interactions between Cytokinin and salicylic acid in plant immunity. *PLoS Genetics* 8: e1002448.

33 Freebairn, H.T. and Buddenhagen, I.W. (1964). Ethylene production by *Pseudomonas solanacearum*. *Nature* 202: 313–314. DOI: 10.1038/202313a0

34 Van de Poel, B., Smet, D., and Van Der Straeten, D. (2015). Ethylene and hormonal cross talk in vegetative growth and development. *Plant Physiology* 169: 61–72. DOI: 10.1104/pp.15.00724

35 Guinel, F.C. (2015). Ethylene, a hormone at the center-stage of nodulation. *Frontiers in Plant Science* 6: 1121. DOI: 10.3389/fpls.2015.01121

36 Tsang, D.L., Edmond, C., Harrington, J.L., and Nühse, T.S. (2011). Cell wall integrity controls root elongation via a general 1-aminocyclopropane-1-carboxylic acid-dependent, ethylene-independent pathway. *Plant Physiology* 156: 596–604. DOI: 10.1104/pp.111.175372

37 Vanderstraeten, L. and Van Der Straeten, D. (2017). Accumulation and transport of 1-aminocyclopropane-1-carboxylic acid (A.C.C.A.C.C.) in plants: Current status, considerations for future research and agronomic applications. *Frontiers in Plant Science* 8: 38. DOI: 10.3389/fpls.2017.00038

38 Digiacomo, F., Girelli, G., Aor, B., Marchioretti, C., Pedrotti, M., Perli, T., ... Del Bianco, C. (2014). Ethylene-producing bacteria that ripen fruit. *ACS Synthetic Biology* 3 (12): 935–938.

39 Ravanbakhsh, M., Sasidharan, R., Voesenek, L., Kowalchuk, G.A., and Jousset, A. (2018). Microbial modulation of plant ethylene signaling: Ecological and evolutionary consequences. *Microbiome* 6 (1): 52. https://doi.org/10.1186/s40168-018-0436-1

40 Glick, B.R. (1995). The enhancement of plant growth by free-living bacteria. *Canadian Journal of Microbiology* 41 (2): 109–117.

41 Sahin, F., Cakmakci, R., and Kanta, F. (2004). Sugar beet and barley yields in relation to inoculation with N_2-fixing and phosphate solubilizing bacteria. *Plant Soil* 265: 123–129. DOI: 10.1007/s11104-005-0334-8

42 Penalver, R., Vicedo, B., Salcedo, C.I., and López, M.M. (1994). *Agrobacterium radiobacter* strains K84, K1026 and K84Agr produce an antibiotic-like substance, active in vitro against *A. tumefaciens* and phytopathogenic *Erwinia* and *Pseudomonas* spp. *Biocontrol Science and Technology* 4 (3): 259–267.

43 Mousa, W.K., Schwan, A., Davidson, J., Strange, P., Liu, H., Zhou, T., and Raizada, M.N. (2015). An endophytic fungus isolated from finger millet (*Eleusine coracana*) produces antifungal natural products. *Frontiers in Microbiology* 6: 1157.

44 Mousa, W.K., Shearer, C., Limay-Rios, V., Ettinger, C.L., Eisen, J.A., and Raizada, M.N. (2016). Root-hair endophyte stacking in finger millet creates a physicochemical barrier to trap the fungal pathogen *Fusarium graminearum*. *Nature Microbiology* 1 (12): 1–12.

45 Tian, L., Chang, C., Ma, L., Nasir, F., Zhang, J., Li, W., and Tian, C. (2019). Comparative study of the mycorrhizal root transcriptomes of wild and cultivated rice in response to the pathogen *Magnaportheoryzae*. *Rice* 12 (1): 35.

46 Prabha, R., Singh, D.P., Gupta, S., Gupta, V.K., El-Enshasy, H.A., and Verma, M.K. (2019). Rhizosphere metagenomics of *Paspalum scrobiculatum* L. (KodoMillet) reveals rhizobiome multifunctionalities. *Microorganisms* 7 (12): 608.

47 Fan, B., Carvalhais, L.C., Becker, A., Fedoseyenko, D., von Wirén, N., and Borriss, R. (2012). Transcriptomic profiling of *Bacillus amyloliquefaciens* FZB42 in response to maize root exudates. *B.M.C.B.M.C. Microbiology* 12 (1): 1–13.

48 Figueroa-López, A.M., Cordero-Ramírez, J.D., Martínez-Álvarez, J.C., López-Meyer, M., Lizárraga-Sánchez, G.J., Félix-Gastélum, R., and Maldonado-Mendoza, I.E. (2016). Rhizospheric bacteria of maize with potential for biocontrol of *Fusarium verticillioides*. *SpringerPlus* 5 (1): 1–12.

49 Chen, Y., Wang, J., Yang, N., Wen, Z., Sun, X., Chai, Y., and Ma, Z. (2018). Wheat microbiome bacteria can reduce virulence of a plant pathogenic fungus by altering histone acetylation. *Nature Communications* 9 (1): 1–14.

50 Colombo, E.M., Kunova, A., Pizzatti, C., Saracchi, M., Cortesi, P., and Pasquali, M. (2019). Selection of an endophytic *Streptomyces* sp. strain DEF09 from wheat roots as a biocontrol agent against *Fusarium graminearum*. *Frontiers in Microbiology* 10: 2356.

51 Zhang, S., Jiang, Q., Liu, X., Liu, L., and Ding, W. (2020). Plant growth promoting rhizobacteria alleviate aluminum toxicity and ginger bacterial wilt in acidic continuous cropping soil. *Frontiers in Microbiology* 11: 2976.

52 Rodríguez, M.A., Cabrera, G., Gozzo, F.C., Eberlin, M.N., and Godeas, A. (2011). Clonostachys rosea BAFC3874 as a *Sclerotinia sclerotiorum* antagonist: Mechanisms involved and potential as a biocontrol agent. *Journal of Applied Microbiology* 110 (5): 1177–1186.

53 Kosawang, C., Karlsson, M., Vélëz, H., Rasmussen, P.H., Collinge, D.B., Jensen, B., and Jensen, D.F. (2014). Zearalenone detoxification by zearalenone hydrolase is important for the antagonistic ability of *Clonostachys rosea* against mycotoxigenic *Fusarium graminearum*. *Fungal Biology* 118 (4): 364–373.

54 Xu, J., Wang, H., Zhu, Z., Ji, F., Yin, X., Hong, Q., and Shi, J. (2016). Isolation and characterization of *Bacillus amyloliquefaciens* ZDS-1: Exploring the degradation of Zearalenone by *Bacillus* spp. *Food Control* 68: 244–250.

55 Jensen, B., Knudsen, I.M., and Jensen, D.F. (2002). Survival of conidia of *Clonostachysrosea* on stored barley seeds and their biocontrol efficacy against seed-borne *Bipolaris sorokiniana*. *Biocontrol Science and Technology* 12 (4): 427–441.

56 Fiorilli, V., Vannini, C., Ortolani, F., Garcia-Seco, D., Chiapello, M., Novero, M., ... Bonfante, P. (2018). Omics approaches revealed how arbuscular mycorrhizal symbiosis enhances yield and resistance to leaf pathogen in wheat. *Scientific Reports* 8 (1): 1–18.

57 Zhang, S., Xu, B., and Gan, Y. (2019). Seed treatment with *Trichoderma longibrachiatum* T6 promotes wheat seedling growth under NaCl stress through activating the enzymatic and nonenzymatic antioxidant defense systems. *International Journal of Molecular Sciences* 20 (15): 3729.

58 Xu, S., Chang, J., Chang, C., Tian, L., Li, X., and Tian, C. (2020). Rhizospheric microbiomes help Dongxiang common wild rice (*Oryza rufipogon* Griff.) rather than *Leersia hexandra* Swartz survive under cold stress. *Archives of Agronomy and Soil Science* 67(14): 1–13.

59 Beirinckx, S., Viaene, T., Haegeman, A., Debode, J., Amery, F., Vandenabeele, S., ... Goormachtig, S. (2020). Tapping into the maize root microbiome to identify bacteria that promote growth under chilling conditions. *Microbiome* 8: 1–13.

60 Abdel Latef, A.A.H., Zaid, A., Abo-Baker, A.B.A.E., Salem, W., and Abu Alhmad, M.F. (2020). Mitigation of Copper Stress in Maize by Inoculation with *Paenibacillus polymyxa* and *Bacillus circulans*. *Plants* 9 (11): 1513.

61 Labanca, E.R.G., Andrade, S.A.L., Kuramae, E.E., and Silveira, A.P.D. (2020). The modulation of sugarcane growth and nutritional profile under aluminum stress is dependent on beneficial endophytic bacteria and plantlet origin. *Applied Soil Ecology* 156: 103715.

62 Masciarelli, O., Llanes, A., and Luna, V. (2014). A new PGPR co-inoculated with *Bradyrhizobium japonicum* enhances soybean nodulation. *Microbiological Research* 169 (7–8): 609–615.

63 Islam, A., Kabir, S., and Khair, A. (2019). Characterization and evaluation of *Bacillus siamensis* isolate for its growth-promoting potential in tomato. *Agriculture* 65 (2): 42–50.

64 Alves, G.C., Videira, S.S., Urquiaga, S., and Reis, V.M. (2015). Differential plant growth promotion and nitrogen fixation in two genotypes of maize by several *Herbaspirillum* inoculants. *Plant and Soil* 387 (1): 307–321.

65 Da Silveira, A.P.D., Iório, R.D.P.F., Marcos, F.C.C., Fernandes, A.O., de Souza, S.A.C.D., Kuramae, E.E., and Cipriano, M.A.P. (2019). Exploitation of new endophytic bacteria and their ability to promote sugarcane growth and nitrogen nutrition. *Antonie Van Leeuwenhoek* 112 (2): 283–295.

66 Ahemad, M. and Khan, M.S. (2012). Effects of pesticides on plant growth-promoting traits of Mesorhizobium strain MRC4. *Journal of the Saudi Society of Agricultural Sciences* 11 (1): 63–71.

67 Bhattacharyya, P.N. and Jha, D.K. (2012). Plant growth-promoting rhizobacteria (PGPR): Emergence in agriculture. *World Journal of Microbiology and Biotechnology* 28 (4): 1327–1350.

68 Dixon, R.O.D. and Wheeler, C.T. (1986). *Nitrogen Fixation in Plants*. Glasgow, United Kingdom: Blackie.

69 Sprent, J.I. and Sprent, P. (1990). *Nitrogen-fixing Organisms. Pure and Applied Aspects*. London, United Kingdom: Chapman & Hall.

70 Reardon, C.L., Gollany, H.T., and Wuest, S.B. (2014). Diazotroph community structure and abundance in wheat-fallow and wheat–pea crop rotations. *Soil Biology and Biochemistry* 69: 406–412.

71 Tairo, E.V. and Ndakidemi, P.A. (2013). Possible benefits of rhizobial inoculation and phosphorus supplementation on nutrition, growth and economic sustainability in grain legumes. *American Journal of Research Communication* 1 (12): 532–556.

72 Gupta, V.V., Zhang, B., Penton, C.R., Yu, J., and Tiedje, J.M. (2019). Diazotroph diversity and nitrogen fixation in summer active perennial grasses in a Mediterranean region agricultural soil. *Frontiers in Molecular Biosciences* 6: 115.

73 Ng, B.H. (1987). The effects of salinity on growth, nodulation and nitrogen fixation of *Casuarina equisetifolia*. *Plant and Soil* 103 (1): 123–125.

74 Ferguson, B.J., Foo, E., Ross, J.J., and Reid, J.B. (2011). Relationship between gibberellin, ethylene and nodulation in *Pisum sativum*. *New Phytologist* 189 (3): 829–842.

75 Fred, E.B., Baldwin, I.L., and McCoy, E. (2002). *Root Nodule Bacteria and Leguminous Plants*, (No. 5). UW-Madison Libraries Parallel Press.

76 Franche, C., Lindström, K., and Elmerich, C. (2009). Nitrogen-fixing bacteria associated with leguminous and non-leguminous plants. *Plant and Soil* 321 (1): 35–59.

77 Burris, R.H. (1969). Progress in the biochemistry of nitrogen fixation. *Proceedings of the Royal Society of London. Series B. Biological Sciences* 172 (1029): 339–355.

78 Postgate, J. (ed.) (2012). *The Chemistry and Biochemistry of Nitrogen Fixation*. Springer Science & Business Media.

79 Wilson, P.W. (1940). The biochemistry of symbiotic nitrogen fixation. *The Biochemistry of Symbiotic Nitrogen Fixation* 34(1):789 p.

80 Bergersen, F.J. (1971). Biochemistry of symbiotic nitrogen fixation in legumes. *Annual Review of Plant Physiology* 22 (1): 121–140.

81 Sagan, M. and Duc, G. (1996). Sym28 and Sym29, two new genes involved in the regulation of nodulation in pea (*Pisum sativum*L.). *Symbiosis* 20 (3): 229-245.

82 Borisov, A.Y., Madsen, L.H., Tsyganov, V.E., Umehara, Y., Voroshilova, V.A., Batagov, A.O., ... Stougaard, J. (2003). The Sym35 gene required for root nodule development in pea is an ortholog of *NIN* from *Lotus japonicus*. *Plant Physiology* 131 (3): 1009–1017.

83 Tsyganov, V.E., Morzhina, E.V., Stefanov, S.Y., Borisov, A.Y., Lebsky, V.K., and Tikhonovich, I.A. (1998). The pea (*Pisum sativum* L.) genes sym33 and sym40 control infection thread formation and root nodule function. *Molecular and General Genetics M.G.G.M.G.G.* 259 (5): 491–503.

84 Kuppusamy, K.T., Endre, G., Prabhu, R., Penmetsa, R.V., Veereshlingam, H., Cook, D.R., and VandenBosch, K.A. (2004). LIN, a *Medicago truncatula* gene required for nodule differentiation and persistence of rhizobial infections. *Plant Physiology* 136 (3): 3682–3691.

85 Yan, Z., Hossain, M.S., Arikit, S., Valdés-López, O., Zhai, J., Wang, J., and Stacey, G. (2015). Identification of micro R.N.A.R.N.A. s and their mRNA targets during soybean nodule development: Functional analysis of the role of miR393j-3p in soybean nodulation. *New Phytologist* 207 (3): 748–759.

86 Yan, Z., Hossain, M.S., Wang, J., Valdés-López, O., Liang, Y., Libault, M., and Stacey, G. (2013). miR172 regulates soybean nodulation. *Molecular Plant-Microbe Interactions* 26 (12): 1371–1377.

87 Li, H., Deng, Y., Wu, T., Subramanian, S., and Yu, O. (2010). Misexpression of miR482, miR1512, and miR1515 increases soybean nodulation. *Plant Physiology* 153 (4): 1759–1770.

88 Schauser, L., Roussis, A., Stiller, J., and Stougaard, J. (1999). A plant regulator controlling the development of symbiotic root nodules. *Nature* 402 (6758): 191–195.

89 Hayashi, S., Gresshoff, P.M., and Ferguson, B.J. (2014). Mechanistic action of gibberellins in legume nodulation. *Journal of Integrative Plant Biology* 56 (10): 971–978.

90 Mazid, M. and Khan, T.A. (2015). Future of bio-fertilizers in Indian agriculture: An overview. *International Journal of Agricultural and Food Research* 3 (3). Mendes, R. et al. (2011) Deciphering the rhizosphere microbiome for disease-suppressive bacteria. *Science* 332: 1097–1100.

91 Berendsen, R.L., Vismans, G., Yu, K., Song, Y., de Jonge, R., Burgman, W.P., Burmølle, M., Herschend, J., Bakker, P.A.H.M., and Pieterse, C.M.J. (2018). Disease-induced assemblage of a plant-beneficial bacterial consortium. *The ISME Journal* 12: 1496–1507.

92 Mishra, J. and Arora, N.K. (2016). Bioformulations for plant growth promotion and combating phytopathogens: A sustainable approach. In: *Bioformulations: For Sustainable Agriculture*, Arora, N.K. editor, Department of Environmental Microbiology School for Environmental Sciences, Babasaheb Bhimrao Ambedkar University, Lucknow, India 3–33.

93 Barman, M., Paul, S., Choudhury, A.G., Roy, P., and Sen, J. (2017). Biofertilizer as prospective input for sustainable agriculture in India. *International Journal of Current Microbiology and Applied Sciences* 6: 1177–1186.

7

Microbiome Engineering and Biotechnology

The Real Finenesses of a Robust Rhizosphere

Barkha Sharma[1], Shalini Tiwari[1], and Kailash Chand Kumawat[2,3]

[1]*Department of Microbiology College of Basic Sciences & Humanities, G. B. Pant University of Agriculture & Technology Pantnagar, Uttarakhand India*
[2]*Department of Soil Science Punjab Agricultural University Ludhiana, Punjab India*
[3]*Department of Industrial Microbiology, Jacob Institute of Biotechnology and Bioengineering Sam Higginbottom University of Agriculture, Technology and Sciences Prayagraj, Uttar Pradesh India*

7.1 Agriculture and Microorganisms

Agriculture depends on the interaction of four natural systems.

(i) **Climate:** The two main variables of climate that can affect crop productivity (farming) are the amount of solar energy (or heat) and the amount of moisture (precipitation level). All forms of agriculture are primarily controlled by temperature [1].

(ii) **Soil:** It contains nutrients (potassium, nitrogen, and phosphorus), organic materials (earthworms, microbes, dead plants, water, and air). Rich soil as plant food is the chief requirement of thriving agriculture [2].

(iii) **Topography:** It affects agriculture in relation to soil erosion, difficulty of tillage, and inadequate transport facilities. In general, leveled land is best for farming [3].

(iv) **Biology:** Some organisms are highly beneficial to agriculture, such as earthworms, bees, and microbes, while some are detrimental, for example, grasshoppers and disease-causing microbes. The growth of unwanted plants, such as weeds, can interfere with the productivity of agriculture [4].

Soil is a paradigm of nutrient sources that we use for crop cultivation; wherefore, it is the most pivotal factor that governs thriving agriculture [5]. Even though it has massive microbial diversity, the population of microorganisms acquires merely 1% of the entire soil volume, aggregated in microhabitats either as colonies or biofilms. These regions are also termed "microbial hotspots," which were distinguished from bulk soil owing to accelerated biogeochemical processes and divided into four active groups by Kuzyakov and Blagodatskaya [6], including (i) rhizosphere, a narrow zone inhabited by soil microbes in the vicinity of plant roots due to the release of root exudates (root microbiome); (ii) detritus, a thin soil layer (< 5 mm) of decomposed and degraded litter of plant materials and other organic matter; (iii) spores, a deep underground root growth zone; and (iv) soil aggregate surface [7]. These regions control the microbial growth limiting factors, such as moisture content of the soil, available nitrogen and oxygen, and recalcitrant organic compounds for bioprocesses [6].

Core Microbiome: Improving Crop Quality and Productivity, First Edition. Edited by Javid A. Parray, Nowsheen Shameem, Elsayed Fathi Abd-Allah, and Mohammad Yaseen Mir.

7.2 Microbiome

The microbiome can be defined as either the total microbial genomes in a particular niche or the entire plant microbial community, particularly protists, bacteria, fungi, microalgae, archaea, and viruses, which live, flourish, and communicate with different roots, shoots, flowers, leaves, and seeds of plants [8–10]. We use *microbiome* in the sense of the final definition presented, as it is also the mostly accepted one.

7.2.1 Rhizosphere Microbiome: Active Microbial Hotspot

The rhizosphere hosts a coupled interaction of plant, soil, and microorganisms, creating an exclusive active environment [11]. A comprehensive understanding of the interaction of these three elements is required to grasp the intricacy of regulation and operation of the rhizosphere system as the conditions established by them regulates the environment. Microorganisms are the most specific elements for manipulation since sustainable agriculture revolves around a healthy and flourishing rhizosphere because of bioactive compounds released by the diverse microbial communities and the microbiome ecosystem directly or indirectly [12].

There are various reasons behind such a flourishing space for microbial growth. The first one is the ready availability of carbon and nutrient sources that govern gas exchange, respiration, nutrients and moisture utilization, and other rhizospheric bioprocesses [13]. The second and significant phenomenon is rhizodeposition, wherein plants release various molecules such as vitamins, sugars, secondary metabolites, amino acids, etc., into the soil. They are also called root exudates and known to alter the rhizosphere and hence soil chemistry and biology [14]. The third is the migration ability of rhizospheric microbes from roots to either phyllosphere (aerial regions) or endosphere (internal tissues) [15].

7.2.2 Significance of Rhizospheric Microbiome

1. The rhizosphere microbiome contributes to the decomposition of organic matter for soil fertility enhancement.
2. Furthermore, the antagonistic microorganisms in the rhizosphere prevent pathogens via the production of hydrolytic enzymes or antibiotics by competing for limiting nutrient sources and niche (space), increasing host immunity. Antimicrobial metabolites produced by microorganisms include ammonia, ethyl acetate extract, flavonol glycosides, oligomycin A, phenazine-1-carboxylic acid (PCA), bacitracin, erythromycin, pyrrolnitrin, and other moieties, while in a few plants, rhizobacteria, specifically *Bacillus* sp. and *Pseudomonas* sp., trigger inducing systemic resistance to exterminate pathogens [16].
3. The genetic and metabolic proficiencies of the rhizosphere microbiome of the genera *Pseudomonas*, *Bacillus*, *Achromobacter*, *Burkholderia*, *Enterobacter*, *Azotobacter*, and *Trichoderma* assist in the improvement of plant growth and development by the alleviation of plant abiotic stress [17,18].
4. Several rhizobacteria and fungi contribute to plant growth promotion by producing phytohormones, such as auxins, cytokinins, gibberellins, ethylene, salicylic acid, and abscisic acid as secondary metabolites. *Gibberella fujikuroi*, a fungal rice plant

pathogen, synthesizes gibberellins. Gibberellin-like substances conspicuously synthesized by sundry microbial species, for example, *Fusarium moniliforme*, *Bacillus* sp., *Azospirillum* sp., *Acetobacter diazotrophicus*, and *Rhizobium* sp. gibberellins upregulate plant growth by regulating germination of seeds, stem prolongation, sex expression, and also the formation of fruit [19].

5. Although most of the rhizospheric microorganisms endow plant growth promotion, a fraction of them impairs plant health and productivity by competing for nutrient sources and habitat, resulting in umpteen economic crop loss and destruction. *Olpidium* sp., *Agrobacterium tumefaciens, Sclerotinias clerotiorum, Verticillium dahlia, Dickeya* sp., *Pythium* sp., *Phytophthora sojae, Fusarium oxysporum, Pratylenchus neglectus, Rhizoctonia* sp., *Gaeumannomyces graminis, Colletotrichum* sp., *Polymyxa* sp., *Plasmodiophora brassicae, Ralstonia solanacearum, Plasmopara halstedii*, and *Rhizoctonia solani* are some predominant microbial plant pathogens in rhizospheric soil [20,21].

7.2.3 Why Do We Modify the Rhizosphere Microbiome?

In the past hundred years, rapid urbanization and industrialization in the global economy have worsened repercussions on the environment. Soil micro-ecosystem is the central province for the interaction of chemical communications and the exchange of compounds between plant and soil microorganisms [22]. Heavy use of fertilizers and urban environment concentrate a plethora of chemicals into the soil, which weakens and contaminates the rhizospheric communities, leading to declining crop yield and quality. From this perspective, reconstruction of the rhizosphere microbiome can be an excellent option for the rhizosphere and plant microbiome improvement [20,23].

Rhizosphere microbiota comprises active microbes that help in improving plant physiology and soil conditions and inducing resistance against biotic (phytopathogens) as well as abiotic stress (including drought, salinity, and contaminated soil) [24,25]. Even so, the entire microbiome does not reflect all such functions due to their positive or negative interaction with other strains. Hence, enhancement of these few microbes can be done to improve the whole rhizosphere. Different plants interacting in the rhizosphere with microorganisms have varied biotic and abiotic stress tolerance capacity, optimum growth range at different altitude or latitudes, and sometimes nutrient-poor conditions allow them to survive in such a harsh environment. Therefore, transferring those potential microorganisms or their genetic materials or gene numbers (e.g. genes coding for *Bacillus thuringiensis* Cry toxins) to the desired rhizosphere or its inhabitant microorganisms that lack those functions can be a good alternative to rhizosphere improvement [26,27].

7.3 Microbiome Engineering

Latest research has been focused on developing agriculture by improving plant functions by altering plant microbiome. This method depends on host-mediated microbial interaction. Artificial selection of a host-specific microbiome aims to enhance plant performance, and it can be applied over several generations; thus, the host's microbiome is reproduced stably to create offspring [28].

7.3.1 Conventional Microbiome Engineering

7.3.1.1 Soil Amendments

Soil amendments contain several organic and inorganic additives supplied to the soil to manipulate plant–microbiome interactions to optimize crop productivity. Inorganic substances such as lime, vermiculite, perlite, and sand reduce the after-effects of soil acidity and salinity on plants [29]. A 30% crop-yield increment was reported when inorganic additives were applied together with organic additives, compared with 8% organic fertilizers [30].

Organic additives serve as a source of energy for plant-beneficial heterotrophic microorganisms. Benefits of these systems are to supplement decimated soil fertility and reduce agro-waste. Prolonged application of conventional organic fertilizers increases microbial biomass and diversity (according to the type of fertilizer applied). Since the mid-19th century, bones were used as a renewable phosphorus (P) supply. It is well established that P in the bone can be solubilized by organic acids releasing soil microbes, e.g., *Bacillus mucilaginous* and *Bacillus megaterium* var. *phosphatic*, to make them available for plant uptake [31].

7.3.1.2 Organic Additives

Biochar, a carbon (C)-rich compound, is formed from organic matter, such as bones, compost, agro-waste, and slaughterhouse waste, by partial or complete pyrolysis, enhancing potential water-holding capacity and available nutrients in the soil. Biochar positively increases crop productivity by providing a secure and flourished habitat for soil microbes. Anti-pathogenic effects of biochar also eliminate several disease-causing microbes, e.g., *Fusarium oxysporum*, *F. solani*, *Galibacterium anatis*, *Podosphaera aphanis*, *Ilyonectria* sp., *Phytophthora cactorum*, and *Rhizoctonia solani* [32,33].

Microbial biopesticides and biofertilizers, commonly termed microbial soil inoculants, are organic additives that benefit agriculture by fostering plant productivity and protecting plant pathogens. Three significant processes boost the population of beneficial rhizosphere microorganisms: (i) biocontrol (diagnosis of physiological alterations and induction of host defenses in plants during biotic stresses); (ii) biofertilization (inoculation of preparation of microbes having nitrogen-fixing, phosphate-solubilizing, siderophore-producing abilities); and (iii) biostimulation (inoculation of phytohormones synthesizing plant-associated microbes to resist abiotic stress and reinforce growth and development of plants) [5,34].

7.3.2 Contemporary Rhizospheric Microbiome Engineering

7.3.2.1 Phytomicrobiome Targeting

The rhizosphere microbiome is positively modified through either adding/activating signal molecules (phytohormones, salicylic acid, hexokinase 1, proline, ions, etc.) or substrates (including root exudates such as sugars, growth regulators, amino acids, malic acid, vitamins, flavonoids, tannins, coumarins, steroids, camalexin, benzoxazinoids, etc.) or soil conditioning (including traditional soil modifications) to attract/stimulate the desired microorganisms. As active consortia, several microbes intercommunicate in the rhizospheric region. For instance, arbuscular mycorrhizal fungi with plant-growth-promoting rhizobacteria, in the form of consortia, mutually complement each other in terms of supply of limited nutrients and their environment efficacy [28]. To completely alter the microbial community structure, core microbes can be introduced into the soil directly – for example, treatment of growing *Capsicum frutescens* plant roots with *Bacillus amyloliquefaciens*, *Acinetobacter* sp. strain UQ202, and *Bacillus velezensis* encouraged growth and also synchronously tolerated soil-borne *Phytophthora capsici* infection [35].

7.3.2.2 Artificial Consortia of Microbes

Artificial (or synthetic) microbial consortia (AMC) corresponds to synthetic biology capable of rebuilding the rhizosphere microbiome in terms of its structure and function because of their multi-functions and resolving the limitations of single microbial strains. Engineering AMC includes selecting core microorganisms from their origin, culturing, compatible optimization of the interaction, and consortia assessment [36]. A recent example of bacterial consortia for better plant growth promotion than the parent strains involves two synthetic microbial communities of bacteria exhibiting ACC deaminase synthesis for improved plant growth on the poor substrate and antimicrobial effect contrary to *F. oxysporum f.* sp. *lycopersici* in tomato plant [37].

7.3.2.3 Microbiome Breeding and Transplantation

The influence of early or late flowering in *Arabidopsis thaliana* was analyzed by microbiome breeding. Based on flowering phenotype, over 10 generations, four adequate microbial communities were selected, and the results displayed the enhanced inflorescence in plants inoculated with "late-flowering microbiome." Hence, microbiome breeding is based on the underlying principle that plant traits support an explicit microbial population through generations. This method includes (i) selecting the microbiome; (ii) allowing the group of microorganisms carrying these microbiomes to interact with the host plant, (iii) screening out unfavorable microorganisms, and then (iv) vertically transferring the selected microbiome to the offspring. For this reason, the one phenotype of the host that affects the microbial population will be propagated [38]. Microbiome transplantation can be achieved outside soil with no breeding system. *Leptospermum scoparium*, a medicinal plant, secretes antimicrobial oils in defense against phytopathogen *Pseudomonas syringae*. Transplantation of the microbiome of *L. scoparium* via wound infection in kiwi fruit mirrors the antimicrobial impact on *P. syringae* [39,40].

7.3.2.4 Host-Mediated Selection and Manipulation

The interaction between plant and microorganisms contributes to establishing the rhizosphere, plant growth, soil textures, and other environmental systems. It is a well-known fact that a particular plant genotype regulates the establishment and functions of the rhizospheric microbiome. For example, inoculation of the *Arabidopsis* plant with *Pseudomonas simiae* WCS417r biofertilizer quadrupled plant biomass than in other genotypes in which the bacterial fertilizer's effect was roughly evident [41]. Ergo, microbe-friendly cultivars can offer higher crop yield with more significant potential. Variation in quantity and type of root exudates of different plant genotypes attract various beneficial and infection-resistant microorganisms. Also, they help in the reconstruction of the whole microbiome. In Columbia, eight bean (*Phaseolus vulgaris*) accessions were tested for genotype-dependent implication on rhizosphere microbiome and results were in agreement; merely 0.7% of operational taxonomic units were shared between them [42]. To achieve plant-specific microbial communities, "designer plants" that release exudates to attract specific beneficial microbiota can be engineered.

7.4 Advance Rhizosphere Microbiome Engineering

Microbiome engineering can alter and improve the rhizosphere in two ways. The first one is the bottom-up approach, which includes the isolation of the desired microorganism from its parent site, followed by its genetic modification and the inoculation of this core microbe as a synthetic microbial community (SynComs) into the plant with vigorous host recolonization, while the second method, the top-down approach, is based on the mechanism involved in horizontal gene transfer (HGT) for *in situ* introduction of desired traits in multiple hosts [43].

7.4.1 Bottom-Up Approach

Synthetic biology of microbiome engineering improves the work on model microbes and non-model microbes. It solves the problem of the use of universal genetic tools, which has been obstructed by organism-specific nuances. Synthetic quorum sensing (QS) circuits on broad-host-range plasmids help in the study of microbe–microbe interaction and plant-benefiting traits of secondary metabolite biosynthetic gene clusters (BGCs) [43,44]. However, the maintenance of stable transfer generation after generation is complex. Microbiome engineering successfully and stably regulates the integration of a > 50kb DNA construct, which means many traits can be simultaneously permitted to integrate into a single step [45].

7.4.1.1 Phage Integrase System

This system regulates the successful recombination phage genome to the host (actinobacteria) attachment sites (e.g., attP–attB) and offers a multifaceted genome engineering method. The essential requirement for this system is the availability of an integrase-specific attachment site in the host, or else these sites can be artificially introduced in chromosomes [46,47].

7.4.1.2 Integrative and Conjugative Elements System

This system integrates MGEs and their active propagation via replication into a host (firmicutes) chromosome because they encode a functional gene for their intercellular transport [48]. Specific ICEs integration sites include conserved sequences (e.g., tRNA genes) that increase their host range. In contrast, a few of them attaching onto less specific areas (AT-rich regions) may disrupt essential genes. *E. coli* and *B. subtilis* can be used as common model microbes for engineering when integrase encoding genes, conjugative machinery, and transposase associated with mobile genetic elements are incorporated into them [49].

7.4.1.3 Chassis-independent Recombinase-assisted Genome Engineering System

Chassis-independent recombinase-assisted genome engineering (CRAGE) technology precisely integrates large and complex genes into non-model microbial chromosomes (proteobacteria) [50,51]. The whole process includes first combining a landing pad containing LoxP site flanking P1 phage cre recombinase, constructing and then assembling them with heterologous genes into accessory vectors [52]. The extensive host range of the CRAGE system offers a better and more straight forward approach for engineering the plant growth-promoting rhizomicrobial genome.

7.4.2 Top-Down Approach

Top-down approaches with nominal commotion introduce/delete target traits into and out of desired microbial communities [53]. *In situ* microbiome engineering uses mobile genetic elements (MGEs) and phages as a vector to transfer desired functional traits into the rhizosphere, which are described as follows [54,55].

7.4.2.1 Mobile Genetic Elements

The use of MGEs to transfer a desired trait into the whole microbial community is a very efficient protocol due to HGT mechanisms followed by many microorganisms. *In situ* genome, engineering tools can redesign these MGEs according to the trait transferred [56]. A conjugate donor microbial strain carrying a target gene moves it into the desired microbial community. An experiment proved that 10kb *nif* gene cluster could be delivered *in situ* through *B. subtilis* donor system (XPORT) in four synthetic soil microbial communities [57]. This mechanism can also regulate integrating other plant growth promotion traits into phytomicrobiome and transfer them into the host plant.

7.4.2.2 Phages

The other approach in HGT is transduction, which includes replication and integration machinery of phages (viruses), used to carry and transfer DNA fragment containing desired traits. Phage-based *in situ* manipulation of phytomicrobial communities has successfully wiped out specific pathogenic microbial strains or has helped transfer PGP genes into them [58].

7.5 Rhizospheric Biotechnological Approaches

Over the past decades, new biochemical and enzymatic technologies have been developed to explore global health and develop modern scientific materials. These methods have been thoroughly tested for their resilience and ability to distinguish between different forms and utilization of soils and their importance to the environment [59].

7.5.1 Fingerprinting Methods Based on Polymerase Chain Reaction

Polymerase chain reaction (PCR) amplification, which furnishes knowledge regarding the community's genetic structure, is used in many genetic fingerprinting techniques. They are classified into two classes suitably depending on electrophoresis relocation either with agarose or with polyacrylamide gels: (i) size movement (includes single-strand conformation polymorphism (SSCP), automated ribosomal intergenic spacer analysis (ARISA)/RISA, random amplified polymorphic DNA (RAPD), and terminal restriction fragment length polymorphism (T-RFLP)) and (ii) sequence movement (similar to denaturing gradient gel electrophoresis, DGGE; and temperature gradient gel electrophoresis, TGGE). Specific primers play a crucial role in generating amplified fragments that can be utilized to determine the structure of microbial communities. Fingerprinting methodology based on PCR is helpful when corresponding bacterial communities are present in the rhizospheric section of plants [60]. However, these methods are liable to depreciate the diversity, which makes them essential for finding domineering representatives of communities in complicated environments like soil. This happens as PCR amplifies tiny fragments and frail secondary structures (such as low guanine–cytosine (GC) contents) and those cells from which DNA can be easily extracted [61].

7.5.1.1 Denaturing Gradient Gel Electrophoresis/Temperature Gradient Gel Electrophoresis

For the division of ribosomal DNA fragments of similar size, which are PCR-amplified except for different nucleotide composition, DGGE and TGGE were formulated [62]. These methods have been adapted over time to analyze bacterial communities. Both ways utilize either a linear gradient of DNA denaturing agents (urea/formamide used in DGGE) or temperature (TGGE) onto polyacrylamide gel that electrophoretically creates an impact on the mobility of partly melted double-stranded DNA. The sequences are linked up to melting temperatures, and the DNA portion ceases mobility when base pairs having the least melting weather reach that temperature. Conformational change from helical to partially melting takes place, and as a result, displacement along the electric field is halted by a special primer, a GC clamp (high GC content sequence) linked to the region of 5′-end and utilized to hold the PCR fragments and counteract them from entirely separating [63]. These methods are still used to assess rhizospheric bacterial dynamics, structure, and diversity, but they have been progressively regained by introducing high-throughput sequencing methodology. Both DGGE and TGGE usually help discover the most significant aspects of the pioneers in the microbial community.

Furthermore, the descriptions are deceptive because a particular band can symbolize heterogeneous species and similar species illustrated by many bands [64]. DGGE/TGGE methods are widely used in a variety of rhizosphere studies to distinguish communities of microbes under various changing conditions, such as shifts in the composition and structure of both bacterial and fungal communities, in response to multiple concentrations of autotoxin [65]. Moreover, microbial communities having molecular fingerprinting revealed variations in environmental circumstances at a volcanic CO_2 vent [66]. Broadly speaking, these gradient gel electrophoresis-based methodologies are more time-consuming than other gel electrophoresis-based fingerprinting methods, such as RISA or RAPD.

7.5.1.2 Terminal Restriction Fragment Length Polymorphism Fingerprinting

The T-RLFP method uses either 5′-PCR primers that have been labeled with a dye that has a fluorescent property [67]. Labeled primers enable the marking of amplified products and are broken down by one or more restriction enzymes. As a result, labeled terminal restriction fragments with various lengths are resolved either by capillary action or sequencing gel [68]. Due to variation in the sequences of the amplified gene, diverse rhizospheric communities of microbes may have discrete composites of restriction sites, "fingerprints" obtained for specific organism assortments. However, different amounts of DNA in a T-RFLP profile can cause variations in abundance and phylotypes. An improved understanding of soil community changes in various environments, including those of soil structure and composition, can be arrived at using the T-RFLP methodology [69].

7.5.1.3 Single-strand Conformation Polymorphism

Similar to DGGE/TGGE, SSCP is a method based on electrophoresis, suitable for examining various communities of microorganisms; nevertheless, partition based on DNA is single-stranded. The secondary structures of single-stranded DNA can distinguish amid products of various phylotypes. In contrast to the T-RFLP/TGGE/DGGE methodology, SSCP does not require pioneer GC clamps or restriction digestion. Although different studies have indicated the variety and allocation of bacteria for the production of the secondary metabolite, i.e. polyhydroxyalkanoate, SSCP was prominently utilized as a culture-independent method. This methodology has assisted in verifying that the rhizosphere is a wanted habitat for a bacterial population that is a significant producer of polyhydroxyalkanoate. However, some inconsistencies amid the culture-dependent and culture-independent (SSCP) methods were revealed at the strain level [70]. SSCP was done to identify the rhizospheric fungal population in the Canary Islands. Moreover, it was found that *Trichoderma*-specific communities are an isolated population that maintain soil health and yield [71].

7.5.1.4 Automated Ribosomal Intergenic Spacer Analysis

Automated ribosomal intergenic spacer analysis is used to examine the alterations in microbial diversity, which are dependent on variations in the size of the unique region known as the intergenic transcribed spacer region, having rRNA gene encoding between the 16S and 23S, in addition to rRNA-encoding genes, i.e. 18S and 28S, for bacteria and eukaryotes (chiefly, fungi) [72,73]. This approach compares the structure of microbial communities along with an estimation of species richness in various samples from different ecosystems like soils. The effect of *Azospirillum lipoferum* strain CRT1 on the constitution of bacterial communities in the rhizosphere in field-grown maize was investigated using ARISA fingerprints. The consequences have shown that this process helped with distinguishing plant-to-plant variability in addition to detecting variability in native communities of rhizobacteria following the inoculation of *Azospirillum* [74]. Two major drawbacks have been identified. First, the spacer length in a particular genome may vary because of a variation in many operons, resulting in richness overestimation. Second, a spacer region of a similar size has been discovered in distinct microbes. An *in silico* methodology was created for a bacterial dataset of 16S–23S spacers and ARISA profiles to assess the precision of richness. Also, ARISA is unsuitable for estimating the richness of enormously diverse ecosystems like soils. Moreover, this methodology magnifies the richness of species, which may have been corrected to analyze the richness of species in the least diversified ecosystem [75].

7.5.1.5 Random Amplified Polymorphic DNA

This method is used for analyzing genetic relationships and diversity [76]. To anneal at several genomes of microbial populations, synthetic oligonucleotide (10 mer nucleotides) primers of any nucleotide sequence are utilized. Following that, a set of amplicons that are exclusive to the template is synthesized. To permit diverse pairing of the primer, shallow annealing temperatures (approximately 36°C) are typically utilized [77]. As a result, profile reproducibility is frequently a concern. Therefore, RAPD is a widely analyzed molecular technique for developing DNA markers. Unlike T-RFLP, it is less time-consuming and less expensive. RAPD has been a prominent tool for analyzing soil microbial communities to gain insight into soil microbial genetic makeup and genetic reliability of micro-propagated plants [78].

7.5.1.6 Amplified Ribosomal DNA Restriction Analysis

Amplified ribosomal DNA restriction analysis renders profiles of restriction fragment from a bacterial population having 16S rRNA gene amplicons or fungal population 18S gene amplicons. Tetra cutter restriction enzymes are utilized for product generation after amplification of a specific region [79]. The products get amplified and used either as fingerprinting pools or libraries of clones to distinguish every sequence in subsequent analysis. Universal primers cannot indicate the presence of definite organisms in the samples; nevertheless, they are utilized to evaluate the structure of the microbial community's crosswise models. Whereas, for phylogenetic analysis, definite primers for a particular microbial genome have been widely used. This technique is utilized for (i) classifying isolates having antagonistic properties against *Phytophthora capsici* [80]; (ii) identifying the clusters that are segregated out of glacier soils [81]; and (iii) confirming various PGP microbial strains inoculated in *Fragaria ananassa* via stolon among mother-plant and daughter-plant [82].

7.6 Non-PCR Methods

7.6.1 Phospholipid Fatty Acid Analysis

Phospholipids are major components of microbial cell membranes that are used to differentiate between different microbial species. The features of phospholipids include distinct fatty acids with varying acyl chains. They are also utilized as prominent microbial population biomarkers. Phospholipid fatty acid (PLFA) analysis is a chemotaxonomic marker of microorganisms that is broadly used in microbial ecology. This methodology is dependent on the assumption that phospholipids get degraded easily so that any remaining phospholipids must fit into living organisms. Recently, the PLFA method has been used to investigate soil organic matter (SOM) formation and the role of cell-envelope fragments of microbial cells in this analysis [83]. The findings of the PLFA study improved our indulgence of SOM formation and the link-up with microbial abundance and activity. Also, PLFA is used to determine diversity indices condemned as the phylogenetic resolution of community characterization, which is relatively scanty [84].

7.6.2 Stable-Isotope Probing

Stable-isotope probing involves incorporating various 13C-labeled, 12C-labeled, or 15N substrates within cells, which represent cellular biomarkers. It is used to detect lively microorganisms devoid of the utilization of radioactive isotopes [85,86]. Moreover, density-gradient centrifugation is used to undo the substrate from the unlabeled nucleic acid [87]. Stable-isotope probing is used to identify the spatial variation of microorganisms active and present in the rhizosphere during carbon flow. Furthermore, SIP has been widely used to examine fungal–bacterial interactions to understand better every microbial community's critical role in the degradation of soil litter [88]. The method also allows for tracking various processes in microbial communities, such as matter fluxes and biochemical activities. Unlike other methodologies, SIP provides much knowledge regarding carbon fluxes along with soil microbial communities.

7.6.3 DNA Arrays

DNA array is a technique in which a solid surface has spots of various DNA samples. DNA, cDNA, or oligonucleotide-containing spots usually depict genes with both recognized and unidentified functions, in addition to non-coding RNAs. The DNA/RNA gets hybridized in these locations, allowing, in turn, many genes to be gathered at the same time. Numerous records on genome transcriptional profiling of microbial communities present in soil use microarrays to assess changes in bacterial population, following disturbance or treatment. Microarray approach on soil microbial population comprises taxonomic group profiling and functional gene or RNA profiling, i.e. cDNA from mRNA/rRNA. Functional gene arrays (FGA) comprise DNA microarrays that contain active microbial genes. An example is GeoChip 3.0, which includes 57,000 genes plus identified functions of abundant species and gene variants from 292 functional gene families [89]. The main goal of profiling mRNA-based expression is to classify and exemplify distinct expressed genes; few are clustered and dependent on the likes of a synchronized gene expression pattern. In microorganisms, "regulons" can be part of particular signaling or biochemical pathways. On the other hand, rather than gene expression, most FGAs are used for microbial DNA profiling and to detect the presence of specific genes (mRNA) [90]. The utilization of microarrays intended for analyzing the communities of microbes present in the soil is restricted to the sequences rendered as probes. The data generated by microarray and added methodologies are typically validated and investigated for specific microbial genome and individual occurrence that express candidate genes [91].

7.6.4 Fluorescence *In Situ* Hybridization

Fluorochrome is a fluorescent molecule linked to an oligonucleotide (10 mer sequence) probe. Because of genetic stability and elevated copy number, 16S rRNA is commonly utilized to probe microbiology. The fluorescent probe usually links to a complementary sequence, allowing fluorescence microscopy to detect it [92]. CARD-fluorescence *in situ* hybridization (FISH), a type of FISH that improves sensitivity by using horseradish peroxidase-labeled probes in amalgamation with catalyzed deposition, was used to detect the incidence of live bacteria in *Arabidopsis thaliana* root segments. This study improved our perception of the function of the soil profile, plant growth and development stage, and genotype in determining the root microbiome [93].

7.6.5 High-Throughput Sequencing Technologies

The high-throughput sequencing approach known as next-generation sequencing (NGS) is progressively employed for culture-independent estimation of the diversity of microbes in composite conditions, i.e. soil. Different technologies for increasing the throughput of DNA/RNA concatenation have become prevalent due to the advancement in nanotechnology and bioinformatics. Such technologies are crucial in metagenomics (DNA-based), and metatranscriptomics (RNA-based), procedures that facilitate the inclusive representation of the capability, along with robust functionality, of microbial communities, respectively.

7.6.6 Illumina Genome Analyzer

This analyzer is the instrument built on the simultaneous, fluorescence-dependent readout of thousands of immobilized forms of libraries whose sequencing is done by adopting reversible terminator chemistry. Illumina involves four sequencers: (i) the HiSeq 2500, (ii) the HiSeq 2000, (iii) the Genome Analyzer IIx, and (iv) the MiSeq platform [94]. The HiSeq 2500 is the most commanding and delivers up to 600 Gb of data with the highest reads per run with a read length of about 2 × 100 bp. Illumina sequencers read short sequences but have more throughput than the Roche 454 FLX pyrosequencer, making them ideal for gene expression studies in multifaceted environments involving soils. It also allows the recognition of a variety of microbial genomes [95]. Using this platform to study microbial diversity involves more output than pyrosequencing. However, their read length is shorter, making taxonomic data least accurate [96].

7.6.7 Ion PGM (Personal Genome Machine)

Semiconductor sequencing technology is used in this platform. A proton is set free every time the addition of a single nucleotide in DNA occurs, along with the altered pH of the resulting product, calculated using a pH-sensitive field-effect transistor. As a result, nucleotides that are not labeled are utilized, and presence is directly detected. When compared with fluorescence detection systems, it has shorter run times. Currently, it can read 400 bp in 4 hours. The PGM has recently been proposed as a microbial ecology sequencing platform for assessing bacterial and archaeal community dynamics [97].

7.6.8 Helioscope Single-Molecule Sequencer

Direct sequencing of DNA/RNA fragments is done using a helioscope single-molecule sequencer, eliminating the need for amplification. The template DNA is fragmented and hybridized on glass flow cells. For sample addition, each of the 25 channels on a standard flow cell can be directed individually. After the setup of flow cells is done, these are then placed in the HeliScope Sequencing system, alongside the entire chemical reagents requisite for sequence analysis and image pattern generation. In this systemic manner, lakhs of reads per run (25–35 bp) can be produced, with the information of more than one gigabyte per day [98]. Helioscope could help with genome biology by performing direct nucleic acid sequencing. The telescope sequencer was used to obtain sequence data to analyze the numerous small RNAs (sRNAs) and discover new sRNAs in samples. This methodology has an elevated inaccuracy rate that ranges between 3% and 4%. In comparison with supplementary methods, it is more expensive than other platforms. Therefore, it is not widely used in the NGS market [99].

7.6.9 Quorum Sensing in Defense of Abiotic Stress

Quorum sensing is an extensive and quick method that changes genome expression arrangements to answer community densities in the bacterial population in a coordinated way. The term "quorum sensing" was used primarily to define the minor threshold level of single-cell masses needed for collective responses of microbial populations [100]. Microbes, especially bacteria using QS, produce autoinducers or N-acyl-homoserine lactones (AHL), which are signals that help in communicating with other bacteria in the environment. They also constitute a receptor that detects the inducer exclusively. When the inducer links to the receptor, specific genes are activated, including those for synthesizing the autoinducer [101]. If fewer bacteria of a similar group are located close by, dispersion will reduce the inducer amount in the environment to almost zero. Therefore, only a minimum quantity of the inducer is synthesized by bacteria. When many bacteria of the same type are located nearby, the concentration of the inducer crosses a threshold and is synthesized in more significant volumes, which activates the receptor fully [102].

A wide range of gram-negative bacteria utilize AHL to organize virulence expressions according to the bacterial population density. Several bacterial phenotypes vital to ensure the organization of eukaryotic hosts of symbiotic, pathogenic, or commensal relations use motility, exopolysaccharide production, biofilm development, and toxin production to regulate the process of QS [103]. Bacterial AHLs are released into the rhizosphere at organically emotional levels [104]. This study depicts how QS is caught up in the occurrence of ailment, where multiple pathogens like *Pseudomonas corrugate, Erwinia carotovora, Xanthomonas campestris*, and *Burkholderia* sp. prevail [104].

According to an investigation on genetically modified tomato plants, the pattern of QS in plant growth-promoting microbe (PGPM) changes. This expands the critical role of QS and enhances the idea of QS interference either to control attacks by plant pathogens or to encourage defense against various stresses [105]. Various AHL signaling molecules could occur between M12 and M14, helping gram-negative PGPMs or rhizobacteria associated with *Burkholderia graminis* thrive. These molecules were investigated for their involvement in promoting plant growth and development along with the induction of salt defense. Thin-layer chromatography, in addition to LC-MS/MS, were used to assess *in vitro* AHL development. Green fluorescent protein-based biosensor fabricated with confocal laser scanning microscopy was used to detect *in situ* AHL output in the rhizosphere of *A. thaliana* plants. Then, these rhizobacteria had been tested over wild-type (wt) plants of tomato to see whether QS was implicated in plant growth promotion and salt stress defense in addition to their YenI (short-chain AHL producers) and LasI-expressing transgenics. Only M12 stimulated plant development in wt tomato plants, and this impact was lost in mutual transgenic lines. The strain M14, on the other hand, did not encourage growth in wt tomatoes but did in LasI. In wt tomato, strain M14 induced salt stress resistance, which was lost in both transgenic lines. Yet, salt resistance was not influenced in wt tomato plants, but it was seen in LasI tomato plants. Therefore, the findings show that AHL signal molecules are involved in both M12 and M14 plant growth-promoting rhizobacteria (PGPR) strains and their ability to promote plant growth and development and defense against salt stress.

7.7 Metagenomics: the Rhizosphere as a Basis of Genes

The metagenomic approach is a relatively new area of study in rhizosphere biotechnology [106]. This method is known as the branch of molecular biology that studies the genomes of entire microorganisms rather than those of a single species. This branch of science, also known as "community genomics," explores the improved genetic aspects of ecological samples from whole microbial communities. Traditional microbiology and genetics are focused on the isolation and sequencing of single microbial genomes to detect and evaluate genes with newer and valuable functions.

Rhizosphere metagenomics investigates the rhizosphere via the extraction of complete DNA and the development of a clonal library from the extracted DNA along with a selection of clones to get the gene of interest from a substantial genetic pool derived from microbial communities present in soil [107]. This methodology has been used to identify newer biomolecules dependent on discovering various genes coded for proteins with novel functions. The identification and isolation of more unique genes that encode degradative enzymes [108], antibiotic resistance [109], and antibiotic production [110] have all supported this strategy. However, linking a gene to its function is complex. Despite the time-consuming and labor-intensive method of gene isolation and detection, rhizosphere metagenomics appears to be a vital area of study in discovering newer, more efficient genes with possibilities of several biotechnological applications, except for scientific restrictions, which must be addressed to ensure future technological advancements.

7.7.1 Metabolic Engineering

Despite the efforts that have been made to produce numerous bioactive compounds in plants on a large scale, increasing their content *in vivo* and obtaining copies of bioactives under field production conditions have remained difficult. Efforts have been made through transgenic and non-transgenic approaches to increase the stages of functional metabolites in plants, which necessitate complex regulatory mechanisms. If a restrictive location is established, bacterial elicitors can be utilized to identify the primary genes that limit the metabolic pathway. We may be able to resolve the low levels of target compounds generated using transgenic approaches. Finally, it is a very appealing and cheering challenge; newer molecules may come into view after the activation of a metabolic pathway following elicitation [111].

7.8 Conclusion and Future Perspectives

Modern agriculture should concentrate on sustainable tools and techniques to minimize the detrimental effects on environment without jeopardizing food security and standards. Microbiome engineering and rhizosphere biotechnology are fast-growing fields with notably successful results in the agriculture sector. This chapter explored recently developed tools for improvement of the rhizosphere by improving the resident microbiome exclusively as well as intrusively. However, future research should focus on customizing rhizospheric microbiota, developing beneficial consortia, and stabilizing them in the field over multiple generations. More research will be required in the field of top-down and bottom-up strategies, PCR and non-PCR methods, and omics techniques, along with studies about their effects on various crops. A combination of soil modifications, plant breeding and genetic manipulation techniques, and targeted engineering of rhizospheric microbiota with microbiome engineering/biotechnology can be used to minimize the use of chemical fertilizers/pesticides and improve the yield and quality of crops. The practical implementation of this integrated approach should be studied, and the possibilities of successfully blending microbiome engineering and biotechnology should be investigated across a range of other applications such as bioremediation, human health, etc.

References

1 Lobell, D.B. and Gourdji, S.M. (2012). The influence of climate change on global crop productivity. *Plant Physiology* 160 (4): 1686–1697.
2 Allison, F.E. (1973). *Soil Organic Matter and its Role in Crop Production*. USA: Elsevier.
3 Renschler, C.S. and Harbor, J. (2002). Soil erosion assessment tools from point to regional scales-the role of geomorphologists in land management research and implementation. *Geomorphology* 47 (2–4): 189–209.
4 Carpenter, J., Felsot, A., Goode, T. et al. (2002). Comparative environmental impacts of biotechnology-derived and traditional soybean, corn, and cotton crops.
5 Bisht, N., Sharma, B., Joshi, N. et al. (2018). Role of microbial siderophore in Iron acquisition and amelioration of plant stress in saline soil. *Biotech Today: An International Journal of Biological Sciences* 8 (2): 7–15.
6 Kuzyakov, Y. and Blagodatskaya, E. (2015). Microbial hotspots and hot moments in soil: Concept & review. *Soil Biology and Biochemistry* 83: 184–199.
7 Mohanram, S. and Kumar, P. (2019). Rhizosphere microbiome: Revisiting the synergy of plant-microbe interactions. *Annals of Microbiology* 69 (4): 307–320.
8 Bulgarelli, D., Schlaeppi, K., Spaepen, S. et al. (2013). Structure and functions of the bacterial microbiota of plants. *Annual Review of Plant Biology* 64: 807–838.
9 Mueller, U.G. and Sachs, J.L. (2015). Engineering microbiomes to improve plant and animal health. *Trends in Microbiology* 23 (10): 606–617.
10 Hardoim, P.R., Van Overbeek, L.S., Berg, G. et al. (2015). The hidden world within plants: Ecological and evolutionary considerations for defining functioning of microbial endophytes. *Microbiology and Molecular Biology Reviews* 79 (3): 293–320.
11 Pii, Y., Mimmo, T., Tomasi, N. et al. (2015). Microbial interactions in the rhizosphere: Beneficial influences of plant growth-promoting rhizobacteria on the nutrient acquisition process. A review. *Biology and Fertility of Soils* 51 (4): 403–415.
12 Rekadwad, B. (ed.) (2020). *Microbial Systematics: Taxonomy, Microbial Ecology, Diversity*. India: C.R.C. Press.
13 Chukwuneme, C.F., Ayangbenro, A.S., Babalola, O.O. et al. (2021). Functional diversity of microbial communities in two contrasting maize rhizosphere soils. *Rhizosphere* 17: 100282.
14 Prashar, P., Kapoor, N., and Sachdeva, S. (2014). Rhizosphere: Its structure, bacterial diversity and significance. *Reviews in Environmental Science and Bio/Technology* 13 (1): 63–77.
15 Compant, S., Clément, C., and Sessitsch, A. (2010). Plant growth-promoting bacteria in the rhizo-and endosphere of plants: Their role, colonization, mechanisms involved and prospects for utilization. *Soil Biology and Biochemistry* 42 (5): 669–678.
16 Parmar, N. and Dufresne, J. (2011). Beneficial interactions of plant growth-promoting rhizosphere microorganisms. In: *Bioaugmentation, Biostimulation and Biocontrol* (eds. A. Singh, N. Parmar, and R.C. Kuhad), 27–42. Berlin: Springer.
17 Gupta, G., Parihar, S.S., Ahirwar, N.K. et al. (2015). Plant growth-promoting rhizobacteria (PGPR): Current and future prospects for development of sustainable agriculture. *Journal of Microbial and Biochemical Technology* 7 (2): 096–102.
18 Meena, K.K., Sorty, A.M., Bitla, U.M. et al. (2017). Abiotic stress responses and microbe-mediated mitigation in plants: The omics strategies. *Frontiers in Plant Science* 8: 172.

19 Pirog, T.P., Iutynska, G.O., and Leonova, N.O. (2018). Microbial synthesis of phytohormones. *Biotechnologia Acta* 11 (1).

20 Mendes, R., Garbeva, P., and Raaijmakers, J.M. (2013). The rhizosphere microbiome: Significance of plant beneficial, plant pathogenic, and human pathogenic microorganisms. *FEMS Microbiology Reviews* 37 (5): 634–663.

21 Glick, B.R. (2020). Biocontrol of Bacteria and Fungi. In: *Beneficial Plant-Bacterial Interactions* (ed. B.R. Glick), 181–230. Cham: Springer.

22 Zhou, Q., Li, K., Jun, X. et al. (2009). Role and functions of beneficial microorganisms in sustainable aquaculture. *Bioresource Technology* 100 (16): 3780–3786.

23 Rao, D.L.N. (2007). Microbial diversity, soil health and sustainability. *Journal-Indian Society of Soil Science* 55 (4): 392.

24 Del Carmen Orozco-mosqueda, M., Del Carmen Rocha-granados, M., Glick, B.R. et al. (2018). Microbiome engineering to improve biocontrol and plant growth-promoting mechanisms. *Microbiological Research* 208: 25–31.

25 Kumar, A. and Dubey, A. (2020). Rhizosphere microbiome: Engineering bacterial competitiveness for enhancing crop production. *Journal of Advanced Research* 24: 337–352.

26 Kolstø, A.B., Tourasse, N.J., and Økstad, O.A. (2009). What sets *Bacillus anthracis* apart from other *Bacillus* species? *Annual Review of Microbiology* 63: 451–476.

27 Santoyo, G., Pacheco, C.H., Salmerón, J.H. et al. (2017). The role of abiotic factors modulating the plant-microbe-soil interactions: Toward sustainable agriculture. A review. *Spanish Journal of Agricultural Research* 15 (1): 13.

28 Arif, I., Batool, M., and Schenk, P.M. (2020). Plant microbiome engineering: Expected benefits for improved crop growth and resilience. *Trends in Biotechnology* 38 (12): 1385–1396.

29 Campbell, S. (1999). *Improving Your Soil (202)*. Storey Publishing, USA.

30 Bass, A.M., Bird, M.I., Kay, G. et al. (2016). Soil properties, greenhouse gas emissions and crop yield under compost, biochar and co-composted biochar in two tropical agronomic systems. *Science of the Total Environment* 550: 459–470.

31 Kremer, R.J. (2017). Biotechnology impacts on soil and environmental services. In: *Soil Health and Intensification of Agroecosystems* (eds. M.M. Al-kaisi and B. Lowery, 353–375. United States: Academic Press.

32 Fischer, D. and Glaser, B. (2012). Synergisms between compost and biochar for sustainable soil amelioration. *Management of Organic Waste* 1: 167–198.

33 Poveda, J., Martínez Gómez, Á., Fenoll, C. et al. (2021). The use of biochar for plant-pathogen control. *Phytopathology* 1–10.

34 Bargaz, A., Lyamlouli, K., Chtouki, M. et al. (2018). Soil microbial resources for improving fertilizers efficiency in an integrated plant nutrient management system. *Frontiers in Microbiology* 9: 1606.

35 Syed-Ab-Rahman, S.F., Xiao, Y., Carvalhais, L.C. et al. (2019). Suppression of *Phytophthora capsici* infection and promotion of tomato growth by soil bacteria. *Rhizosphere* 9: 72–75.

36 Ghosh, S., Chowdhury, R., and Bhattacharya, P. (2016). Mixed consortia in bioprocesses: Role of microbial interactions. *Applied Microbiology and Biotechnology* 100 (10): 4283–4295.

37 Loganathan, M., Garg, R., Venkataravanappa, V. et al. (2014). Plant growth-promoting rhizobacteria (PGPR) induces resistance against *Fusarium* wilt and improves lycopene content and texture in tomato. *African Journal of Microbiological Research* 8 (11): 1105–1111.

38 Wagner, M.R., Lundberg, D.S., Coleman-Derr, D. et al. (2014). Natural soil microbes alter flowering phenology and the intensity of selection on flowering time in a wild *Arabidopsis* relative. *Ecology Letters* 17 (6): 717–726.

39 Jones, R.J., Johnson, R., Maclean, P. et al. (2017). More than just plants: A study of biotic stress impacts on the root microbiomes of *Trifolium repens*, *Proceedings of the NZMS Annual conference*, Auckland, New Zealand, 65.

40 Wicaksono, W.A. (2016). Inner Plant Space-bacterial Endophytes of Leptospermum Scoparium (Mānuka) (Doctoral dissertation, Lincoln University).

41 Desrut, A., Moumen, B., Thibault, F. et al. (2020). Beneficial rhizobacteria *Pseudomonas simiae* WCS417 induce major transcriptional changes in plant sugar transport. *Journal of Experimental Botany* 71 (22): 7301–7315.

42 Pérez-Jaramillo, J.E., de Hollander, M., Ramírez, C.A. et al. (2019). Deciphering rhizosphere microbiome assembly of wild and modern common bean (*Phaseolus vulgaris*) in native and agricultural soils from Colombia. *Microbiome* 7 (1): 1–16.

43 Ke, J., Wang, B., and Yoshikuni, Y. (2020). Microbiome engineering: Synthetic biology of plant-associated microbiomes in sustainable agriculture. *Trends in Biotechnology* 39 (3): 244–261.

44 Martínez-García, E., Benedetti, I., Hueso, A. et al. (2015). Mining environmental plasmids for synthetic biology parts and devices. Plasmids: Biology and Impact in Biotechnology and Discovery, 633–649.

45 Zou, X., Wang, L., Li, Z. et al. (2018). Genome engineering and modification toward synthetic biology for the production of antibiotics. *Medicinal Research Reviews* 38 (1): 229–260.

46 Herisse, M., Porter, J.L., Guerillot, R. et al. (2018). The ΦBT1 large serine recombinase catalyzes D.N.A. integration at pseudo-attB sites in the genus. *Nocardia. Peer J* 6: e4784.

47 Li, L., Liu, X., Wei, K. et al. (2019). Synthetic biology approaches for chromosomal integration of genes and pathways in industrial microbial systems. *Biotechnology Advances* 37 (5): 730–745.

48 Johnson, C.M. and Grossman, A.D. (2015). Integrative and conjugative elements (I.C.E.s): What they do and how they work. *Annual Review of Genetics* 49: 577–601.

49 Delavat, F., Miyazaki, R., Carraro, N. et al. (2017). The hidden life of integrative and conjugative elements. *FEMS Microbiology Reviews* 41 (4): 512–537.

50 Wang, G., Zhao, Z., Ke, J. et al. (2019). C.R.A.G.E. enables rapid activation of biosynthetic gene clusters in undomesticated bacteria. *Nature Microbiology* 4 (12): 2498–2510.

51 Wang, B., Zhao, Z., Jabusch, L.K. et al. (2020). CRAGE-Duet facilitates modular assembly of biological systems for studying plant–microbe interactions. *A.C.S. Synthetic Biology* 9 (9): 2610–2615.

52 Ohtsuka, M., Kimura, M., Tanaka, M. et al. (2009). Recombinant D.N.A. technologies for construction of precisely designed transgene constructs. *Current Pharmaceutical Biotechnology* 10 (2): 244–251.

53 Gilmore, S.P. and O'Malley, M.A. (2016). Microbial communities for bioprocessing: Lessons learned from nature. *Current Opinion in Chemical Engineering* 14: 103–109.

54 Miguel, A.B.R., Jetten, M.S., and Welte, C.U. (2020). The role of mobile genetic elements in organic micropollutant degradation during biological wastewater treatment. *Water Research* 9: 100065.

55 Arbas, S.M., Narayanasamy, S., Herold, M. et al. (2021). Roles of bacteriophages, plasmids and CRISPR immunity in microbial community dynamics revealed using time-series integrated meta-omics. *Nature Microbiology* 6 (1): 123–135.

56 Pinilla-Redondo, R., Cyriaque, V., Jacquiod, S. et al. (2018). Monitoring plasmid-mediated horizontal gene transfer in microbiomes: Recent advances and future perspectives. *Plasmid* 99: 56–67.

57 Whitfill, T. and Oh, J. (2019). Recoding the metagenome: Microbiome engineering in situ. *Current Opinion in Microbiology* 50: 28–34.

58 Brigulla, M. and Wackernagel, W. (2010). Molecular aspects of gene transfer and foreign D.N.A. acquisition in prokaryotes with regard to safety issues. *Applied Microbiology and Biotechnology* 86 (4): 1027–1041.

59 Killham, K. (2011). Integrated soil management–moving towards globally sustainable agriculture. *Journal of Agricultural Sciences-London* 149: 29–36.

60 Ranjard, L., Poly, F., and Nazaret, S. (2000). Monitoring complex bacterial communities using culture-independent molecular techniques: Application to soil environment. *Research in Microbiology* 151: 167–177.

61 McGrath, K.C., Thomas-Hall, S.R., Cheng, C.T. et al. (2008). Isolation and analysis of mRNA from environmental microbial communities. *Journal of Microbiological Methods* 75: 172–176.

62 Maoloni, A., Milanović, V., Cardinali, F. et al. (2020). Bacterial and fungal communities of Gioddu as revealed by PCR–D.G.G.E. analysis. *Indian Journal of Microbiology* 60 (1): 119–123.

63 Dowd, S.E., Sun, Y., Secor, P.R. et al. (2008). Survey of bacterial diversity in chronic wounds using pyrosequencing, D.G.G.E., and full ribosome shotgun sequencing. *B.M.C. Microbiology* 8: 43.

64 Nübel, U., Garcia-Pichel, F., and Muyzer, G. (1997). PCR primers to amplify 16S rRNA genes from cyanobacteria. *Applied and Environmental Microbiology* 63: 3327–3332.

65 Zhou, X. and Wu, F. (2012). P-coumaric acid influenced cucumber rhizosphere soil microbial communities and the growth of *Fusarium oxysporum* f. Sp. *Cucumerinum owen. PLoSOne* 7: e48288.

66 Frerichs, J., Oppermann, B.I., Gwosdz, S. et al. (2012). Microbial community changes at a terrestrial volcanic CO_2 vent induced by soil acidification and anaerobic microhabitats within the soil column. *FEMS Microbiology Ecology* 84: 60–74.

67 Kari, A., Nagymáté, Z., Romsics, C. et al. (2019). Monitoring of soil microbial inoculants and their impact on maize (*Zea mays* L.) rhizosphere using T-RFLP molecular fingerprint method. *Applied Soil Ecology* 138: 233–244.

68 Osborn, A.M., Moore, E.R., and Timmis, K.N. (2000). An evaluation of terminal-restriction fragment length polymorphism (T-RFLP) analysis for the study of microbial community structure and dynamics. *Environmental Microbiology* 2: 39–50.

69 Aiken, J.T. (2011). Terminal restriction fragment length polymorphism for soil microbial community fingerprinting. *Soil Science Society of American Journal* 75: 102–111.

70 Gasser, I., Müller, H., and Berg, G. (2009). Ecology and characterization of polyhydroxyalkanoate-producing microorganisms on and in plants. *FEMS Microbiology Ecology* 70: 142–150.

71 Zachow, C., Berg, C., Müller, H. et al. (2009). Fungal diversity in the rhizosphere of endemic plant species of Tenerife (Canary Islands): Relationship to vegetation zones and environmental factors. *Internal Society for Microbial Ecology Journal* 3: 79–92.

72 Fuhrman, J.A., Steele, J.A., Hewson, I. et al. (2008). A latitudinal diversity gradient in planktonic marine bacteria. *Proceedings of the National Academy of Sciences of the United States of America* 105: 7774–7778.

73 Srivastava, M., Kaushik, M.S., and Mishra, A.K. (2016). Linking the physicochemical properties with the abundance and diversity of Rhizospheric bacterial population inhabiting Paddy soil based on a concerted multivariate analysis of PCR-DGGE and R.I.S.A. *Geomicrobiology Journal* 33 (10): 894–905.

74 Baudoin, E., Nazaret, S., Mougel, C. et al. (2009). Impact of inoculation with the phytostimulatory PGPR *Azospirillum lipoferum* crt1 on the genetic structure of the rhizobacterial community of field-grown maize. *Soil Biology and Biochemistry* 41: 409–413.

75 Kovacs, A., Yacoby, K., and Gophna, U. (2010). A systematic assessment of automated ribosomal intergenic spacer analysis (A.R.I.S.A.) as a tool for estimating bacterial richness. *Research in Microbiology* 161: 192–197.

76 Hadrys, H., Balick, M., and Schierwater, B. (1992). Applications of random amplified polymorphic D.N.A. (R.A.P.D.) in molecular ecology. *Molecular Ecology* 1: 55–63.

77 Amorim, J., Vidal, R., Lacerda-Junior, G. et al. (2012). A simple boiling-based D.N.A. extraction for R.A.P.D. profiling of landfarm soil to provide representative metagenomic content. *Genetics and Molecular Research* 11: 182–189.

78 Singh, S.K., Rai, M.K., and Sahoo, L. (2012). An improved and efficient micropropagation of *Eclipta alba* through transverse thin cell layer culture and assessment of clonal fidelity using R.A.P.D. analysis. *Industrial Crops and Products* 37: 328–333.

79 Nocker, A., Burr, M., and Camper, A.K. (2007). Genotypic microbial community profiling: A critical technical review. *Microbial Ecology* 54: 276–289.

80 Yang, M.M., Xu, L.P., Xue, Q.Y. et al. (2012). Screening potential bacterial biocontrol agents towards *Phytophthora capsici* in pepper. *European Journal of Plant Pathology* 134: 811–820.

81 Liu, G.-X., Hu, P., Zhang, W. et al. (2012). Variations in soil culturable bacteria communities and biochemical characteristics in the Dongkemadi glacier forefield along a chronosequence. *Folia Microbiologica* 57: 485–494.

82 Guerrero-Molina, M.F., Winik, B.C., and Pedraza, R.O. (2012). More than rhizosphere colonization of strawberry plants by *Azospirillum brasilense*. *Applied Soil Ecology* 61: 205–212.

83 Schurig, C., Smittenberg, R.H., Berger, J. et al. (2013). Microbial cell-envelope fragments and the formation of soil organic matter: A case study from a glacier forefield. *Biogeochemistry* 113: 595–612.

84 Frostegard, A., Tunlid, A., and Baath, E. (2011). Use and misuse of P.L.F.A. measurements in soils. *Soil Biology and Biochemistry* 43: 1621–1625.

85 Dumont, M.G. and Murrell, J.C. (2005). Stable isotope probing-linking microbial identity to function. *Nature Reviews Microbiology* 3: 499–504.

86 Radajewski, S., Ineson, P., Parekh, N.R. et al. (2000). Stable-isotope probing as a tool in microbial ecology. *Nature* 403: 646–649.

87 Lu, Y., Abraham, W.R., and Conrad, R. (2007). Spatial variation of active microbiota in the rice rhizosphere revealed by in situ stable isotope probing of phospholipid fatty acids. *Environmental Microbiology* 29: 474–481.

88 Bodé, S., Fancy, R., and Boeckx, P. (2013). Stable isotope probing of amino sugars – a promising tool to assess microbial interactions in soils. *Rapid Commun. Mass Spectrometry* 27: 1367–1379.

89 He, Z., Deng, Y., van Nostrand, J.D. et al. (2010). Geochip 3.0 as a high-throughput tool for analyzing microbial community composition, structure and functional activity. *International Society of Microbial Ecology* 4: 1167–1179.

90 Waldron, P.J., Wu, L., van Nostrand, J.O.Y.D. et al. (2009). Functional gene array-based analysis of microbial community structure in groundwaters with a gradient of contaminant levels. *Environmental Science and Technology* 43: 3529–3534.

91 Wu, L., Kellogg, L., Devol, A.H. et al. (2008). Microarray-based characterization of microbial community functional structure and heterogeneity in marine sediments from the Gulf of Mexico. *Applied and Environmental Microbiology* 74: 4516–4529.

92 Moter, A. and Göbel, U.B. (2000). Fluorescence *in situ* hybridization (FISH) for direct visualization of microorganisms. *Journal of Microbiological Methods* 41: 85–112.

93 Lundberg, D.S., Lebeis, S.L., Paredes, S.H. et al. (2012). Defining the core *Arabidopsis thaliana* root microbiome. *Nature* 488: 86–90.

94 Sequencing systems. Available online: http://www.illumina.com/systems/sequencing.ilmn (accessed on 6 May 2013).

95 Fierer, N., Leff, J.W., Adams, B.J. et al. (2012). Cross-biome metagenomic analyses of soil microbial communities and their functional attributes. *Proceedings of the National Academy of Sciences of the United States of America* 109: 21390–21395.

96 Bartram, A.K., Lynch, M.D.J., Stearns, J.C. et al. (2011). Generation of multimillion-sequence 16S rRNA gene libraries from complex microbial communities by assembling paired-end Illumina reads. *Applied and Environmental Microbiology* 77: 3846–3852.

97 Whiteley, A.S., Jenkins, S., Waite, I. et al. (2012). Microbial 16S rRNA ion tag and community metagenome sequencing using the Ion Torrent (P.G.M.) platform. *Journal of Microbiological Methods* 91: 80–88.

98 Shokralla, S., Spall, J.L., Gibson, J.F. et al. (2012). Next-generation sequencing technologies for environmental D.N.A. research. *Molecular Ecology* 21: 1794–1805.

99 Kapranov, P., Ozsolak, F., and Milos, P.M. (2012). Profiling of short R.N.A.s using helicos single-molecule sequencing. In: *Next-Generation microRNA Expression Profiling Technology* (ed. J. Fan), 219–232. UK: Springer.

100 Fuqua, W.C. and Winans, S.C. (1994). A LuxR-LuxI type regulatory system activates *Agrobacterium* Ti plasmid conjugal transfer in the presence of a plant tumor metabolite. *Journal of Bacteriology* 176: 2796–2806.

101 Chamkhi, I., El Omari, N., Benali, T. et al. (2020). Quorum Sensing and Plant–Bacteria Interaction: Role of Quorum Sensing in the Rhizobacterial Community Colonization in the Rhizosphere. In: *Quorum Sensing: Microbial Rules of Life* (ed. S.S. Dhiman), 139–153. Washington, DC: American Chemical Society.

102 Whitehead, N.A., Barnard, A.M.L., Slater, H. et al. (2001). Quorum sensing in gram-negative bacteria. *FEMS Microbiology. Reviews* 25: 65–404.

103 Gonzalez, J.E. and Keshavan, N.D. (2006). Messing with bacterial quorum sensing. *Microbiology and Molecular Biology Reviews* 70: 859–875.

104 Barriuso, J., Ramos Solano, B., Fray, R.G. et al. (2008). Transgenic tomato plants alter quorum sensing in plant growth-promoting rhizobacteria. *Plant Biotechnology Journal* 6: 442–452.

105 Ahmad, I., Farrukh, A., Ahmad, F. et al. (2008). Quorum sensing in bacteria. Potential in plant health protection. In: *Plant–bacteria Interactions: Strategies and Techniques to Promote Plant Growth*, (eds. I. Ahmad, J. Pichtel and S. Hayat), Weinheim: Wiley-VCH.

106 Rolf, D. (2005). The metagenomics of soil. *Nature Reviews of Microbiology* 3: 470–478.

107 Enebe, M.C. and Babalola, O.O. (2020). Effects of inorganic and organic treatments on the microbial community of maize rhizosphere by a shotgun metagenomics approach. *Annals of Microbiology* 70 (1): 1–10.

108 Henne, A., Daniel, R., Schmitz, R.A. et al. (1999). Construction of environmental D.N.A. libraries in *Escherichia coli* and screening for the presence of genes conferring utilization of 4-hydroxybutyrate. *Applied Environmental Microbiology* 65: 3901–3907.

109 Riesenfeld, C.S., Goodman, R.M., and Handelsman, J. (2004). Uncultured soil bacteria are a reservoir of new antibiotic resistance genes. *Environmental Microbiology* 6: 981–989.

110 Wang, G.Y., Graziani, E., Waters, B. et al. (2000). Novel natural products from soil D.N.A. libraries in a *Streptomycete* host. *Organic Letters* 2: 2401–2404.

111 Solano, B.R., Maicas, J.B., and Mañero, J.G. (2009). Biotechnology of the rhizosphere. In: *Recent Advances in Plant Biotechnology* (eds. A. Kirakosyan and P.B. Kaufman), 137–162. USA: Springer.

8

Role of Rhizospheric Microbiome in Enhancing Plant Attributes and Soil Health for Sustainable Agriculture

Sandeep Sharma[1] *and Kailash Chand Kumawat*[1,2]

[1]*Department of Soil Science, Punjab Agricultural University, Ludhiana – 141004, Punjab*
[2]*Department of Industrial Microbiology, Jacob Institute of Biotechnology and Bio-engineering, Sam Higginbottom University of Agriculture, Technology and Sciences, Prayagraj – 211007, Uttar Pradesh, India.*

8.1 Introduction

As the global human population continues exponentially, arable agricultural land has been decreased by urbanization and industrialization. The demand for food and supply has severely affected these changes, while the ongoing growth in the global population demands a significant increase in agricultural products to meet current requirements [1]. Conventional agriculture, however, faces production declines and increases in input and agrochemical prices. Furthermore, loss of productivity in agriculture due to natural and anthropogenic activities results in soil depletion and decreased crop productivity. Modernization and urbanization also differ in land-use pattern, which is why arable land is reduced. Farmers often abandon this profession due to the low-cost advantages and the introduction of numerous seed variants as they are less informed of the new technologies. The higher costs of farm income affect the interests of farmers and deprive them of farming. These two factors significantly impact agricultural development because agriculture plays a significant essential role in low-income developing countries and gross domestic product. Above 50% of Indians rely on agriculture. Heavy use of agrochemicals often detracts from soil and food quality for higher crop production [2].

Adverse climatic conditions are a threat to agriculture production, considering their anthropogenic causes. Various abiotic and biotic modifications are exposed to plants. The demands for agricultural product supply will demand developing agricultural nations, as the world's human population continues to rise. Indian farmers inappropriately use agrochemicals to enhance agricultural produce to meet as per explosive growing population demand. Food security is the product of the successful realization of the Green Revolution in India and a large part of the Indian economy. Rising agricultural production has played an essential role in chemical fertilizers. However, these substances and chemicals adversely influence soil physico-chemical properties and soil health [3]. A technique to enhance agricultural production sustainably is the biggest challenge in agriculture. In the past few years, several studies have concentrated on seeking potential methods for mitigating these

Core Microbiome: Improving Crop Quality and Productivity, First Edition. Edited by Javid A. Parray, Nowsheen Shameem, Elsayed Fathi Abd-Allah, and Mohammad Yaseen Mir.

adverse effects due to the significant negative impact of agrochemicals. The "ecological awareness" idea leads us to search for alternative approaches to enhance agricultural productivity and soil health. Using an ever-growing knowledge base and understanding of population "needs" and human–Nature interactions, some planners, managers, and researchers have developed guidelines for managing social-ecological structures based on the philosophy of ecology [4].

Ecological information may direct adaptive integrated management, which is then applied to soil restructuring and ecosystem. Ecologically sustainable engineering projects are built to the environmental advantage of the current ecosystems [5]. Microbial inoculants can also be grown while preserving sustainability to the advantage of soil in the form of bio-fertilizers. The use of beneficial microbiota may be a possible alternative to hazardous agrochemicals. The essential role of microbes in the enhancement of agriculture production and soil management is critical. In improving plant growth characteristics, the rhizospheric microbiome plays a crucial role, including absorbing nutrients and improving soil fertility and quality of crop production [6]. Soil microbial dynamics establish the capacity of soil productivity. Although plant–microbial interactions are necessary to control the functioning of the ecosystem, these interactions with plant microbes rely significantly on the nutrients available. Primarily, PGPRs are used to improve plant growth-promoting attributes through direct or indirect approaches such as phytohormone, siderophore, nutrient solubilization, nitrogen fixation (N_2 fixation), and phyto-pathogen antagonism [7].

Moreover, PGPRs have been documented to enhance plant growth-promoting traits under adverse climatic conditions, such as high temperature, drought, and salinity. The rhizosphere microbiota induces plant growth promotion through cell division, differentiation of root, proliferation, and enzyme activities through plant–microbial interaction. However, complex relationships in plants at various levels include genetics, physiology, ecology, and morphology [8]. These plant–microbial interactions differ from one plant to the other. Today, an alternative approach must be implemented to make agriculture more sustainable and effective. A new and potential sustainable strategy to boost agriculture is necessary to feed the rising world population. Food protection is a grave challenge as the world's human population has risen indiscriminately. Despite this, advanced technology and the use of agrochemicals in agriculture contribute to soil fertility degradation.

Furthermore, pathogens adversely affect crop productivity. Therefore, a biological control unit is required to advance our agricultural system to resolve or mitigate the above effects. This strategy is both sustainable and economical for the future cropping system with regard to exponential population growth. Improvement of agricultural production is achieved by adding PGPRs that encourage sustainable agriculture as they are beneficial for soil health, consumers, and farmers without any harmful impact on the environment.

8.2 Beneficial Effects of Plant Growth-Promoting Rhizobacteria

To meet the needs of agriculture and future food security, sustainable farming is necessary in today's world. Because of numerous issues, our agricultural heritage is unable to do so. We urgently need to establish a system for doing this, which is sustainable and reliable. Sustainable agriculture should satisfy our agricultural needs, and it is not possible to do so through our conventional methods. Special farming techniques are employed by this form of farming activity in which the environment is used without compromise. Biological methods, a unique formation aspect, may be an effective alternative to filling up the conventional gap. This method of farming is advantageous, and natural resources are being used without affecting future generations. The plant's root is colonized by the diversity of the dense microbic community, including bacteria, fungi, and actinomycetes. These microorganisms are beneficial microbes used as inoculants to improve plant growth-promoting attributes of crop plants [7]. These microbial groups have many valuable plant growth-promoting (PGP) traits that attract agricultural scientists and policymakers.

Moreover, these microbial communities enhance crop productivity as well as soil environment and soil health. In the form of root exudates, organic matter attracts a wide range of rhizospheric microbes and habitats. Rhizobacteria are found in the rhizospheric area of the plant root in response to root exudates by chemo-tactic mechanism and competent rhizobacteria. Some microbes live near plants and interact in different ways [9]. At the molecular level, contact occurs through a particular signaling molecule. It can provide a favorable environment for the microbes in the nodules to fix atmospheric elemental nitrogen, depending on the plant–microbial interactions [10]. However, the rhizospheric microbiota is made up of selected microbial groups with plant growth attributes. The microbes can be found in the rhizosphere and encourage PGP traits. Rhizospheric microbiome also helps to ensure the sustainability of all ecosystems in a wide variety of applications. These microbes regulate the cycling of nutrients, regulate soil organic dynamics, and increase the efficiency of acquiring nutrients. The microbial symbiosis also improves the productivity in plant nutrient acquisition. These microbial associations also regulate decomposition, mineralization, and nutrient flow.

8.3 Improvement of Soil Health and Agricultural Production in Sustainable Agriculture

Soil rhizosphere is an essential biological component for plant–microbial interactions. It is a natural habitat for the rhizospheric microbiome's diversity, and estimated from previous research that rhizospheric soil contains up to 10 billion microbial population densities per gram of soil. Rhizospheric microbes contribute to a wide range of physiological function in maintaining soil fertility or soil health and crop productivity [11]. Beneficial microbes manage soil health or soil fertility by altering soil physico-chemical properties directly or indirectly. Soil–plant–microbial interaction is one of the essential alternative aspects of a sustainable agricultural system. This type of microbial association may help to achieve the main objective of future sustainable agriculture. Soil rhizosphere holds a variety of microbial diversity, such as bacteria, fungi, and actinomycetes. Bacteria, fungi, and actinomycetes are three main groups of rhizospheric soil microbiome that form significant soil microbial biomass for soil–plant interaction. Among these, rhizospheric microbes create a strong interaction with plants and improve nutrient-use efficiency and water-uptake supply and ameliorate diverse types of adverse climate-changing stresses.

Rhizospheric microbiomes are a biological indicator of soil enzyme activities and regulate the physico-chemical properties of soil. The diversity of indigenous microbes is a fundamental component of soil for all biogeochemical cycles that ultimately improve plant nutrient uptake efficiency and soil health. Low water availability, temperature variation, salinity, drought, anthropogenic factors, and grazing cause adverse effects on indigenous microbial diversity and soil biogeochemical processes. Soil–plant–microbes interaction forms a comparatively stable and beneficial association for PGP attributes or traits. Some potentially beneficial microbes can also harm soil rhizosphere zone and plant growth promotion [7]. Due to the intensive agricultural cropping system and toxic effect of agrochemicals, this relation deteriorated soil health or soil fertility and beneficial potential microbial diversity.

8.4 Microbial Mechanism for Plant Growth Attributing Characteristics

Rhizospheric microbiota can be used differently for the plant growth attributes required for beneficial potential microbial diversity. There are two main mechanisms for promoting plant growth (direct and indirect). Natural approaches are provided by the accessibility of various beneficial metabolites synthesized by the rhizospheric microbial diversity. Similarly, they facilitate crop plant growth by essential nutrients solubilizing/mineralizing, phytohormones production, availability of essential nutrients, and siderophore production [12]. The indirect approach includes the decrease of some harmful effects of phytopathogens by various pathways. Such catecholate-type siderophores secreted by microbes reduces the availability of iron by phytopathogen and reduces their survival efficacy due to iron competition. In addition, induced systemic resistance provides defense against phytopathogens. Beneficial rhizospheric microbes induce PGP processes such as (1) production of phytohormones (indole acetic acid (IAA), cytokinin, and gibberellins), (2) solubilization and increase in the availability of nutrients, (3) synthesis of secondary metabolites, i.e. exo-polysaccharide (EPS) production, hydrogen cyanide (HCN), catecholate-type siderophore, 1-amino cyclopropane-1-carboxylate (ACC) deaminase, fluorescent pigment, and intrinsic antibiotic spectra, (4) stimulation of systemic resistance [13–17]. For example, *Pseudomonas aeruginosa* LSE-2 strain of soybean has multifarious PGP traits, i.e. synthesis of IAA, biofilm, P and Zn solubilization, siderophore, EPS, fluorescent pigment, and antibiotic production [18,19]. Despite these multifunctional PGP traits, PGPRs also increase the growth of plants through ACC deaminase enzyme, biofilm, antibiotics, cell degradation enzymes, and EPSs. ACC deaminase enzyme produced by the rhizospheric microbiomes converts ethylene into ammonia (NH_3) and α-ketobutyrate and better survival of host plants under abiotic stress [20–22]. Microbes that produce ACC deaminase improve plant survival and enhance crop and soil fertility or soil health [23,24]. Exo-polysaccharides protect host plants through biofilm formation and soil aggregation, maintain structural integrity, promote cation exchange and water-holding capacity, protect against environmental fluctuation, enhance epiphytic colonization and nutrient uptake efficiency (Figure 8.1) [25,26].

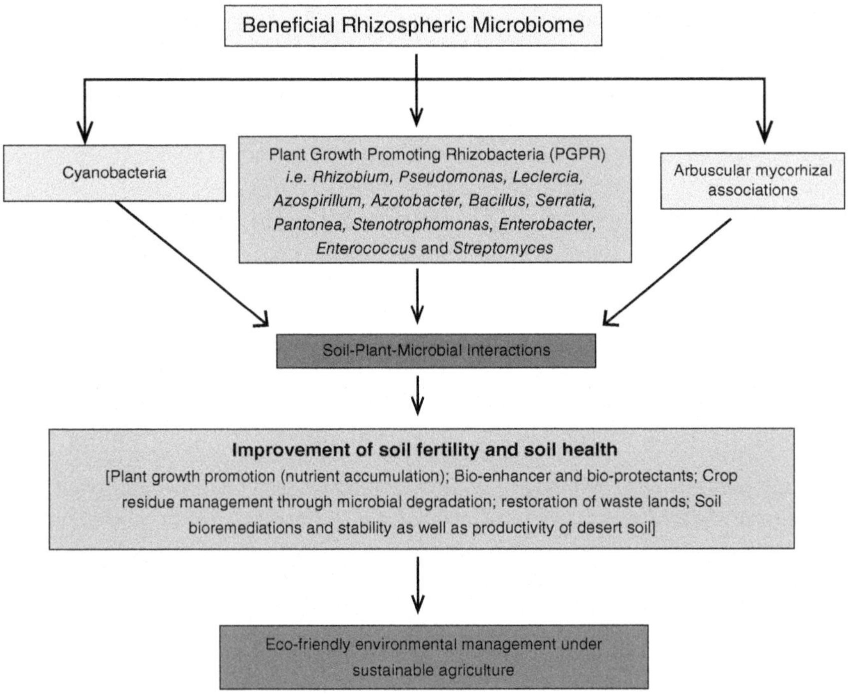

Figure 8.1 Role of the rhizospheric microbiome and their interactions for eco-friendly environmental management under sustainable agriculture.

8.5 Direct Mechanism

8.5.1 Biological Nitrogen Fixation

Nitrogen (N) is one of the essential macro-nutrients for growth promotion in plants, which enhances crop production. Approximately 78% elemental N is present in the atmosphere but still restricted for uptake by the growing of crops due to the heavy losses caused by emission or leaching in the intensive cropping systems. Although it exists either in elemental or molecular form and is not directly available for plant nutrition, it is only available to legumes via biological nitrogen fixation (BNF) to convert into ammonia, which is an essential metabolic activity of BNF carried out by oxygen-sensitive, nitrogenase enzyme complex [74]. Nitrogenase enzyme also has two most crucial metalloenzyme complex component systems comprising di-nitrogenase reductase as iron protein and di-nitrogenase as a metal cofactor. On their basis, three different nitrogenase enzyme-fixing systems have been reported namely, Mo- and Fe-nitrogenase and the rarely present V-nitrogenase [75]. BNF is better performed by Mo-nitrogenase in the most rhizospheric microbes carrying N_2 fixation in non-leguminous plants by establishing non-obligate interaction [20,76]. Beneficial rhizospheric microbes involved in BNF can be broadly divided into involving three groups based on their association, i.e. symbiotic association with (a) legumes and (b) non-legume plants and non-symbiotic association with (c) endophytes as well as free-living ones, i.e. *Acetobacter, Pseudomonas, Bacillus, Enterobacter, Azospirillum, Klebsiella*, etc., which fix a minor portion of elemental nitrogen from the atmosphere. Most of the primary or molecular form of nitrogen is set via biological nitrogen fixer microbes, such as *Rhizobium, Mesorhizobium, Sinorhizobium*, and *Bradyrhizobium* in legumes and *Frankia* in non-leguminous plants [20,54,77]. Several researchers have also documented that free-living N_2-fixing bacteria and endophytic rhizobia can also induce the growth promotion in plants of non-legumes and contributed to reduced dependence on inorganic nitrogen agrochemicals [78].

Although PGPRs are present everywhere in plant tissue, their potential to fix atmospheric nitrogen through BNF is not fully exploited. In addition to endophytic rhizobia, non-symbiotic microbes with beneficial PGP traits have been found in diverse agricultural crop plants (Table 8.1) and also habitats of several other diazotrophic microbiota, including *Pantoea, Herbaspirillum, Brevundimonas, Ideonella*, and *Azospirillum* in maize [79] and *Enterobacter, Pseudomonas, Rheinheimera, Pantoea, Bacillus*, and *Enterococcus* in rice [80]. The considerable involvement of diazotrophic rhizospheric microbes of N_2 fixation and N as an essential nutrient in plant growth might be easily assessed *in vitro* and *in vivo* studies with variable mechanisms. Dobbelaere et al. [81] reported that a significant amount of N could be easily provided by the rhizospheric microbiome in the legumes and cereal crops. However, bio-inoculation of a symbiotic association of *Azospirillum* in sorghum, maize, and wheat contributed 5–10 kg N/ha/year. This amount is non-significant compared with 150–200 kg N/ha/year, which is conventional practice in the intensive agriculture system. The contribution of a free-living or symbiotic association within the rhizospheric microbiome for various crop plants in Australia is less than 10 kg N/ha/year [82] [83]. Also documented a varied range of N_2 fixation from 0 to 15 kg N/ha/year. Similarly, recommended amount of BNF was estimated at 10 kg N/ha/year in the rhizosphere by Bottomley and Myrold [84]. The capacity of beneficial rhizospheric microbes to fix molecular N_2 is no longer considered as essential criteria to classify them as bio-fertilizers/bio-enhancers or bio-protectants.

8.5.2 Phytohormone Production

Auxins are the natural compounds that act as chemical signals generated through the diverse physiological and metabolic processes that are dispersed all over the plant tissue and bacterial system. They are considered target-specific for appropriate plant growth promotion under different environments and are also termed "plant growth hormones" or "regulators." Plant growth regulators have been classified into five major classes: auxins, cytokinins, abscisic acid, gibberellins, and ethylene (stress hormone). Among these, auxins constitute IAA, which is synthesized by many rhizospheric microbes in the vicinity of root that evoke many physiological changes, leading to enhanced plant biomass and surface area. It aids in improved essential nutrient and water uptake even under abiotic and biotic stresses [85,86]. Previous researchers have demonstrated the secretion and release of phytohormones by approximately 80% of rhizospheric microbiomes as their secondary metabolite, which may alter the indigenous synthesis of phytohormones as they change the cell membrane permeability of plants for increased secretion of root exudates [20].

Table 8.1 Role of beneficial rhizospheric microbiome for plant growth promotion in different crops.

Crops	Beneficial rhizospheric bacteria	Plant growth promotion mechanisms	References
Wheat	*Chryseobacterium gleum*, *Enterobacter cloacae*, *Serratia*, and *Dietzia natronolimnaea*	Phosphate solubilization, IAA, HCN, and cell wall degradation enzyme production	[23,27–31]
Maize	*Gracilibacillus*, *Virgibacillus*, *Salinicoccus*, *Staphylococcus*, *Bacillus*, *Brevibacterium*, *Exiguobacterium*, *Pseudomonas*, *Oceanobacillus*, *Arthrobacter*, *Zhihengliuella*, and *Halomonas* spp.	IAA production, phosphate solubilization, EPS, and biofilm formation	[15,32–36]
Rice	*Pantoea agglomerans*, *Curtobacterium albidum*, *Halobacillus dabanensis*, *B. stratosphericus*, *B. pumilus*, *P. pseudoalcaligenes*, and *Citrobacter*	Exo-polysaccharide production, IAA, ammonia production, antioxidant enzyme, ACC deaminase, and phosphate solubilization activity	[37–40]
Cottonseed	*B. amyloliquefaciens*, *Curtobacterium oceanosedimentum*, *P. oryzihabitans*, *P. putida*, *P. chlororaphis*, *Paenibacillus xylanexeden*, *Klebsiella oxytoca*, and *Enterobacter cloacae*	IAA production, antioxidant enzymes, and photosynthetic pigment content enhancement	[41–43]
Rapeseed	*Rhizobium* sp. and *P. putida*	IAA, ACC deaminase, P-solubilization, chlorophyll synthesis, an antioxidant enzyme, membrane transportation and pathogenesis-related responses	[14,44]
Sunflower	*Pseudomonas* sp.	Siderophore, N_2 fixation, IAA, HCN, chitinase, β-1-3 glucanase activity, P-solubilization, biofilm, and EPS production	[45,46]
Chickpea	*Mesorhizobium* sp., *M. cicero*, *P. pseudoalcaligens*, *H. variabilis*, and *P. rifletoensis*	Phosphate solubilization, siderophore, IAA production, biofilm formation, EPS, ACC deaminase enzyme enhanced nodulation status by reducing ethylene level	[47–50]
Soybean	*Bacillus firmus*; *Bradyrhizobium* sp., *P. putida*, *P. simiae*, *P. fluorescens*, and *L. adecarboxylata*	IAA, phosphate solubilization, ACC deaminase, siderophore production, EPS, biofilm formation, salicylic acid, and gibberellins production	[18,19,51–53]

(*Continued*)

Table 8.1 (Continued)

Crops	Beneficial rhizospheric bacteria	Plant growth promotion mechanisms	References
Groundnut	*Klebsiella, Pseudomonas, Agrobacterium, Ochrobactrum, B. saurashtrense, B. casei*, and *Haererohalobacter*	IAA, NH_3, P-solubilization, HCN, production secondary metabolites, i.e. phenol, osmolites, proline, and free amino acids	[54,55]
Mung bean	*E. cloacae, B. cereus, Rhizobium* sp., *Enterococcus mundtii*, and *P. fluorescens*	Phosphate solubilization, IAA, siderophore, ammonia, HCN, EPS production, enhanced photosynthetic rate, chlorophyll content, and water-use efficiency	[56–59]
Lentil	*Rhizobium* sp., *P. jessenii*, and *R. leguminosarum*	IAA production, phosphate solubilization, and enhanced nodulation efficiency	[60,61]
Pea	*G. mosseae, Arthrobacter protophormiae*, and *R. leguminosarum*	Reduction of ethylene concentration through ACC deaminase	[62]
Black gram	*P. fluorescens*	Improvement in water and nutrient uptake efficiency and photosynthetic content	[63]
Faba bean	*P. anguilliseptica, P. putida, P. fluorescens*, and *B. subtilis*	Biofilm formation, EPS production, and enhanced plant growth promotion activities	[64,65]
Pigeon pea	*Bradyrhizobium* sp. and *Burkholderia cepacia*	Improvement of nodulation status and N_2 fixation	[66]
Sugarcane	*Enterobacter* sp.	ACC deaminase activity	[67]
Cucumber	*R. palustris, P. fluorescens, B. megaterium*, and *V. paradoxus*	BNF, P and K-solubilization, IAA, 5-aminolevulinic acid synthesis, and EPS production	[68,69]
Okra	*Enterobacter* sp.	Reactive oxygen species (ROS)-scavenging enzymes	[70]
Potato	*Bacillus pumilus*	P-solubilization, IAA, and siderophore synthesis	[71]
Tomato	*P. fluorescens, P. migulae*, and *Streptomyces* sp.	Phosphate solubilization and IAA production	[72,73]

Apart from plant growth and developmental stages, IAA is also concerned with plant–microbe interaction and defense mechanisms [87]. Many symbiotic bacterial species viz, *Stenotrophomonas, Rhizobium, Nostoc, Azotobacter, Bradyrhizobium, Azospirillum, Rahnella, Pseudomonas*, and *Arthrobacter*, etc. are known to produce auxins (IAA) and influence the diverse physiological processes of host plants, ranging from photo-stimulation to pathogenesis [88,89]. Rhizospheric microbiomes with auxin hormone secretion abilities can be applied as bio-fertilizers and bio-enhancers. They elevate root differentiation through lateral and adventitious root formation, thereby enhancing the surface area to improve nutrient and water utilization efficiency. Recently, Kang et al. [90] demonstrated that auxins produced by the salt-tolerating *Leclercia adecarboxylata* strain, MO1 have strong interactions with carbohydrate, chlorophyll fluorescence improvement (Fv/Fm), *ipdc* gene expression, and organic acid production in tomato. Furthermore, IAA secretion by rhizospheric PGPR-induced ACC deaminase enzyme activity through signaling cascade that hydrolyzes ACC into NH_3 and α-ketobutyrate, helping plant growth even under salt stress by reducing ethylene levels.

Gibberellins (Gbs) are secondary physiological metabolites that can play an essential role as signaling cascades toward the crop plant; PGPRs also accumulate cytokinins and gibberellins that were concerned in root initiation cell division, cell enlargement, and differentiation ultimately enhance root biomass [91]. The higher endogenous level of genetic algorithms in salt-tolerating microbes, such as *Promicromonospora* spp. SE-188, *Burkholderia cepacia* SE-4, and *Acinetobacter calcoaceticus* SE-370 improved plant growth attributes in cucumber under salinity with increased proteins, reducing sugars, ribonuclease, and antioxidant enzymes [17]. Patel and Saraf [92] demonstrated the effectiveness of potential microbes such as *Pseudomonas putida*, *Pseudomonas stutzeri*, and *Stenotrophomonas maltophilia* from the rhizosphere of coleus with enhanced cytokinins (CK), gibberellin, and IAA under salt stress conditions. Cytokinin involved in tissue differentiation and proliferation of cells are produced in many salt-tolerating microbes, such as *Bacillus, Arthrobacter, Pseudomonas, Azospirrilum, Azotobacter, Arthrobacter, Halomonas*, and *Stenotrophomonas* species [93]. Accelerated cytokinin production decreases ethylene stress hormone causing lesser leaf senescence in cereal and legumes supporting plant growth [94,95]. Cytokines functional as master regulators in mitigating salinity stress in crop plants [96].

8.5.3 Phosphate Solubilization

Phosphorus (P) is the next most essential macro-nutrient after nitrogen for plant growth promotion and crop production. It is also present in plant rhizospheric microorganisms and various organic and inorganic components. However, the total P content in average soil is 0.05%, and only 0.1% P content from the soil is available as a nutrient to crop plants. The minimal availability of P to crop plants is due to low solubility and high fixation ability. Approximately 80% of applied inorganic P-fertilizers as di-ammonium phosphate are immobilized due to the formation of a complex with Al or Fe in acidic soils or Ca in sodic or calcareous soil. The best alternative way for phosphate-solubilizing rhizospheric microbes is phosphate mobilizing/solubilizing microorganisms (PSMs) that provide necessary P to crop plants by converting the fixed organic and inorganic phosphates into two soluble stages, the monobasic (H_2PO^{4-} and HPO_2^{-4}) ions [77]. Among the diversity of the rhizospheric microbiome, PSMs, such as *Bacillus*, *Enterobacter*, *Serratia*, *Rhizobium*, *Enterobacter*, *Pseudomonas*, *Enterococcus*, *Burkholderia*, *Microbacterium*, *Proteus*, and *Flavobacterium*, could be helpful as bio-inoculants/bio-enhancers for sustainable agriculture since they were able to convert the unavailable form of phosphorus to available format through the various producing types of organic acids (acetate, gluconate, lactate, citrate, succinate, etc.) made by PSM [97,98]. Different proto-cooperative effects such as solubilization or mineralization improved BNF efficacy through modulation, and enhanced uptake of essential macro- and micronutrients have been reported in various crop plants inoculated with single or dual combination of bio-inoculants with phosphate-solubilizing/mineralizing ability [18,19,99,100]. Bhise and Dandge [17] also reported that high salt-tolerating rhizospheric microbes possessing the ability of phosphate, zinc, and potassium solubilization; N_2 fixation; and ACC deaminase and phytohormone production aid in nutrient mobilization, making them accessible to plants even under abiotic (drought and salt) stress conditions. Presently, a significant limitation is the inconsistent effect on PSM by PGPRs *in vivo*, which are rigorously affected because of the various environmental stresses and competition with indigenous micro-flora [101]. It appears that the field evaluation of the ability to solubilize P *in vitro* does not contribute as efficiently under field conditions [102]. Thus, it can be concluded that the bio-protectant/bio-enhancer/bio-fertilizers with dual action might have been mediated by directly solubilizing inorganic sources of P, mineralizing organic P, other PGP activities, and stimulating root growth or the formation of solid mycorrhizal association.

8.5.4 Siderophore Production

Iron (Fe), the fourth most abundant element in the earth's crust, is essential for cytochrome, ribonucleotide reductase and acts as a cofactor of about 140 enzymes. Crop plants require available ferrous (Fe^{+2}) in small quantities, which is essential for various metabolic pathways, viz chlorophyll synthesis, photosynthesis, Kreb's cycle, and nitrogenase activity BNF and also act as cofactors of many intermediate metabolic synthetic pathways [103]. Due to the scarcity of ferrous ions (Fe^{+2}), vegetative plant parts turn yellow due to less photosynthetic activity. Ultimately, plants die, causing a loss in crop production and deteriorating the product's quality [104]. In the rhizospheric area under aerobic environmental conditions, the ferric (Fe^{+3}) form of iron gets transformed into insoluble hydroxides and oxy-hydroxides, thereby raising a severe problem of iron deficiency [105]. In an iron-limiting or competitive ecosystem, diverse rhizospheric microbes synthesize intra and extracellular water-soluble peptide ferric iron chelators, i.e., siderophore (catecholate-type) with different functional groups on side-chain behaving as legends with a specific affinity [106,107]. Other changing climatic factors, such as concentration and various forms of iron, moisture content, pH and availability of essential macro and micronutrients, and organic acids can regulate the secretion of siderophore [100]. Siderophore molecules contain more than 500 compounds identified based on their legend architecture and classified into three major groups: catecholates and hydroxyl-carboxylic acid hydroxamates [108].

Siderophore efficiency mainly depends on various complex compound formations with ferric (Fe^{+3}) ions, and siderophore upholds the high association content, thereby generating high iron-deficient conditions and phytopathogenic agents. Siderophores also function as nutrient-solubilizing agents under iron-limiting conditions by reducing Fe^{+3} ions to Fe^{+2}, which are further transported to the interior part of the cell through ATP-binding cassette transporter and ultimately recycled or destroyed [105,109,110]. Along with Fe sequestration, siderophores uphold the capability to form stable complexes with xenobiotic compounds or bio-hazardous heavy metal, i.e. Pb, Zn, Cu, Al, Cd, and Ga and radionuclide, i.e. U and Np, etc., which are alarming signals of a polluted environment [100]. Siderophore-producing rhizobia and rhizospheric bacteria positively affect plant height, enhancing nutrient acquisition, preventing phytopathogens by antibiotics, and iron-scavenging activity [111,112]. Pseudochelin A, a new siderophore of *Pseudoalteromonas piscicida* strain S2040, shows suppressing activity against the *Aspergillus fumigatus* [110]. It is also reported that several mycorrhizal associations can result in secreting various low molecular weight organic chelating compounds, viz oxalic acid, citric acid, siderophores, and proton for mobilization of available Fe present in soil rhizosphere and might have been involved in the uptake of Fe in crop plants [113–115].

8.5.5 ACC Deaminase Activity

Ethylene (ET), a vital stress hormone, is accumulated in crop plants under stress, which regulates mainly metabolic processes (seed emergence, senescence, and organ abscission), concerned in conventional plant growth promotion of host plant and up-regulates crucial responses for environmental stress tolerance [116]. Low levels of ethylene have been documented as a defense when the plant is under abiotic stress [117]. At the same time, high ethylene production may cause ethylene stress, hindering plant growth [118]. Ethylene is produced in a biphasic model representing two peaks of production in plants [119], where the first small peak is responsible for the up-regulation of gene-encoding plant defense (protective proteins) system and the second more prominent peak known as "stress ethylene" has harmful impacts on plant growth-causing senescence, leaf abscission, and chlorosis [20]. Chemicals such as $CoCl_2$ and amino-ethoxy-vinyl-glycine are used as inhibitors to overcome the problem associated with salinity but negatively impact them [117].

Rhizospheric beneficial microbes can assimilate ACC, converting it into ammonia and α-ketobutyrate, providing nitrogen and energy under salt stress conditions [120] [15]. Documented ACC deaminase enzyme activity in halo-tolerant potential microbial genera of *Brevibacterium, Arthrobacter, Pseudomonas, Virgibacillus, Rhizobium meliloti, Gracilibacillus, Stenotrophomonas, Bacillus, Salinicoccus, Exiguobacterium, Enterobacter, Klebsiella, Methylobacterium,* Planococcus, *Acidovorax,* and *Variovorax.* The beneficial rhizospheric microbes with ACC deaminase enzyme activity and other PGP traits have been studied to alleviate salt stress and improve growth in tomato and rice [22]. Due to the induced ACC deaminase activity of PGPRs, the internal concentration of stress hormone (ethylene) reduces, which, in turn provides resistance or tolerance against several abiotic and biotic stresses, i.e., salinity, water logging conditions, drought, heavy metals, aromatic hydrocarbons, high temperature, high radiations, insect predation, and phytopathogens [20,21,107]. Germination percentage, shoot and root biomass, enhanced uptake of nitrogen, phosphorus, and potassium, and nodulation status with ecto- and endo-mycorrhizal associations are some of the detectable physiological changes seen in crop plants inoculated with potential indigenous PGPRs [20].

8.5.6 Antagonistic Activities, Extracellular Polymeric Substances, and Antioxidant Enzymes for Managing Biotic and Abiotic Stresses

Soil-borne pathogens decrease agricultural productivity. Rhizospheric microbiome function in controlling these pathogens can be improved by providing competition for colonization sites, antimicrobial metabolites (antibiosis), secretion of iron, scavenging protein, and induced systemic resistance. The synthesis of cell wall degrading enzymes (cellulase, proteases, lipase, chitinase, pectinase, and hydrolase) by rhizospheric microbes against phytopathogens leads to leakage of cell metabolites from cell membrane, and, ultimately, collapse of the evading pathogen [120]. Due to antibiosis and cell wall degradation, enzyme production from diverse rhizospheric microbes plays a significant role in plant growth promotion by suppressing pathogens viz, *Sclerotium rolfsii*, *Rhizoctonia solani*, *Botrytis cinerea*, *Phytophthora* sp., *Fusarium oxysporum*, and *Pythium ultimum* [121].

Beneficial microbes in the rhizosphere can produce varied polysaccharides that promote biofilm formation and root colonization, which aids in a better interaction with the rhizoplanes. Early colonization of root by EPS-producing PGPRs makes phosphorus available to plants with nutrient transport, growth, development, and protects phytopathogen [122]. To compensate for abiotic stresses, PGPRs form a biofilm where EPS causes EPS aggregation of soil particles, maintains structural integrity, promotes cation exchange and water-holding capacity, and protects against environmental fluctuation, nutrient retention, and enhances epiphytic colonization [25,123].

Plants under abiotic and biotic stresses produce reactive oxygen species (ROS), damaging major cell components viz, proteins, lipids, and nucleic acids [124]. Exposure to plant stress (salt and drought) decreases the level of ROS-scavenging enzymes, i.e. superoxide dismutase (SOD), catalase (CAT), and ascorbate peroxidase (APX) by increasing the peroxidation of lipids [17,125]. Rhizospheric microbiota enables the augmentation of the inherent ability of the plant to defend against phytopathogens associated with different crops, apart from being a better substitute against agrochemicals [126,127]. Several types of research demonstrated that plant growth-promoting attributes and improved disease resistance against phytopathogens with the potential of indigenous PGPRs, such as *Pseudomonas* sp., *Bacillus* sp., and *Trichoderma* sp., etc., on diverse host plants, such as soybean, chickpea, mung bean, pea, wheat, rice, and rockcress [107,128,129].

The principal mechanism of induction innate defense responses on crops by rhizospheric microbes is through stimulation of induced systemic resistance, operating under the response to a microbial elicitor [130,131]. In this context, Jain et al. [132] demonstrated the stimulus of various defense enzyme activities, such as peroxidase, polyphenol oxidase, superoxide dismutase, glucanase, chitinase, phenol peroxidase production, and lignin deposition, in response to priming with a consortium mixture of PGPRs as bio-enhancers/bio-protectants. Thus, all of the above-mentioned reports/studies indicated that rhizospheric microbiota induces an augmented form of defense response in the crop plant and enables the quenching of factors responsible for the induction of oxidative stress response with a specific host [110,133]. It is an economic and sustainable approach to overcome phytopathogens in most cereals and legume crops.

8.6 Mycorrhizae

Mycorrhizal symbioses are proto-cooperative association between fungal mycelium. The root of the higher plant can be characterized agronomically by enhanced plant growth promotion or ecologically by increasing the strength or reproductive capability of the host plant. Vesicular arbuscular mycorrhiza (VAM) have an extensive ramified hyphal network spread in internal and external tissues in the higher plant roots and have access to a higher amount of water and essential minerals. Aborted microspores can improve the supply of N, P, Cu, Zn, Fe, Ca, B, Mn, Ni, K, etc. [54]. Multi-inoculated (VAM, *Rhizobium* sp., and PGPRs) combinations have also been very beneficial for enhancing plant growth-promoting attributes, nutrient acquisitions, BNF, and improving root architectures under normal and salt stress conditions [134]. Another most important beneficial effect on the plant is the production of secondary metabolites (high molecular weight glycoprotein – glomalin), produced in abundance by arbuscular mycorrhizal fungi. These materials accumulate higher in the rhizosphere and are positively correlated with soil aggregation stability. In addition, several studies have registered that the mycorrhizal fungi also affect the physiological process of the plant through osmoregulation, comprising proline content enhancement, higher water and nutrient efficiency, osmolytes, and anti-oxidative enzymes under abiotic or biotic conditions [135].

8.7 Application of Microbial Inoculants and Future Prospects

Application of rhizospheric microbes, such as *Bacillus, Pseudomonas, Rhizobium, Mesorhizobium, Enterobacter, Enterococcus, Pantone, Serratia, Bradyrhizobium, Azospirillum, Flavobacterium*, and *Azotobacter*, has resulted in significant enhancement of grain yield, plant biomass, nodulation status, soil enzyme activities and nutrient as well as water-use efficiency in various crops, such as moonbeam, groundnut, soybean, common bean, wheat, rice, chickpea, cotton, and oilseed crops [56,100]. Different PGP mechanisms, such as phytohormone production (IAA, cytokinin, and gibberellin), N_2 fixation, nutrient (P, K, and Zn) solubilization, ACC deaminase, biofilm, and EPS production are essential PGP attributes of plant growth promotion and improvement of crop production and soil health in a sustainable manner.

Integrated disease management of plants mediated by beneficial microbes will minimize agro-chemicals concentration and reduce plant pathogens in an eco-friendly manner, mainly via competition for essential nutrients, cell wall degrading enzymes, antibiotic production, and induced systemic resistance [107]. Some potential PGPRs can stimulate plant growth under environmental stresses, i.e. salinity, drought, waterlogged condition, temperature fluctuations, xenobiotic compounds, organic contaminants and ameliorate biotic stresses predation of insect, nematodes, fungi, bacteria, and viruses [20]. Thus, all of the above-discussed properties of PGPRs equip it as potential bio-fertilizers/bio-enhancer/bio-protectants, phyto-stimulators, and phyto-remediators.

The exceedingly growing world population, continuously increasing demand for food production, the simultaneous buildup of agrochemical residue in the food chain have led to higher environmental contamination and health issues. The shift toward an intensive cropping system involving environmental-friendly approaches of integrated disease management has thus addressed the immediate need of the hour. Formulation of a microbial consortium with multifunctional PGP traits might be a cost-effective approach over the single bio-inoculants in plant growth promotion and amelioration of abiotic and biotic stresses. In this context, Tewari and Arora [136] suggested that future research need to be directed toward bio-engineering of rhizospheric biology to accomplish the optimum level of crop production by genetically manipulating beneficial indigenous microbes and their habitats, such as optimizing shelf life, quality control, tolerance to unfavorable environmental conditions, enhanced crop production, and development of cost-effective bio-inoculants affordable to the consumer at the field level. The molecular and biotechnological approaches need to explore rhizospheric microbiota and attain the desired defense mechanisms against phytopathogens. These will help us identify, screen, and develop efficient bio-formulation with multifarious PGP traits to combat phytopathogens and maintain a sustainable agroecosystem [137].

8.8 Conclusion

After looking at the application and future prospects, we can conclude that rhizospheric PGPRs have a multifunctional PGP traits/approach favoring rhizospheric microbes and their environment. The main aim of a sustainable agricultural system is to maintain soil health, benefit from good produce for farmers, and protect human health and the ecological environment. Rhizospheric microbiomes are the potential alternative tools for sustainable agriculture and trend for the future. For this reason, recent research efforts should be focused on the development of biocompatible synergistic consortium of PGPR inoculants with agrochemical and soil organic matter under changing climatic conditions. Studies at the molecular level can provide us with next-generation sequencing/meta-genomics/proteomics studies through forward or reverse genetics. On a special note, beneficial rhizospheric microbes could be a better and safer alternative to chemical fertilizers under sustainable agriculture either in combination or alone.

References

1 Tilman, D., Balzer, C., Hill, J., and Befort, B.L. (2011). Global food demand and the sustainable intensification of agriculture. *Proceedings of the National Academy of Sciences* 108 (50): 20260–20264.

2 Kang, S.M., Radhakrishnan, R., Khan, A.L., Kim, M.J., Park, J.M., Kim, B.R., Shin, D.H., and Lee, I.J. (2014). Gibberellin secreting rhizobacterium, *Pseudomonas putida* H-2-3 modulates the hormonal and stress physiology of soybean to improve the plant growth under saline and drought conditions. *Plant Physiology and Biochemistry* 84: 115–124.

3 Singh, J.S. (2015). Microbes: The chief ecological engineers in reinstating equilibrium in degraded ecosystems. *Agriculture, Ecosystems & Environment* 203: 80–82.

4 Patten, D.T. (2016). The role of ecological wisdom in managing for sustainable interdependent urban and natural ecosystems. *Landscape and Urban Planning* 155: 3–10.

5 Achal, V., Mukherjee, A., and Zhang, Q. (2016). Unearthing ecological wisdom from natural habitats and its ramifications on development of bio-cement and sustainable cities. *Landscape and Urban Planning* 155: 61–68.

6 Yan, N., Marschner, P., Cao, W., Zuo, C., and Qin, W. (2015). Influence of salinity and water content on soil microorganisms. *International Soil and Water Conservation Research* 3 (4): 316–323.

7 Ahmad, F., Ahmad, I., and Khan, M.S. (2008). Screening of free-living rhizospheric bacteria for their multiple plant growth-promoting activities. *Microbiological Research* 163 (2): 173–181.

8 Islam, F., Yasmeen, T., Ali, Q., Ali, S., Arif, M.S., Hussain, S., and Rizvi, H. (2014). Influence of *Pseudomonas aeruginosa* as PGPR on oxidative stress tolerance in wheat under Zn stress. *Ecotoxicology and Environmental Safety* 104: 285–293.

9 Singh, J.S., Pandey, V.C., and Singh, D.P. (2011). Efficient soil microorganisms: A new dimension for sustainable agriculture and environmental development. *Agriculture, Ecosystems & Environment* 140 (3): 339–353.

10 Masciarelli, O., Llanes, A., and Luna, V. (2014). A new PGPR co-inoculated with *Bradyrhizobium japonicum* enhances soybean nodulation. *Microbiological Research* 169 (7): 609–615.

11 Sahoo, R.K., Ansari, M.W., Tuteja, R., and Tuteja, N. (2015). Salt tolerant SUV3 overexpressing transgenic rice plants conserve physicochemical properties and microbial communities of the rhizosphere. *Chemosphere* 119: 1040–1047.

12 Złoch, M., Thiem, D., Gadzała-Kopciuch, R., and Hrynkiewicz, K. (2016). Synthesis of siderophores by plant-associated metallotolerant bacteria under exposure to Cd^{2+}. *Chemosphere* 156: 312–325.

13 Duan, J., Jiang, W., Cheng, Z., Heikkila, J.J., and Glick, B.R. (2013). The complete genome sequence of the plant growth-promoting bacterium *Pseudomonas* sp. UW4. *PLoS ONE* 8 (3): e58640.

14 Saghafi, D., Ghorbanpour, M., and Lajayer, B.A. (2018). Efficiency of *Rhizobium* trains as plant growth-promoting rhizobacteria on morpho-physiological properties of *Brassica napus* L. under salinity stress. *Journal of Soil Science and Plant Nutrition* 18: 253–268.

15 Aslam, F. and Ali, B. (2018). Halotolerant bacterial diversity associated with *Suaeda fruticose* (L.) Forssk. improved growth of maize under salinity stress. *Agron* 8: 131.

16 Vimal, S.R., Patel, V.K., and Singh, J.S. (2019). Plant growth-promoting *Curtobacterium albidum* strain SRV4: An agriculturally important microbe to alleviate salinity stress in paddy plants. *Ecological Indicators* 105: 553–562.

17 Bhise, K.K. and Dandge, P.B. (2019). Mitigation of salinity stress in plants using plant growth-promoting bacteria. *Symbiosis* 1–14. https://doi.org/10.1007/s13199-019-00638-y

18 Kumawat, K.C., Sharma, P., Singh, I., Sirari, A., and Gill, B.S. (2019a). Co-existence of *Leclercia adecarboxylata* (LSE-1) and *Bradyrhizobium* sp. (LSBR-3) in nodule niche for multifaceted effects and profitability in soybean production. *World Journal of Microbiology and Biotechnology* 35 (11): 1–17.

19 Kumawat, K.C., Sharma, P., Sirari, A., Singh, I., Gill, B.S., Singh, U., and Saharan, K. (2019b). Synergism of *Pseudomonas aeruginosa* (LSE-2) nodule endophyte with *Bradyrhizobium* sp. (LSBR-3) for improving plant growth, nutrient acquisition and soil health in soybean. *World Journal of Microbiology and Biotechnology* 35 (3): 1–17.

20 Glick, B.R. (2014). Bacteria with ACC deaminase can promote plant growth and help to feed the world. *Microbiological Research* 169: 30–39.

21 Glick, B.R. (2012). Plant growth-promoting bacteria: Mechanisms and applications. *Scientifica* (2012: 1–15.

22 Bal, H.B., Nayak, L., Das, S., and Adhya, T.K. (2013). Isolation of ACC deaminase producing PGPR from rice rhizosphere and evaluating their plant growth-promoting activity under salt stress. *Plant Soil* 366: 93–105.

23 Bhise, K.K., Bhagwat, P.K., and Dandge, P.B. (2017). Plant growth-promoting characteristics of salt-tolerant *Enterobacter cloacae* strain KBPD and its efficacy in amelioration of salt stress in *Vigna radiata* L. *Journal of Plant Growth Regulation* 36: 215–226.

24 Tiwari, G., Duraivadivel, P., Sharma, S., and Hariprasad, P. (2018). 1-Aminocyclopropane-1-carboxylic acid deaminase producing beneficial rhizobacteria ameliorate the biomass characters of *Panicum maximum* Jacq. by mitigating drought and salt stress. *Scientific Reports* 8: 17513. 10.1038/s41598-018-35565-3

25 Balsanelli, E., Baura, V.A., Pedrosa, F.O., Souza, E.M., and Monteiro, R.A. (2014). Exo-polysaccharide biosynthesis enables mature biofilm formation on abiotic surfaces by *Herbaspirillum seropedicae*. *PloS One* 9: 1–10.

26 Kasim, W.A., Osman, M.E., Omar, M.N., Abd El-Daim, I.A., Bejai, S., and Meijer, J. (2013). Control of drought stress in wheat using plant growth-promoting bacteria. *Journal of Plant Growth Regulation* 32: 122–130.

27 Singh, R.P., Jha, P., and Jha, P.N. (2017). Bio-inoculation of plant growth-promoting rhizobacterium *Enterobacter cloacae* ZNP-3 increased resistance against salt and temperature stresses in wheat plant (*Triticum aestivum* L.). *Journal of Plant Growth Regulation* 36 (3): 783–798.

28 Kamran, S., Shahid, I., Baig, D.N., Rizwan, M., Malik, K.A., and Mehnaz, S. (2017). Contribution of zinc solubilizing bacteria in growth promotion and zinc content of wheat. *Frontiers in Microbiology* 8: 2593.

29 Khande, R., Sushil, K.S., Ramesh, A., and Mahaveer, P.S. (2017). Zinc solubilizing *Bacillus* strains that modulate growth, yield and zinc biofortification of soybean and wheat. *Rhizosphere* 4: 126–138.

30 Habib, S.H., Kausar, H., and Saud, H.M. (2016). Plant growth-promoting rhizobacteria enhance salinity stress tolerance in okra through ROS-scavenging enzymes. *BioMed Research International* (2016: 6284547.

31 Bharti, N., Pandey, S.S., Deepti, B., Patel, V.K., and Kalra, A. (2016). Plant growth-promoting rhizobacteria *Dietzia natronolimnaea* modulates the expression of stress-responsive genes providing protection of wheat from salinity stress. *Scientific Reports* 6: 34768. https://doi.org/10.1038/srep34768

32 Li, H.Q. and Jiang, X.W. (2017). Inoculation with plant growth-promoting bacteria (PGPB) improves salt tolerance of maize seedling. *Russian Journal of Plant Physiology* 64: 235–241.

33 Chen, L., Liu, Y., Wu, G., Veronican, N.K., Shen, Q., and Zhang, N. (2016). Induced maize salt tolerance by rhizosphere inoculation of *Bacillus amyloliquefaciens* SQR9. *Physiologia Plantarum* 158: 34–44.

34 Akram, M.S., Shahid, M., Tariq, M., Azeem, M., Javed, M.T., and Saleem, S. (2016). Deciphering *Staphylococcus sciuri* SAT-17 mediated anti-oxidative defense mechanisms and growth modulations in salt-stressed maize (*Zea mays* L.). *Frontiers in Microbiology* 7: 867.

35 Zerrouk, I.Z., Benchabane, M., Khelifi, L., Yokawa, K., Ludwig-Muller, J., and Baluska, F. (2016). A *Pseudomonas* strain isolated from date palm rhizospheres improves root growth and promotes root formation in maize exposed to salt and aluminum stress. *Journal of Plant Physiology* 191: 111–119.

36 Gond, S.K., Torres, M.S., Bergen, M.S., Helsel, Z., and White, J.F. (2015). Induction of salt tolerance and up-regulation of aquaporin genes in tropical corn by rhizobacterium *Pantoea agglomerans*. *Letters in Applied Microbiology* 60: 392–399.

37 Sultana, S., Paul, S.C., and Karim, M.M. (2018). Salinity intrusion and coastal agriculture: Adaptation strategies using salt-tolerant plant growth-promoting rhizobacteria for sustainable food security. *Regional Problems* 21: 58–61.

38 Rima, F.S., Biswas, S., Sarker, P.K., Islam, M.R., and Seraj, Z.I. (2018). Bacteria endemic to saline coastal belt and their ability to mitigate the effects of salt stress on rice growth and yields. *Annals of Microbiology* 68: 525–535.

39 Misra, S., Dixit, V.K., Khan, M.H., Mishra, S., Dviwedi, G., and Yadav, S. (2017). Exploitation of agro-climatic environment for selection of 1-aminocyclopropane-1-carboxylic acid (ACC) deaminase producing salt-tolerant indigenous plant growth-promoting rhizobacteria. *Microbiological Research* 205: 25–34.

40 Khan, A., Zhao, X.Q., Javed, M.T., Khan, K.S., Bano, A., and Shen, R.F. (2016). *Bacillus pumilus* enhances tolerance in rice (*Oryza sativa* L.) to combined stresses of NaCl and high boron due to limited uptake of Na^{+}. *Environmental and Experimental Botany* 124: 120–129.

41 Irizarry, I. and White, J. (2017). Application of bacteria from non-cultivated plants to promote growth, alter root architecture and alleviate salt stress of cotton. *Journal of Applied Microbiology* 122: 1110–1120.

42 Egamberdieva, D., Jabborova, D., and Hashem, A. (2015). *Pseudomonas* induces salinity tolerance in cotton (*Gossypium hirsutum*) and resistance to *Fusarium* root rot through the modulation of indole 3 acetic acid. *Saudi Journal of Biological Sciences* 22: 773–779.

43 Yaish, M.W., Antony, I., and Glick, B.R. (2015). Isolation and characterization of endophytic plant growth-promoting bacteria from date palm tree (*Phoenix dactylifera* L.) and their potential role in salinity tolerance. *Avan Leeuw J Microb* 107: 1519–1532.

44 Cheng, Z., Woody, O.Z., Mcconkey, B.J., and Glick, B.R. (2012). Combined effects of the plant growth-promoting bacterium *Pseudomonas putida* UW4 and salinity stress on the *Brassica napus* proteome. *Applied Soil Ecology* 61: 255–263.

45 Tewari, S. and Arora, N.K. (2016). Fluorescent *Pseudomonas* sp. PF17 as an efficient plant growth regulator and bio-control agent for sunflower crop under saline conditions. *Symbiosis* 68: 99–108.

46 Shilev, S., Sancho, E.D., and Benlloch-González, M. (2012). Rhizospheric bacteria alleviate salt produced stress in sunflower. *Journal of Environmental Management* 95: 37–41.

47 Egamberdieva, D., Shurigin, V., Gopalakrishnan, S., and Sharma, R. (2014). Growth and symbiotic performance of chickpea (*Cicer arietinum*) cultivars under saline soil conditions. *Journal of Biological and Chemical Research* 56: 1–10.

48 Patel, D., Jha, C.K., Tank, N., and Saraf, M. (2012). Growth enhancement of chickpea in saline soils using plant growth-promoting rhizobacteria. *Journal of Plant Growth Regulation* 31: 53–62.

49 Qurashi, A.W. and Sabri, A.N. (2012). Bacterial exo-polysaccharide and biofilm formation stimulate chickpea growth and soil aggregation under salt stress. *Brazilian Journal of Microbiology* 43: 1183–1191.

50 Nagpal, S., Sharma, P., and Kumawat, K.C. (2019). Assessment of native single and dual inoculants of *Mesorhizobium* sp. and endophytic rhizobacteria for plant growth promotion in chickpea. *Agricultural Research Journal* 56 (4): 746–751.

51 El-Esawi, M.A., Alaraidh, I.A., Alsahli, A.A., Alamri, S.A., Ali, H.M., and Alayafi, A.A. (2018a). *Bacillus Firmus* (SW5) augments salt tolerance in soybean (*Glycine max* L.) by modulating root system architecture, antioxidant defense systems and stress-responsive genes expression. *Plant Physiology and Biochemistry* 132: 375–384.

52 Egamberdieva, D., Davranov, K., Wirth, S., Hashem, A., and Abd-Allah, E.F. (2017). Impact of soil salinity on the plant-growth-promoting and biological control abilities of root-associated bacteria. *Saudi Journal of Biological Sciences* 24: 1601–1608.

53 Vaishnav, A., Kumari, S., Jain, S., Verma, A., Tuteja, N., and Choudhary, D.K. (2016). PGPR mediated expression of salt tolerance gene in soybean through volatiles under sodium nitroprusside. *Journal of Basic Microbiology* 56: 1274–1288.

54 Sharma, S., Kulkarni, J., and Jha, B. (2016). Halotolerant rhizobacteria promote growth and enhance salinity tolerance in peanut. *Frontiers in Microbiology* 7: 1600. 10.3389/fmicb.2016.01600

55 Ghorai, S., Pal, K.K., and Dey, R. (2015). Alleviation of salinity stress in groundnut by application of PGPR. *International Journal of Research in Engineering Technology* 2: 742–750.

56 Kumawat, K.C., Sharma, P., Nagpal, S., Gupta, R.K., Sirari, A., Nair, R.M., Bindumadhava, H., and Singh, S. (2021). Dual Microbial Inoculation, a Game Changer? Bacterial Biostimulants with Multifunctional Growth Promoting Traits to Mitigate Salinity Stress in Spring Mungbean. *Frontiers in Microbiology* 11: 600576. 10.3389/fmicb.2020.600576

57 Mahmood, S., Daur, I., Al-Solaimani, S.G., Ahmad, S., Madkour, M.H., Yasir, M., Hirt, H., Ali, S., and Ali, Z. (2016). Plant growth-promoting rhizobacteria and silicon synergistically enhance salinity tolerance of mungbean. *Frontiers in Plant Science* 7: 876. https://doi.org/10.3389/fpls.2016.00876

58 Islam, F., Yasmeen, T., Arif, M.S., Ali, S., Ali, B., Hameed, S., and Zhou, W. (2016). Plant growth-promoting bacteria confer salt tolerance in *Vigna radiata* by up-regulating antioxidant defense and biological soil fertility. *Plant Growth Regulation* 80: 23–36.

59 Aamir, M., Aslam, A., Khan, M.Y., and Usman, M. (2013). Co-inoculation with *Rhizobium* and plant growth-promoting rhizobacteria (PGPR) for inducing salinity tolerance in mungbean under field condition of semi-arid climate. *Asian Journal of Agriculture and Biology* 1: 7.

60 Halder, A., Banerjee, J., Bhattacharyya, P., Pramanik, K., and Debnath, A. (2016). Isolation of lentil specific salt-tolerant nitrogen-fixing bacteria from Murshidabad district of West Bengal. *Journal of crop and weed* 12: 14–19.

61 Zahir, Z.A., Zafar-ul-hye, M., Sajjad, S., and Naveed, M. (2011). Comparative effectiveness of *Pseudomonas* and *Serratia* sp. containing ACC-deaminase for co-inoculation with *Rhizobium leguminosarum* to improve growth, nodulation and yield of lentil. *Biology and Fertility of Soils* 47: 457–465.

62 Barnawal, D., Bharti, N., Maji, D., Chanotiya, C.S., and Kalra, A. (2014). ACC deaminase containing *Arthrobacter protophormiae* induces NaCl stress tolerance through reduced ACC oxidase activity and ethylene production resulting in improved nodulation and mycorrhization in *Pisum sativum*. *Journal of Plant Physiology* 171: 884–894.

63 Yasin, N.A., Khan, W.U., Ahmad, S.R., Ali, A., Ahmad, A., and Akram, W. (2018). Imperative roles of halotolerant plant growth-promoting rhizobacteria and kinetin in improving salt tolerance and growth of black gram (*Phaseolus mungo*). *Environmental Science and Pollution Research* 25: 4491–4505.

64 Mohammed, A.F. (2018). Effectiveness of exo-polysaccharides and biofilm-forming plant growth-promoting rhizobacteria on salinity tolerance of faba bean (*Vicia faba* L.). *African Journal of Microbiology Research* 12: 399–404.

65 Metwali, E.M., Abdelmoneim, T.S., Bakheit, M.A., and Kadasa, N. (2015). Alleviation of salinity stress in faba bean (*Vicia faba* L.) plants by inoculation with plant growth-promoting rhizobacteria (PGPR). *Plant Omics* 8: 449.

66 Bano, D.A., Singh, R.K., Waza, S.A., and Singh, N.P. (2015). Effect of cow pea *Bradyrhizobium* (RA-5) and *Burkholderia cepacian* (RRE-5) on growth parameters of pigeon pea under salt stress conditions. *Journal of Pure and Applied Microbiology* 9: 2539–2546.

67 Kruasuwan, W. and Thamchaipenet, A. (2018). 1-Aminocyclopropane-1- carboxylate (ACC) deaminase producing endophytic diazotrophic *Enterobacter* sp. EN-21 modulates salt stress response in sugarcane. *Journal of Plant Growth Regulation* 37: 849–858.

68 Ge, H. and Zhang, F. (2019). Growth promoting ability of *Rhodopseudomonas palustris* G5 and its effect on induced resistance in cucumber against salt stress. *Journal of Plant Growth Regulation* 38: 180–188.

69 Nadeem, S.M., Ahmad, M., Naveed, M., Imran, M., Zahir, Z.A., and Crowley, D.E. (2016). Relationship between *in vitro* characterization and comparative efficacy of plant growth-promoting rhizobacteria for improving cucumber salt tolerance. *Archives of Microbiology* 198: 379–387.

70 Sheikh, H.H., Hossain, K., and Halimi, M.S. (2016). Plant growth-promoting rhizobacteria enhances salinity stress tolerance in okra through ROS-scavenging enzymes. *BioMed Research International* 2016 (6284547): 10. https://doi.org/10.1155/2016/6284547

71 Gururani, M.A., Upadhyaya, C.P., Baskar, V., Venkatesh, J., Nookaraju, A., and Park, S.W. (2013). Plant growth-promoting rhizobacteria enhance abiotic stress tolerance in *Solanum tuberosum* through inducing changes in the expression of ROS scavenging enzymes and improved photosynthetic performance. *Journal of Plant Growth Regulation* 32: 245–258.

72 Ali, S., Charles, T.C., and Glick, B.R. (2014). Amelioration of high salinity stress damage by plant growth-promoting bacterial endophytes that contain ACC deaminase. *Plant Physiology and Biochemistry* 80: 160–167.

73 Palaniyandi, S.A., Damodharan, K., Yang, S.H., and Suh, J.W. (2014). *Streptomyces* sp. strain PGPA39 alleviates salt stress and promotes growth of 'micro tom' tomato plants. *Journal of Applied Microbiology* 117: 766–773.

74 Kim, J. and Rees, D.C. (1994). Nitrogenase and biological nitrogen fixation. *Biochemistry* 33 (2): 389–397.

75 Dean, D.R. and Jacobson, M.R. (1992). Biochemical genetics of nitrogenase. In: *Biological Nitrogen Fixation* (eds. G. Stacey, H. Burris and H.J. Evans), 763–834. New York: Chapman and Hall.

76 Glick, B.R., Patten, C.L., Holguin, G., and Penrose, G.M. (1999). *Biochemical and Genetic Mechanisms Used by Plant Growth-promoting Bacteria*. London: Imperial College Press.

77 Bhattacharyya, P.N. and Jha, D.K. (2012). Plant growth-promoting rhizobacteria (PGPR): Emergence in agriculture. *World Journal of Microbiology and Biotechnology* 28: 1327–1350.

78 Bhattacharjee, R.B., Singh, A., and Mukhopadhyay, S.N. (2008). Use of nitrogen-fixing bacteria as biofertilizer for non-legumes: Prospects and challenges. *Applied Microbiology and Biotechnology* 80: 199–209.

79 Roesch, L.F.W., Camargo, F.A.O., Bento, F.M., and Triplett, E.W. (2008). Biodiversity of diazotrophic bacteria within the soil, root and stem of field-grown maize. *Plant Soil* 302: 91–104.

80 Mano, H. and Morisaki, H. (2008). Endophytic bacteria in the rice plant. *Microbes and Environments* 23: 109–117.

81 Dobbelaere, S., Vanderleyden, J., and Okon, Y. (2003). Plant growth-promoting effects of diazotrophs in the rhizosphere. *Critical Reviews in Plant Sciences* 22: 107–149.

82 Unkovich, M. and Baldock, J. (2008). Measurement of a symbiotic N_2 fixation in Australian agriculture. *Soil Biology and Biochemistry* 40: 2915–2921.

83 Peoples, M.B., Giller, K.E., Herridge, D.F., and Vessey, J.K. (2002). Limitations to biological nitrogen fixation as a renewable source of nitrogen for agriculture. In: *Nitrogen Fixation: Global Perspectives* (eds. T.M. Finan, M.R. O'Brian, D.B. Layzell, J.K. Vessey and W. Newton), 356–360. Wallingford: CAB International.

84 Bottomley, P.J. and Myrold, D.D. (2007). Biological N inputs. In: *Soil Microbiology, Ecology and Biochemistry* (ed. E. Paul), 365–387. Oxford: Academic.

85 Egamberdieva, D. and Kucharova, Z. (2009). Selection for root-colonizing bacteria stimulating wheat growth in saline soils. *Biology and Fertility of Soils* 45: 561–573.

86 Patel, R.R., Patel, D.D., Thakor, P., Patel, B., and Thakkar, V.R. (2015). Alleviation of salt stress in germination of *Vigna radiata* L. by two halotolerant *Bacilli* sp. isolated from saline habitats of Gujarat. *Journal of Plant Growth Regulation* 76: 51–60.

87 Spaepen, S. and Vanderleyden, J. (2011). Auxin and plant-microbe interactions. *Cold Spring Harbor Perspectives in Biology* 3: 4.

88 Abd-Allah, E.F., Alqarawi, A.A., Hashem, A., Radhakrishnan, R., Al-Huqail, A.A., and Al-Otibi, F.A. (2017). Endophytic bacterium *Bacillus subtilis* (BERA71) improves salt tolerance in chickpea plants by regulating the plant defense mechanisms. *Journal of Plant Interactions* 3: 37–44.

89 Egamberdieva, D., Jabborova, D., Wirth, S., Alam, P., Alyemeni, M.N., and Ahmad, P. (2018). Interactive effects of nutrients and *Bradyrhizobium japonicum* on the growth and root architecture of soybean (*Glycine max* L.). *Frontiers in Microbiology* 9: 1000. 10.3389/fmicb.2018.01000

90 Kang, S.M., Shahzad, R., Bilal, S., Khan, A.L., Park, Y.G., and Lee, K.E. (2019). Indole-3-acetic-acid and ACC deaminase producing *Leclercia adecarboxylata* MO1 improves *Solanum Lycopersicum* L. growth and salinity stress tolerance by endogenous secondary metabolites regulation. *BMC Microbiology* 19: 80. 10.1186/s12866-019-1450-6

91 Salamone, I.E.G., Hynes, R.K., and Nelson, L.M. (2005). Role of cytokinins in plant growth promotion by rhizosphere bacteria. In: *PGPR: Biocontrol and Bio Fertilization* (ed. Z.A. Siddiqui), 173–195. Amsterdam: Springer.

92 Patel, T. and Saraf, M. (2017). Biosynthesis of phytohormones from novel rhizobacterial isolates and their *in vitro* plant growth-promoting efficacy. *Journal of Plant Interactions* 12: 480–487.

93 TrParray, A.P., Jan, S., Kamili, A.N., Qadri, R.A., Egamberdieva, D., and Ahmad, P. (2016). Current perspectives on plant growth-promoting rhizobacteria. *Journal of Plant Growth Regulation* 35: 877–902.

94 Pallai, R., Hynes, R.K., Verma, B., and Nelson, L.M. (2012). Phytohormone production and colonization of canola (*Brassica napus* L.) roots by *Pseudomonas fluorescens* 6-8 under gnotobiotic conditions. *Canadian Journal of Microbiology* 58: 170–178.

95 Kapoor, R. and Kaur, M. (2016). Cytokinins production by fluorescent *Pseudomonas* isolated from rhizospheric soils of Malus and Pyrus. *African Journal of Microbiology Research* 10: 1274–1279.

96 Fahad, S., Hussain, S., Bano, A., Saud, S., Hassan, S., Shan, D., and Tabassum, M.A. (2015). Potential role of phytohormones and plant growth-promoting rhizobacteria in abiotic stresses: Consequences for changing environment. *Environmental Science and Pollution Research* 22: 4907–4921.

97 Zaidi, A., Khan, M.S., Ahemad, M., and Oves, M. (2009). Plant growth promotion by phosphate solubilizing bacteria. *Acta Microbiologica Et Immunologica Hungarica* 56: 263–284.

98 Srinivasan, R., Yandigeri, M.S., Kashyap, S., and Alagawadi, A.R. (2012). Effect of salt on survival and P-solubilization potential of phosphate solubilizing microorganisms from salt-affected soils. *Saudi Journal of Biological Sciences* 19: 427–437.

99 Ahemad, M. and Khan, M.S. (2012). Effect of fungicides on plant growth-promoting activities of phosphate solubilizing *Pseudomonas putida* isolated from mustard (*Brassica compestris*) rhizosphere. *Chemosphere* 86: 945–950.

100 Ahemad, M. and Kibret, M. (2014). Mechanisms and applications of plant growth-promoting rhizobacteria: Current perspective. *Journal of King Saud University-Science* 26: 1–20.

101 Ahemad, M. and Khan, M.S. (2010). Phosphate-solubilizing and plant-growth-promoting *Pseudomonas aeruginosa* PS1 improves green gram performance in quizalafop-p-ethyl and clodinafop amended soil. *Archives of Environmental Contamination and Toxicology* 58: 361–372.

102 Rengel, Z. (2008). Bioavailability of phosphorus and micronutrients in the soil plant-microbe continuum. *Journal of Soil Science and Plant Nutrition* 8: 84–91.

103 Kobayashi, T. and Nishizawa, N.K. (2012). Iron uptake, translocation and regulation in higher plants. *Annual Review of Plant Physiology and Plant Molecular Biology* 63: 131–152.

104 Murgia, I., Arosio, P., Tarantino, D., and Soave, C. (2012). Crops bio-fortification for combating 'hidden hunger' for iron. *Trends in Plant Science* 17: 47–55.

105 Rajkumar, M., Ae, N., Prasad, M.N.V., and Freitas, H. (2010). Potential of siderophore producing bacteria for improving heavy metal phytoextraction. *Trends in Biotechnology* 28: 142–149.

106 Crosa, J.H. and Walsh, C.T. (2002). Genetics and assembly line enzymology of siderophore biosynthesis in bacteria. *Microbiology and Molecular Biology Reviews* 66: 223–249.

107 Singh, J., Singh, P., Ray, S., Rajput, R.S., and Singh, H.B. (2019). Plant Growth Promoting Rhizobacteria: Benign and Useful Substitute for Mitigation of Biotic and Abiotic Stresses. In: *Plant Growth Promoting Rhizobacteria for Sustainable Stress Management, Microorganisms for Sustainability* (ed. R.Z. Sayyed et al.). https://doi.org/10.1007/978-981-13-6536-2_5

108 Bonnefoy, V. and Holmes, D.S. (2012). Genomic insights into microbial iron oxidation and iron uptake strategies in extremely acidic environments. *Environmental Microbiology* 14: 1597–1611.

109 Braun, V. and Hantke, K. (2011). Recent insights into iron import by bacteria. *Current Opinion in Chemical Biology* 15 (2): 328–334.

110 Kumar, A. and Verma, J.P. (2019). The Role of Microbes to Improve Crop Productivity and Soil Health. In: *Ecological Wisdom Inspired Restoration Engineering, EcoWISE* (eds. V. Achal and A. Mukherjee). Singapore: Springer Nature Singapore Pte Ltd. 2019. https://doi.org/10.1007/978-981-13-0149-0

111 Shen, X., Hu, H., Peng, H., Wang, W., and Zhang, X. (2013). Comparative genomic analysis of four representative plant growth-promoting rhizobacteria in *Pseudomonas*. *BMC Genomics* 14: 1–20.

112 Parray, J.A., Jan, S., Kamili, A.N., Qadri, R.A., Egamberdieva, D., and Ahmad, P. (2016). Current perspectives on plant growth-promoting rhizobacteria. *Journal of Plant Growth Regulation* 35: 877–902.

113 Bharadwaj, D.P., Alstrom, S., and Lundquist, P.O. (2012). Interactions among *Glomus* irregular, arbuscular mycorrhizal spore-associated bacteria and plant pathogens under *in vitro* conditions. *Mycorrhiza* 22: 437–447.

114 Labidi, S., Jeddi, F.B., Tisserant, B., Debiane, D., Rezgui, S., Ferjani, A.G., and Sahraoui, A.L.H. (2012). Role of arbuscular mycorrhizal symbiosis in root mineral uptake under $CaCO_3$ stress. *Mycorrhiza* 22: 337–345.

115 Amanullah, M.M., Archana, J., Manoharan, S., and Subramanian, K.S. (2012). Influence of iron and AM inoculation on metabolically active iron, chlorophyll content and yield of hybrid maize in calcareous soil. *Journal of Agronomy* 11: 27–30.

116 Etesami, H. and Beattie, G.A. (2017). Plant-microbe interactions in adaptation of agricultural crops to abiotic stress conditions. In: *Probiotics and Plant Health* (eds. V. Kumar, M. Kumar, S. Sharma, and R. Prasad), 163–200. Singapore: Springer. https://doi.org/10.1007/978-981-10-3473-2.

117 Dodd, I.C., Zinovkina, N.Y., Safronova, V.I., and Belimov, A.A. (2010). Rhizobacterial mediation of plant hormone status. *Annals of Applied Biology* 157: 361–379.

118 Yoon, G.M. and Kieber, J.J. (2013). 14-3-3 regulates 1- aminocyclopropane-1-carboxylate synthase protein turnover in *Arabidopsis*. *Plant Cell* 25: 1016–1028.

119 Vanloon, L.C., Rep, M., and Pieterse, C.M.J. (2006). Significance of inducible defence-related proteins in infected plants. *Annual Review of Phytopathology* 44: 135–162.

120 Paul, D. and Lade, H. (2014). Plant growth-promoting rhizobacteria to improve crop growth in saline soils: A review. *Agronomy for Sustainable Development* 34: 737–752.

121 Nadeem, S.M., Zaheer, Z.A., Naveed, M., and Nawaz, S. (2013). Mitigation of salinity induced a negative impact on the growth and yield of wheat by plant growth-promoting rhizobacteria in naturally saline conditions. *Annals of Microbiology* 63: 225–232.

122 Tewari, S. and Arora, N.K. (2014). Talc based exo-polysaccharides formulation enhancing growth and production of *Helianthus Annuus* under saline conditions. *Cellular and Molecular Biology* 60: 73–81.

123 Kasim, W.A., Gaafar, R.M., Abou-Ali, R.M., Omar, M.N., and Hewait, H.M. (2016). Effect of biofilm-forming plant growth-promoting rhizobacteria on salinity tolerance in barley. *Annals of Agricultural Sciences* 61: 217–227.

124 Das, K. and Roy Choudhury, A. (2014). Reactive oxygen species (ROS) and response of antioxidants as ROS - scavengers during environmental stress in plants. *Frontiers in Environmental Science* 2: 53.

125 Halo, B.A., Khan, A.L., Waqas, M., Al-Harrasi, A., Hussain, J., Ali, L., and Lee, I.J. (2015). Endophytic bacteria (*Sphingomonas* sp. LK11) and gibberellin can improve *Solanum lycopersicum* growth and oxidative stress under salinity. *Journal of Plant Interactions* 10: 117–125.

126 Sarma, B.K., Yadav, S.K., Singh, S., and Singh, H.B. (2015). Microbial consortium mediated plant defense against phytopathogens: Readdressing for enhancing efficacy. *Soil Biology and Biochemistry* 87: 25–33.

127 Spence, C., Alff, E., and Johnson, C. (2014). Natural rice rhizospheric microbes suppress rice blast infections. *BMC Plant Biology* 14: 130.

128 Singh, A., Jain, A., Sarma, B.K., Upadhyay, R.S., and Singh, H.B. (2013). Rhizosphere microbes facilitate redox homeostasis in *Cicer arietinum* against biotic stress. *Annals of Applied Biology* 163 (1): 33–46.

129 Chauhan, H. and Bagyaraj, D.J. (2015). Inoculation with selected microbial consortia not only enhances the growth and yield of French bean but also reduces fertilizer application under field condition. *Scientia Horticulturae* 197: 441–446.

130 Shoresh, M., Harman, G.E., and Mastouri, F. (2010). Induced systemic resistance and plant responses to fungal biocontrol agents. *Annual Review Of Phytopathology* 48: 21–43.

131 Jain, A., Singh, A., Singh, S., and Singh, H.B. (2015). Bio-control agents-mediated suppression of oxalic acid-induced cell death during *Sclerotinia sclerotiorum*–pea interaction. *Journal of Basic Microbiology* 55 (5): 601–606.

132 Jain, A., Singh, S., Sarma, B.K., and Singh, H.B. (2012). Microbial consortium mediated reprogramming of defense network in pea to enhance tolerance against *Sclerotinia sclerotiorum*. *Journal of Applied Microbiology* 112: 537–550.

133 Ghazalibiglar, H., Hampton, J.G., de Jong, E.V.Z., and Holyoake, A. (2016). Is induced systemic resistance the mechanism for control of black rot in *Brassica oleracea* by a *Paenibacillus* sp.? *Biological Control* 92: 195–201.

134 Calvo-Polanco, M., Sánchez-Romera, B., Aroca, R., Asins, M.J., Declerck, S., Dodd, I.C., Martínez-Andújar, C., Albacete, A., and Ruiz-Lozano, J.M. (2016). Exploring the use of recombinant inbred lines in combination with beneficial microbial inoculants (AM fungus and PGPR) to improve drought stress tolerance in tomato. *Environmental and Experimental Botany* 131: 47–57.

135 Hashem, A., Abd-Allah, E.F., Alqarawi, A.A., Al-Huqail, A.A., Egamberdieva, D., and Wirth, S. (2016). Alleviation of cadmium stress in *Solanum lycopersicum* L. by arbuscular mycorrhizal fungi via induction of acquired systemic tolerance. *Saudi Journal of Biological Sciences* 23 (2): 272–281.

136 Tewari, S. and Arora, N.K. (2013). Transactions among microorganisms and plant in the composite rhizosphere. In: *Plant-microbe Symbiosis: Fundamentals and Advances* (ed. N.K. Arora), 1–50. New Delhi: Springer.

137 Gupta, G., Parihar, S.S., Ahirwar, N.K., Snehi, S.K., and Singh, V. (2015). Plant growth-promoting rhizobacteria (PGPR): Current and future prospects for development of sustainable agriculture. *Journal of Microbial & Biochemical Technology* 7 (2): 096–102.

9

Toxic Effects of Some Herbicides on the Fatty Acid Profile of Wheat Varieties

A Phytomicrobiome Study

Fadime Karabulut and Songul Çanakcı-Gulengul

Department of Biology Firat University Elazıg 23119 Turkey

9.1 Introduction

Environmental pollution occurs when organic and inorganic pollutants occur in and around living beings and their environment, mixing with air, water, and soil. It changes depending on the stress agents caused by the surroundings, the plant species, the process of exposure to stress, and the structure of the tissue and organs affected by the stress. Plants undergo multiple metabolic events, such as photosynthesis, respiration, growth and development, and any disruption in the food chain significantly affects plant, animal, and human health [1]. Chemical fertilizers, herbicides, and soil regulators used to produce the highest agricultural output by the unit area pollute agrarian lands. For this reason, yield from agricultural land has been decreasing [2,3]. For plants grown in agricultural areas to be of quality, it is necessary for plants to struggle against weeds that are ground and food partners. Since weeds can germinate or grow faster and sooner than wheat, weeds can use the growing area of the crop, consuming plant nutrients and water from the soil, and negatively affecting plant growth by preventing the daylight exposure to wheat. As a result of this interaction, weeds among wheat cause losses between 10% and 24% [4]. The chemical struggle is one of the most frequently used and economical methods in this struggle. Besides the many benefits of the herbicides used in this struggle, there are also damages. Herbicides that remain in the soil for a short duration may not be effective in controlling weeds. The presence of herbicides in the soil for a longer duration may result in dispersion, causing environmental pollution. On the other hand, herbicide residue may cause phytotoxicity if it remains in the soil until the following year [5].

Herbicides are significant pollutants commonly used to minimize crop loss and not reduce productivity. The use of herbicides is essential in agriculture [6]. In recent years, the random use of herbicides that cause soil and water pollution leads to ecological imbalance [7]. Because of the accumulation capacity in plants, herbicides accumulate in living tissues. A herbicide can be effectively cleaned by adopting environmental-friendly biological treatment technologies [8].

In this chapter, we determine the toxic effect of herbicides contaminating soil and its fatty acid profile. For this, it is necessary to apply phytoremediation techniques to clean the soil.

Core Microbiome: Improving Crop Quality and Productivity, First Edition. Edited by Javid A. Parray, Nowsheen Shameem, Elsayed Fathi Abd-Allah, and Mohammad Yaseen Mir.

9.2 Herbicides

Herbicide is a Latin word. Latin consists of the words "herbal" meaning "plant" and "cida" meaning "destroy." Soil content effectively retains the herbicides once administered to the soil by clay or organic matter as it increases the density of clay or organic matter in the soil and the adherence of herbicides to the soil increases. In this case, the herbicides are ineffective, as it is difficult for weeds to take herbicides from the soil. Weeds can be controlled using more herbicides in soils with this feature. Thus, the immunity of weeds to pesticides and environmental pollution increases [9].

Herbicides have been divided into systemic and contact, depending on movement within plants. While systemic herbicides are harmful herbicides in motion, contact herbicides are those that do not spread, which cause damage at their point of contact. Contact herbicides need excess water or pressure to reach the whole plant. Systemic herbicides are carried from leaves to roots and quickly destroy weeds in one year. Accordingly, herbicide application to perennial weeds should be applied when the nutrients that have been moved from the root parts are dense. Generally, this application is made when weeds start growing. Because weeds grow before the onset of cold weather and the flowering season, spring is when the movement of nutrients in plants is decreased in the phloem. Since weeds carry nutrients accumulated in roots to leaves, herbicides moving from leaves to roots will not significantly affect weeds.

Herbicides can reach plants without harming the environment by dispersion. Herbicides are distributed in two ways as part and steam scattering. In part scattering, herbicides are distributed by the wind when applied to the plant as small particles. Some manufacturers prevent part scattering by producing herbicides in a sticky formula, which is formed due to the evaporation of the herbicide over an hour or two or a few days after herbicide application in steam scattering [10]. Even if herbicides are not involved in the season, herbicide residues in the soil in the previous seasons can be absorbed by the plant. Atrazine herbicide is more effective with increasing soil pH and can become very effective without being firmly attached to the soil [10]. As atrazine has a longer half-life than other pesticides, most of it remains in the environment [11]. Metolachlor is more permanent in sub-soils than on surface soils. It has been demonstrated that water can be cured of these permanent herbicides by phytoremediation in surface waters treated with metolachlor and atrazine. It is clearly shown that *Ceratophyllum demersum* and *Elodea canadensis* can cure water contaminated with atrazine and metolachlor [12].

9.2.1 Uptake and Transport of Herbicides

After the application of herbicides to roots or leaves, the existent form of herbicides in the soil and how they disperse within the soil are significant. Herbicides can be transported to other regions by the wind when they become vapor due to hot weather conditions and are soluble in water in slopy soil and the flow of soil. Herbicides can disperse by infiltrating underground waters formed by the effect of precipitation. Herbicides, which begin to dissolve in the soil, are adsorbed with the help of organic substances to be decomposed microbiologically. After herbicides desorb with increasing evaporation, capillaries return to the soil with water. Herbicides can accumulate after which, they can be transmitted to aquatic creatures. With precipitation, it can be seen that herbicides come back to the earth and undergo chemical, phytochemical, and microbial disintegration in the environment [13]. The transfer of herbicides contaminated to the soil to the plant takes place in the dissolved liquid phase. Solubility, temperature, steam pressure, soil pH, moisture content, molecular weight, and organic matter content are influential factors in transporting herbicides to the plant. In addition, the root structure and the type of enzyme secreted can be effective, depending on the nature of the plant used [13]. Material removal from leaf xylem to leaf cells involves another membrane transport step [14]. Most studies show that most of the areas where phytoremediation is applied are made for metals (loid), which are inorganic compounds. Specific membrane carrier proteins take up inorganic compounds. Herbicides pass from stem xylem to leaf symplast by simple diffusion. Herbicides are transported to a particular tissue or cellular areas after entering the leaf symplast. It can generally be neutralized where it can be less harmful to fundamental cellular processes [14].

OA: organic acids, GSH: glutathione, NA: nicotianamine, MT: metallothioneins, Glu: glucose, PC: phytochelatins [15].

Pollutants are first taken into the root by binding to the root surface in the form of ions and passed into the stem by diffusion with the mass flow (Figure 9.1). Intake of metals attached to soil particles, involving chelation of molecules secreted from the roots to the rhizosphere

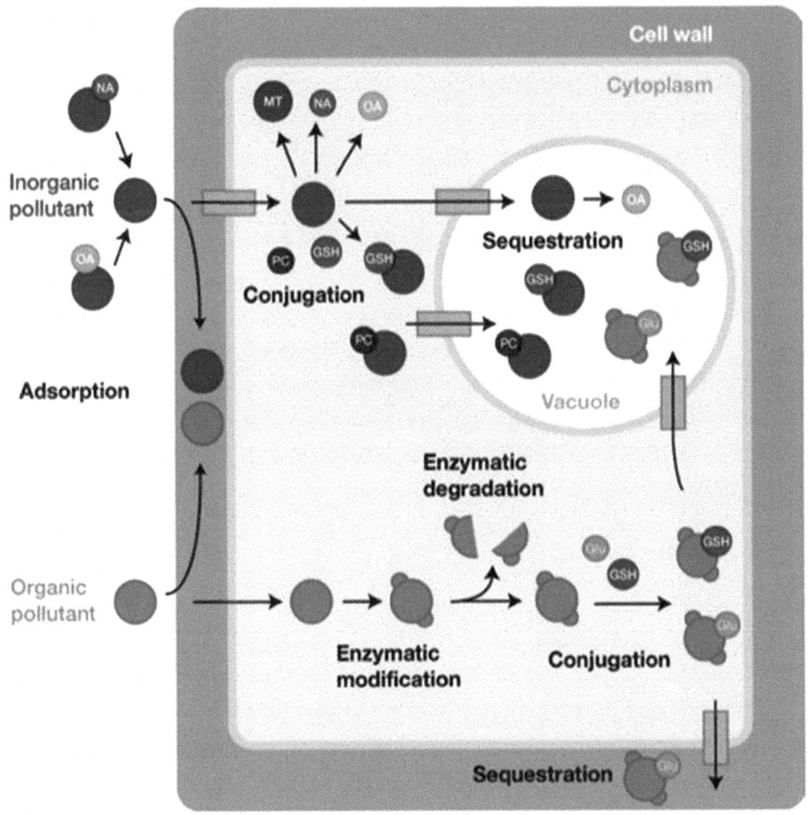

Figure 9.1 Mechanisms of inorganic and organic pollutants in plants. OA: organic acids, GSH: glutathione, NA: nicotianamine, MT: metallothioneins, Glu: glucose, PC: phytochelatins [15]..

is provided by reductase and proton release depending on the plasma membrane [16]. According to plant species and pollutants, ions are taken up by the roots either in a symplast (intercellular) or apoplast (extracellular) route. The cation exchange capacity in the cell wall occurs through apoplastic transport [17]. Herbicides usually accumulate in the vacuole or cell wall in the cell. In addition, it can get into trichomes and epidermis in tissues [14].

9.2.2 Oxidative Stress and Antioxidative Defense System of Herbicides on Plants

Oxygen is required to produce energy in aerobic organisms and plants. Reactive oxygen species emerge with the reduction of O_2 to water. Reactive oxygen types, stress conditions, drought, ozone, and low or high temperature are reported to occur in electron transfer and lipid catabolism. Reactive oxygen species in plant cells comprising superoxide dismutase (SOD) radicals are formed at peroxisomes during metabolism. It has been determined that the reactive oxygen types that are included can neutralize the antioxidant defense system, such as SOD and GR. It is reported that reactive oxygen species at low concentrations respond to defenses in the plant. In contrast, high concentrations can damage plant cells. Reactive oxygen species that cause oxidative damage are involved in the activation of defense enzymes such as SOD and the formation of GSSG.

It has been determined that plant cells turn into H_2O_2 with the help of SOD formed in different organelles. Ascorbate, the density of which increases in chloroplasts, is rapidly reduced to O_2. It is H_2O_2, which rapidly oxidizes thiol groups and is a powerful oxidant. CAT converts H_2O_2 into water and molecular oxygen so that H_2O_2 does not accumulate in the chloroplast together with thiol-regulating enzymes during photosynthesis. CAT has two active zones that bind H_2O_2. H_2O_2 destruction occurs by the activation of peroxidases in cells. So, it reduces H_2O_2 to water [18,19].

Glutathione is a powerful antioxidant that destroys peroxides and includes compounds such as peptide and thiol. In addition, it is compounded with low molecular weight. Glutathione is included in protecting cells from reactive oxygen species, metabolic events, catabolism, and transport. These metabolic events play a role in reactions occurring during protein and nucleic acid synthesis, detoxification of free radicals and peroxides, and cofactors in some enzymes, passage, and cysteine storage. Glutathione is synthesized in the intracellular space and begins to break up on the outer surface of the cell membranes. Glutathione increases against oxidative damage on the cell surface. Glutathione has been identified to protect proteins along with the cell membrane in peroxidases. The other peroxidase is GSH-S-transferase that participates in the ascorbate–GSH cycle to reduce organic peroxides and remove H_2O_2. Ascorbate is oxidized in leaves and tissues. In addition, it converts monodehydroascorbate (MDHA) into dehydroascorbate (DHA). DHA reduces GSH to ascorbate by the activity of dehydroascorbate reductase, and GSSG occurs. GSSG is turned into GSH by NADPH. It is reported that H_2O_2 is reduced to water with electrons coming out of NADPH. Superoxide dismutase acts as the enzyme that catalyzes the transformation of superoxide to hydrogen peroxide [18,19]. H_2O_2 is formed by the dismutation of superoxide molecules and is detoxified by mixing it with water with the help of GHS peroxidase and catalase enzymes. Catalase affects the increase of hydrogen peroxide formation. In addition to

SOD, GHS peroxidase, and catalase enzymes, vitamins E and C act as antioxidants in the cell. These antioxidants are reported to break the lipid peroxidation chain in the cell membrane [18,19].

9.2.3 The Effect of Herbicides on Cereals Growth and Yield

For the plants grown in agricultural areas to be of high quality, it is necessary to fight against weeds that are ground and food partners. Since wheat is a commonly cultivated plant, it is challenging to fight weeds, especially after emergence. Generally, the fight against wheat cannot be done mechanically, and it is done culturally and chemically. Besides the many benefits of the herbicides used in this struggle, there are also damages. For example, when herbicides are given to leaves after the germination of the plant, some of it may fall into the soil. In this case, weed control is more effective since the plant takes chemicals from both leaf and soil. In terms of weeds, the fact that the herbicides have little time to stay in the soil may pose difficulty in controlling weeds. This long time may also cause herbicides to spread to the environment. On the other hand, herbicide residue may cause phytotoxicity if it remains in the soil until the following year [20].

Herbicides form many mechanisms of action on cereals. Acetyl coenzyme A carboxylase (ACCase) inhibitors destroy plants and inhibit cell wall formation of meristem tissue in plant cells. The ACCase enzyme initiates lipid synthesis in plants. The acetohydroxyacid synthase (AHAS) enzyme and acetolactate synthase enzyme, ALS inhibitors, are the first synthesis enzymes of branched valine, leucine, and isoleucine amino acids. Herbicides are effective in lawns and dicotyls. Enolpyruvylshikimate-3-phosphate (EPSP) synthase enzyme inhibitors are used to synthesize tryptophan, phenylalanine, and tyrosine amino acids. Synthetic auxin is an organic herbicide and is known to mimic plant hormone. Photosystem inhibitors reduce electron transfer from water to NADPH during photosynthesis. By binding to the QB (a molecule involved in the ETS) field in the D2 protein, it prevents quinone binding here and accumulates electrons in the chlorophyll molecule. Amino acid synthesis inhibitors, sulfonyl ureas, imidazolinones, triazolopyrimidines, EPSP synthase inhibitors, and glutamine synthase inhibitors specifically affect protein production. Thus, the plant was found to be weakened. It is reported that seedling growth inhibitors inhibit the first stage of growth. In these, powder herbicides should be used as an aqueous solution, and the root part should be targeted. Photosynthetic inhibitors are used as growth-regulating herbicides. In this group, triazines, uracil, phenyl ureas, benzothiadiazole, nitriles, and pyridazines are used. Commercial names used are Atrazine, Sencor, Hyvar, Karmex, Basagran, and Buctril. Lipid synthesis inhibitors cause the plant's cell wall to develop and die. Herbicides provide the formation of superoxide and hydroxyl radicals. They destroy the plant's cell wall. In this herbicide group, diphenyl ethers, aryl triazolinone, phenylphthalamides, and bipyridylium are used. Pigment inhibitors prevent the formation of chlorophyll, causing loss of green color in the plant. The fact that the mechanism of some herbicides is not known indicates that their mode of action is also unknown. So, it cannot be divided into classes [21]. Oxidative stress causes an increase in the activity of O-production due to fatty acyl-CoA oxidase, xanthine oxidase, and NADH, and a decrease in the activity of antioxidant enzymes. As a result, excessive production of oxygen radicals (H_2O_2 and O_2^-), Mn-SOD and catalase cause cell damage [22].

In this study, we investigated the impact of three concentrations of wheat on fatty acids due to the toxic effect of different concentrations of synthetic herbicides called atrazine and metolachlor. As plant material, seeds of Bayraktar, İkizce, and Tosunbey cv. belonging to the species *Triticum aestivum* L. were used. For three different wheat varieties, 15-day-old wheat seedlings were separated into 4 groups and 4 different concentrations (0, 100 μM, 300 μM, and 1000 μM) of atrazine and metolachlor were applied separately. Hydroponic environment from the root was preferred in all applications to seedlings. Seedlings were used at various concentrations (0, 100 μM, 300 μM, and 1000 μM) prepared from atrazine and metolachlor herbicides for one day. All these different herbicide applications were made simultaneously to other seedlings. Fatty acid analysis was done on leaves. For the analysis of fatty acids, leaf tissues were used and investigated by gas chromatography (GC) [23,24]. Results were determined in terms of weight percent of the total. Three replicates were maintained for each treatment. In each analysis, 3 g of leaf tissue were used.

Compared with the control, about 14:0 atrazine herbicide was sprayed on the leaves of seedlings; in Bayraktar wheat, a decrease of 17.88% and 27.99% was determined at concentrations of 300 μM and 1000 μM, respectively. In Ikizce wheat, an increase of 18.63% and 34.78% was detected at 300 μM and 1000 μM, respectively. In Tosunbey wheat, a reduction of 12.51% and 24.38%, respectively, were found at concentrations of 300 μM and 1000 μM ($p \leqslant 0.05$) (Table 9.1). In the ratio of 14:1 compared with the control, an increase of 46.95% and 52.47% was detected in Bayraktar wheat at concentrations of 300 μM and 1000 μM, respectively. In Ikizce wheat, an increase of 26.35% at 1000 μM concentration was detected. In Tosunbey wheat, an increase of 17.21% and 27.66%, respectively, was detected at concentrations of 300 μM and 1000 μM ($p \leqslant 0.05$) (Table 9.1). In the ratio of 16:0 compared with the control, in Tosunbey wheat, an increase of 20.37% at 1000 μM concentration was detected ($p \leqslant 0.05$) [25,26]. In the ratio of 16:1 compared with the control, in Bayraktar wheat, a decrease of 14.33%, 30.96%, and 36.56% was detected in concentrations of 100 μM, 300 μM, and 1000 μM, respectively [27]. In Ikizce wheat, an increase of 42.51% and 51.36%, respectively, was observed at concentrations of 300 μM and 1000 μM [26]. In Tosunbey wheat, an increase of 38.23% and 48.40% was detected at concentrations of 300 μM and 1000 μM, respectively ($p \leqslant 0.05$) [27]. In the ratio of 18:0 compared with the control, an increase of 76.42% and 89.73%, respectively, was observed in Bayraktar wheat at concentrations of 300 μM and 1000 μM [27]. In Ikizce wheat, an increase of 24.30% for 300 μM and a decrease of 29.27% for 1000 μM concentration was detected. In Tosunbey wheat, a reduction of 25.11%, 38.97%, and 36.86%, respectively, was found in concentrations of 100 μM, 300 μM, and 1000 μM ($p \leqslant 0.05$). In the ratio of 18:1 compared with the control, in Bayraktar wheat, a decrease of 18.21% and 24.28% was detected in concentrations of 100 μM and 300 μM, respectively. In Ikizce wheat, an increase of 16.69% was detected at 100 μM concentration, respectively [26]. In Tosunbey wheat, a decrease of 25.72%, 27.98%, and 38.24% was found at concentrations of 100 μM, 300 μM, and 1000 μM ($p \leqslant 0.05$). For a ratio of 18:2 compared with the control, in Bayraktar wheat, a decrease in the concentrations of 100 μM, 300 μM, and 1000 μM was found at 7.36%, 9.35%, and 17.43%, respectively. In İkizce wheat, a decrease of 14.82% and 20.58% was detected at concentrations of 300 μM and 1000 μM, respectively. In Tosunbey wheat, an increase of 6.29%, 10.96%, and 14.14%, were found at concentrations of 100 μM, 300 μM, and 1000 μM respectively ($p \leqslant 0.05$). In the ratio of 18:3 compared with the control, a decrease of 22.16% at 1000 μM concentration was detected in

Table 9.2 Fatty acid profile effects in the *Triticum aestivum* L. Bayraktar cv., İkizce cv., and Tosunbey cv. Wheat seedlings applied to metolachlor herbicide.

Groups (metolachlor)	14:0	14:1	16:0	16:1	18:0	18:1	18:2	18:3
B-Control	0,673 ± 0,065	4,990 ± 0,336	33,811 ± 0,554	2,062 ± 0,363	5,055 ± 0,157	5,440 ± 0,163	10,904 ± 0,084	37,231 ± 2,023
B-100	0,565 ± 0,054*	3,660 ± 0,165*	31,568 ± 0,955	1,924 ± 0,186	5,688 ± 0,443	5,573 ± 0,162	10,337 ± 0,132	41,209 ± 1,672
B-300	0,515 ± 0,036*	2,517 ± 0,174*	29,546 ± 1,162*	1,926 ± 0,522	8,630 ± 0,494*	4,448 ± 0,170*	9,645 ± 0,174*	44,130 ± 0,250*
B-1000	0,457 ± 0,021*	2,227 ± 0,186*	24,229 ± 1,414*	1,779 ± 0,503	5,780 ± 0,197	3,856 ± 0,178*	7,559 ± 0,229*	37,309 ± 0,727
İ-Control	0,755 ± 0,014	3,656 ± 0,075	46,742 ± 1,427	4,337 ± 0,384	5,120 ± 0,210	7,318 ± 0,150	7,066 ± 0,364	25,874 ± 1,981
İ-100	0,827 ± 0,018	3,284 ± 0,222	41,609 ± 1,028*	3,879 ± 0,362	5,922 ± 0,154	6,793 ± 0,228	6,655 ± 0,205	25,239 ± 2,125
İ-300	0,885 ± 0,026*	3,109 ± 0,227	39,831 ± 1,870*	3,213 ± 0,235*	6,875 ± 0,052*	5,139 ± 0,141*	6,304 ± 0,339*	23,868 ± 2,464
İ-1000	0,970 ± 0,014*	2,680 ± 0,242*	33,085 ± 2,675*	3,142 ± 0,245*	7,495 ± 0,221*	4,941 ± 0,180*	5,681 ± 0,101*	21,499 ± 3,074
T-Control	0,540 ± 0,059	4,314 ± 0,089	36,216 ± 2,070	2,925 ± 0,254	4,492 ± 0,644	5,074 ± 0,337	7,115 ± 0,239	38,136 ± 1,664
T-100	0,616 ± 0,022	4,262 ± 0,128	32,684 ± 0,784	3,466 ± 0,377	3,955 ± 0,371	3,888 ± 0,233*	6,902 ± 0,242	41,263 ± 2,824
T-300	0,672 ± 0,012*	4,048 ± 0,257	32,045 ± 1,089	3,015 ± 0,243	3,700 ± 0,336	3,914 ± 0,080*	6,885 ± 0,157	34,570 ± 1,280
T-1000	0,71 ± 0,006*	3,725 ± 0,141	30,452 ± 0,327*	2,381 ± 0,157	2,901 ± 0,251*	3,713 ± 0,100*	6,579 ± 0,209	28,053 ± 0,602*

*Compared with the control; significant at probability levels of $p \leq 0.05$. Mean of the data ± SE (n: 3), B: Bayraktar, İ: İkizce, T: Tosunbey.

Bayraktar wheat. In Ikizce wheat, a decrease of 20.37%, 23.94%, and 34.99%, respectively, was found at concentrations of 100 μM, 300 μM, and 1000 μM [25]. In Tosunbey wheat, an increase of 18.95% was detected in the concentration of 300 μM ($p \leqslant 0.05$) (Table 9.1). Seedlings applied with metolachlor herbicide in the ratio of 14:0 compared with the control; in Bayraktar wheat, a decrease of 16.05%, 23.40%, and 32.11%, respectively, was detected at concentrations of 100 μM, 300 μM, and 1000 μM [25]. In Ikizce wheat, an increase of 17.17% and 28.47% was detected at 300 μM and 1000 μM concentrations, respectively. In Tosunbey wheat, an increase of 24.43% and 31.41% were detected at concentrations of 300 μM and 1000 μM, respectively ($p \leqslant 0.05$) [26] (Table 9.2). In the ratio of 14:1 compared with the control, in Bayraktar wheat, a decrease of 26.64%, 49.56%, and 55.37%, respectively, was detected at concentrations of 100 μM, 300 μM, and 1000 μM. In İkizce wheat, a decrease of 26.69% at 1000 μM concentration was detected ($p \leqslant 0.05$). At 16:0 compared with the control, in Bayraktar wheat, a decrease of 12.62% and 28.34% was detected at concentrations of 300 μM and 1000 μM, respectively. In Ikizce wheat, a decrease of 10.98%, 14.78%, and 29.22% was detected at concentrations of 100 μM, 300 μM, and 1000 μM, respectively. In Tosunbey wheat, a decrease of 15.92% in 1000 μM concentration was detected ($p \leqslant 0.05$) [27]. At 16:1 compared with the control, a decrease of 25.91% and 27.56% in the İkizce wheat at 300 μM and 1000 μM concentrations, respectively, was determined [26] ($p \leqslant 0.05$). At 18:0 compared with the control, an increase of 70.72% in 300 μM concentration was detected in Bayraktar wheat. In Ikizce wheat, an increase of 34.29% and 46.41% was detected at concentrations of 300 μM and 1000 μM, respectively [26]. In Tosunbey wheat, a 35.43% reduction in 1000 μM concentration was detected [27] ($p \leqslant 0.05$). At 18:1 compared with the control; in Bayraktar wheat, a decrease of 18.22% and 29.11%, respectively, was detected at 300 μM and 1000 μM concentrations. In Ikizce wheat, a decrease of 29.78% and 32.48%, respectively, was detected at 300 μM and 1000 μM concentrations [26]. In Tosunbey wheat, a decrease of 23.38%, 22.86%, and 26.83% was found at concentrations of 100 μM, 300 μM and 1000 μM ($p \leqslant 0.05$). At 18:2 compared with the control, in Bayraktar wheat, a decrease of 11.55% and 30.68% was determined at concentrations of 300 μM and 1000 μM, respectively. In Ikizce wheat, a decrease of 10.78% and 19.60%, respectively, was detected at concentrations of 300 μM and 1000 μM ($p \leqslant 0.05$). At 18:3, compared with the control, an increase of 18.53% in 300 μM concentration was detected in Bayraktar wheat. In Tosunbey wheat, a 26.44% decrease in 1000 μM concentration was detected [25] ($p \leqslant 0.05$) (Table 9.2).

9.3 Phytoremediation

Having derived from the words "phyto" meaning plant and "remediation" meaning reclamation, phytoremediation is also referred to as "bioremediation," "botanical remediation," and "green remediation." This method removes water and soil contaminated with inorganic and organic pollutants using plants [28]. Most of the organic pollutants in contaminated areas are human-made as well as toxic and carcinogenic. For organisms, they are xenobiotics. Depending on their properties, xenobiotics can be broken down in the root area of the plants or taken by the plant and broken down in the plant. These organic pollutants can be kept or evaporated from either the vacuole or cell wall areas by

Table 9.1 Fatty acid profile effects in the *Triticum aestivum* L. Bayraktar cv., İkizce cv., and Tosunbey cv. wheat seedlings to which were applied atrazine herbicide.

Groups (atrazine)	14:0	14:1	16:0	16:1	18:0	18:1	18:2	18:3
B-Control	0,527 ± 0,016	2,241 ± 0,061	39,573 ± 0,216	3,870 ± 0,207	3,153 ± 0,477	5,840 ± 0,146	10,506 ± 0,213	37,212 ± 1,536
B-100	0,473 ± 0,009	2,681 ± 0,027	39,642 ± 0,139	3,315 ± 0,150*	3,605 ± 0,535	4,776 ± 0,135*	9,732 ± 0,065*	35,404 ± 2,645
B-300	0,433 ± 0,009*	3,293 ± 0,260*	40,987 ± 0,351	2,672 ± 0,220*	5,563 ± 0,286*	4,422 ± 0,340*	9,523 ± 0,103*	33,563 ± 1,704
B-1000	0,380 ± 0,012*	3,416 ± 0,253*	42,811 ± 0,444	2,455 ± 0,066*	5,983 ± 0,443*	5,701 ± 0,105	8,675 ± 0,152*	28,967 ± 0,624*
İ-Control	0,492 ± 0,035	3,263 ± 0,290	41,637 ± 0,890	1,795 ± 0,175	4,079 ± 0,294	6,623 ± 0,335	7,692 ± 0,142	27,523 ± 1,304
İ-100	0,507 ± 0,028	3,514 ± 0,170	42,867 ± 0,923	1,911 ± 0,120	4,467 ± 0,315	7,728 ± 0,249*	7,258 ± 0,073	21,918 ± 0,825*
İ-300	0,583 ± 0,015*	3,756 ± 0,216	45,247 ± 0,692	2,558 ± 0,224*	5,070 ± 0,327*	6,399 ± 0,272	6,552 ± 0,100*	20,934 ± 0,439*
İ-1000	0,663 ± 0,021*	4,122 ± 0,213*	45,865 ± 0,495	2,717 ± 0,188*	2,885 ± 0,112*	6,288 ± 0,129	6,109 ± 0,030*	17,894 ± 2,564*
T-Control	0,556 ± 0,014	3,938 ± 0,078	33,269 ± 3,446	2,041 ± 0,199	3,929 ± 0,131	6,065 ± 0,205	7,144 ± 0,237	38,740 ± 0,575
T-100	0,516 ± 0,024	4,317 ± 0,132	35,520 ± 3,569	2,303 ± 0,173	2,942 ± 0,100*	4,505 ± 0,254*	7,594 ± 0,108*	42,804 ± 1,348
T-300	0,487 ± 0,021*	4,615 ± 0,094*	38,774 ± 2,887	2,822 ± 0,263*	2,399 ± 0,181*	4,368 ± 0,326*	7,927 ± 0,226*	46,080 ± 1,405*
T-1000	0,421 ± 0,024*	5,027 ± 0,103*	40,044 ± 3,393*	3,029 ± 0,123*	2,481 ± 0,188*	3,746 ± 0,252*	8,154 ± 0,197*	41,704 ± 1,186

*Compared with the control; significant at probability levels of $p \leq 0.05$. Mean of the data ± SE (n: 3), B:Bayraktar, İ: İkizce, T: Tosunbey.

conjugation [15]. Herbicides such as organic pollutants, atrazine, and metolachlor [29] can be successfully removed from the environment by phytoremediation [15]. Furthermore, the phytoremediation technique does not harm natural resources and is highly accepted by public opinion.

9.4 Phytoremediation Techniques

In the methods to be used, factors such as the intake and removal mechanisms of the pollutants by plants, the physical and chemical properties of the pollutant environment, the suitability of the method to be applied to the contaminant, the concentration of the pollutant, the depth of the pollutant in the soil, and the climatic conditions should be taken into consideration (Table 9.3).

9.4.1 Phytoextraction (Herbal Extraction)

It involves the harvesting and destroying of plants following the uptake of contaminants by plant roots and subsequent accumulation in aboveground organs. This technique removes actively taken nutrients (Cu and Zn) and non-nutrient metals (Cd, Ni, and Pb). Phytoextraction technology is applied only for areas where metal pollution is low or moderate. This is because plant growth cannot be maintained in heavily contaminated sites [36]. Natural hyperaccumulator plants are used in this technology [37]. However, chelating agents can be added to increase the solubility of metals with low solubility in soil solution [38]. Successful phytoextraction depends on the ability of plants to rapidly produce biomass and the ability to accumulate the collected metals in high amounts in body tissues [39]. Phytoextraction is a method where inorganic pollutants are transported to the aboveground parts of the plant via the root, and many of the contaminants accumulate there. Therefore, plants get more inorganic substances than other plant species. With this method, the pollutants taken into the plant can be removed with plant harvest, pruning, or removal [28].

9.4.2 Phytodegradation (Vegetal Degradation)

"Phytodegradation," also named "phytotransformation," is the metabolism of herbicides in plant tissues. This method breaks down contaminants with the rhizospheric association between the metabolic functions in plants and soil microorganisms (a thin layer of loose soil attached to the root system). In this method, the primary mechanism is the uptake of contaminants by plants and those metabolized in plants. This process is limited in the root zone, especially to the extreme points of the root zone. There are many factors for the absorption of pollutants within the plant. The plant species, the age of the plant, the duration of the pollution in the soil, and the physical and chemical structure of the soil ought to be studied. It is challenging to instantly obtain melting compounds from plants. Phytodegradation can be used for cleaning contaminated areas of soil, sediment, mud, and groundwater. This method has been used to treat surface waters [28]. In a study about this

Table 9.3 Studies on phytoremediation of herbicide varieties with phytoremediation techniques.

Subject no.	Herbicide	Affected area	Phytoremediation techniques	Results	Reference
1	Atrazine, alachlor, metolachlor, and pendimethalin	Lettuce seedling growth, earthworm accumulation	For the vegetative cover technique, a prairie grass mixture was used	Decrease in mobility and bioavailability indicates that herbicide residues may stabilize to increasing distributing rates	[30]
2	2,4-D and propanil	Soil samples	Potential of *Melissa officinalis* in soil phytoremediation	This study showed that *Melissa officinalis* cultivation could be a fast, effective, and low-cost strategy for phytoremediation of soils contaminated by 2,4-D herbicides and propanil.	[31]
3	For fertilizers, NO_3^--N, NH_3-N, and $PO^4{}_3$ —-P and glyphosate contamination was analyzed.	Soil and water	For vegetative cover technique; *Cyperus textilis, Phragmites australis, Typha capensis, Arundo donax, Canna indica, Cynodon dactylon, Prionium serratum*, and *Pennisetum clandestinum*	In the laboratory, the plant breeding system measured the soil improvement effect and the agricultural pollutant removal performance of the plants.	[3]
4	Fomesafen and sulfentrazone herbicides	Soil	Phytoremediation potential (black oats, vetch, turnip, and lotus)	It was observed that the vetch presented the best results for all evaluated variables.	[32]
5	Atrazine, chloroacetanilides, isoproturon, metolachlor	Soil	Phytoextraction	It was concluded that it is possible to remove metabolic compounds by phytoremediation technique.	[9]

(*Continued*)

Table 9.3 (Continued)

Subject no.	Herbicide	Affected area	Phytoremediation techniques	Results	Reference
6	Atrazine, metolachlor, trifluralin, and pendimethalin	Soil	Rhizodegradation; Kochia plants	Phytoremediation is a suitable treatment method for soils contaminated with a mixture of different herbicides.	[33]
7	Atrazine and metolachlor	Soil	Rhizodegradation	The complete metabolism and binding profile of atrazine and metolachlor were determined in phytoremediation studies.	[34]
8	Atrazine and metolachlor	Macrophyte-cultured ponds and artificial wetlands for the treatment of sewage, industrial effluents, and agricultural drainage water	Rhizodegradation; *Ceratophyllum demersum*, *Lemna minor*, and *Elodea canadensis*	The existence of herbicide-tolerant aquatic plants can accelerate the removal of metolachlor and atrazine with contaminated waters.	[35]

technique, plants absorb atrazine in the soil and later hydrolyze it, which is alkylated in the root, stem, and leaf. As a result of atrazine metabolite toxicity studies, metabolites recognized in plant tissue were less toxic than atrazine [29]. The most important advantage of this method is that it is decomposed or decreased according to physiological events in the plant and is not dependent on microorganisms. The disadvantage of the technique is that toxic intermediate and end products can be formed during the deterioration, which is very difficult to detect [29].

9.4.3 Phytostabilization (Root Fixing)

Phytostabilization, a process of much use to plants, stabilizes herbicides in the soil, reduces their absorbability, and takes the form of immobilizing pollutants accumulating by the roots, adhering to the surface of the roots, or settling in the root area of the plant. Plant roots also define preventing possible movement of pollutants through wind and water erosion, infiltration, and soil dispersion. It is an essential advantage that it does not require soil transport.

Also, as the area is planted, restoration in the ecosystem is enriched. The most crucial disadvantage is that the pollution factors can remain in place and can be transported with changes over a long period mixed with groundwater by washing. It is required for fertilization, and soil improvement works in the system to provide maximum stabilization efficiency. Root depth is significant for the effectiveness of the system. The phytostabilization method is also used in the treatment of soil, sediment, and mud [28]. In the root-fixing method, what is essential is that the herbicides are covered with aboveground plants to prevent penetration of groundwater and erosion. Plant roots immobilize herbicides. With this method, plants must grow in contaminated soils and change the properties of the soil to convert toxic metals into more minor poisonous forms. Plants to be used for phytostabilization should have an extensive root system, produce high biomass in the presence of high concentrations of herbicides, and translocated herbicides to the body at the minimum level [40].

9.4.4 Phytovolatilization (Herbal Evaporation)

Phytovolatilization occurs in plants by taking organic and inorganic pollutant factors into the body and changing their structure to release them into the atmosphere [41]. Phytovolatilization may also include the photodegradation process. The most important advantage of the method is the conversion of highly toxic compounds to less toxic forms. However, the fact that very harmful poisonous materials can be released into the atmosphere is also seen as a disadvantage. Root depth is significant in this system. If groundwater is concerned, plant roots must be profound. Polluted groundwater can be brought up to the surface with pumps, allowing water to be taken up by shallower plant roots. The phytovolatilization method can also be applied above ground, as well as in groundwater. Among the plants used for this purpose, Se element is removed with Indian mustard and canola, and selenium in the form of selenite is released into the atmosphere after conversion into less toxic dimethyl selenite gas [28].

The use of plants to separate organic compounds in a certain amount in the soil and give them to the atmosphere is seen as a method. In this technology, plants become a little volatile after they absorb metals through the root. Thus, plants release metals into the atmosphere through perspiration [42].

9.4.5 Rhizodegradation (Root Degradation)

If degradation is caused by microorganisms or by the effect of plant roots, this event is called photodegradation. Rhizodegradation is the decomposition of organic pollutants due to microorganism activities in the root zone in the soil. Sugar, amino acids, organic acid, fatty acids, sterol, growth factors, nucleotide, flavanone, and enzymes affect microbial activities and are released from the roots. Organic compounds that cause pollution also occur in this environment. The most important benefit of root degradation is that the organic compounds disappear from their natural habitat. However, these are transported to the plant or atmosphere to a small extent, among the pollutants removed by the retrogradation method, total petroleum hydrocarbons, multi-ring aromatic hydrocarbons (PAH), BTEX benzene, toluene, ethylbenzene, xylene (BTEX), pesticides (herbicide, insecticide), chlorinated solvents (TCE, TCA), pentacloro (PCP), polychlorinated biphenyls (PCB), and surfactants (LAS, LAE). The most commonly used plants in biodegradation are *Morus rubra* L., *Mentha spicata*, *Medicago sativa*, and *Typha latifolia* [28].

9.4.6 Rhizofiltration (Root Filtration)

Due to hemofiltration, biotic and abiotic processes, herbicides are taken into the roots or the plant sticks on the roots (adsorption). During the formation of these processes, pollutants can be taken to the plant and transported. It is essential here that pollutants provide immobility on or inside the plant. Then contaminants are removed from the plant in a variety of ways. The advantage of the system is that it allows the use of terrestrial and aquatic plants. In addition to being applicable in natural environments, the system is applied in artificial areas, such as pools, tanks, and ponds. In other words, pollution can be eliminated at the source or far from it [28].

9.4.7 Hydraulic Control

Terms as "phytohydraulic control" or "hydraulic smoke control" in hydraulic control are in use. By this method, plants are used to prevent herbicides from being transported to and accumulated in groundwater and to control them. In this system, more than one of the previously mentioned green breeding categories coexist. The most important advantage is that there is no need to install an artificial system, and the roots spread over more areas than the pumps, so the growing area is expanding. The most crucial disadvantage is that depending on the season and climate, it changes the plant's water intake. However, deciduous trees cannot perform the desired task during the winter. Willow and Eucalyptus species are also used for these purposes. A five-year-old Populus tree receives 100–200 liters of water per day, and the sweating amount of a single willow tree is equivalent to 5000 gallons of water a day, emphasizing the water usability of these plants and its importance [43].

9.4.8 Vegetative Cover Systems

Vegetative cover is a method of controlling herbicides with a long-term and self-growing plant system on the soil surface. Vegetative cover involves water evaporated from the soil and comprises prevention of water loss or remediation. First, while the plant minimizes the water loss from the soil, it also maximizes the ability to hold water, and the herbicides cannot be reduced or moved to the wash formation. In the second type of cover for breeding purposes, the plant minimizes the infiltration of water as a cover and aims to disrupt the pollution in the substrate. The mechanism has water intake, root environment microbiology, and plant metabolism, and involves phytodegradation, photodegradation, phytovolatilization, and phytoextraction categories of hydraulic control. In applications, the vegetative cover is usually formed as a barrier to prevent the dispersion of pollution. This cover can be installed at a low cost as an ecosystem that minimizes the surface erosion that renews itself and has low maintenance requirements. Its most crucial disadvantage is providing long-term care and control, which may be necessary to guarantee proper vegetation. Because some plant species become more dominant over time, this method can be applied in soil, sediment, and mud. For this purpose, poplar trees and lawns are used commercially [28].

9.4.9 Coastal Buffer Strips

Coastal buffer strips are the process of planting suitable plants in strips along the stream to remove the herbicides present underground or in surface water, which generally flow toward streams. With this reclamation, herbicides do not spread to the environment and do not mix with the groundwater. The system also controls erosion and reduces sediment. It has been determined that buffer strips applications minimize soil erosion by 90% and herbicide flow by 42–70% in Canada. In addition, the sediment in the water can decrease by 71–91%, nitrogen by 67–96%, phosphorus by 27–97%, pesticides by 8–100% and fecal coliforms by 70–74% [44]. Most fertilizers and herbicides are examined using this method. Poplars are among the most frequently used plants for this purpose [28].

9.5 Evaluation of Ideal Plants and Harvested Plants for Phytoremediation

Diagnosis of hyperaccumulator plants shows that these plants have a genetic potential in cleaning contaminated areas. Small size and low growth rates of hyperaccumulator plants negatively affect the phytoremediation potential and limit agricultural practices such as mechanical harvesting. To overcome these disadvantages, classical breeding methods are used based on the development of slow-growing and high biomass varieties from low biomass plants [45].

Understanding the genetic make-up and biochemistry of herbicide residue intake, transport, and accumulation can lead to the development of transgenic plants with high phytoremediation capacity (see Figure 9.2) [46].

Approximately 450 plant species (only 0.2% of angiosperms) are defined as hyperaccumulators [47]. Many types of research evaluate the phytoremediation potentials of different plants in terms of various pollutants, such as *Thlaspi caerulescens, Arabidopsis thaliana, Brassica juncea, Lycopersicon esculentum, Zea mays, Hordeum vulgare, Oryza sativa, Pisum sativum*, and *Sedum alfredii*. However, the phytoremediation potential of many hyperaccumulator species is limited due to slow growth rate, low biomass, and often a tight relationship with a particular habitat [48]. For this reason, plants with the necessary properties for phytoremediation, such as deep root system, rapid growth rate, and heavy metal tolerance should be developed [49]. Various cultural and weed species are used very efficiently for phytoremediation. These plants, called hyperaccumulators, are generally found in the plant families Brassicaceae, Euphorbiaceae, Asteraceae, Lamiaceae, and Scrophulariaceae. For example, Indian mustard, canola, Alpine pennycress, sunflower, tobacco, and hybrid poplar trees are the highest fertile plants used as fertility is a priority [45].

The purpose of removing herbicide residue from the soil is to convert the herbicides, which are held by the soil, into a more controllable and portable form by using plants.

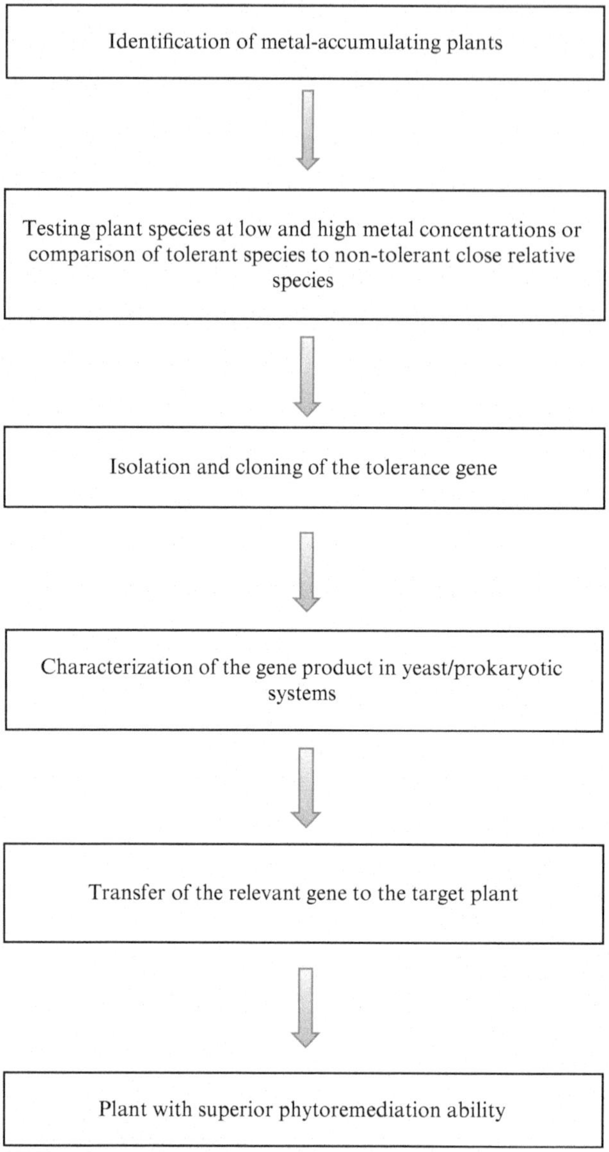

Figure 9.2 Strategy required for the development of plants that accumulate heavy metal.

Therefore, the phytoremediation method is not considered a final removal or elimination method. Final removal or elimination can be accomplished by burning the plants, resulting from phytoremediation or storing them in a suitable storage area.

Some herbicides taken by plants accumulate in the leaves and stems of the plant without any degradation and move away from the environment with the plant's harvest. The residues in the harvested plant resulting from the phytoextraction method can be isolated by drying, burning and turning into ash, limiting it to compost, and recycling it into a biological metal mine [50].

9.6 Conclusion

The problem of herbicide residues, which is increasing day by day in agricultural lands, makes agricultural fields inefficient on a large scale and has significant effects on the functions of the ecosystem and animal and human health through the food chain. One of the most critical environmental problems that scientists working in this field have been focusing on is herbicide residues in agricultural land. Today, people have realized that this problem needs to be addressed urgently and effectively. However, it requires the allocation of a big financial budget for the adoption and practice of classical and traditional physical and chemical methods. Otherwise, the yield of grain products for daily consumption is very low. In our study, the decrease in the fatty acid components of wheat, a critical grain product, shows that the yield rate is low. In these agricultural areas, herbicide residues need to be reduced or destroyed.

Phytoremediation is a technique used to improve soil and water resources with herbicide residues, which can be up to 100 times cheaper and effective compared with traditional methods. Plants are used as ideal tools in remediation work, as they have unique structures and properties. However, there are limitations in using this technology as it requires a good understanding of the mechanisms (molecular and genetic) of most of the hyperaccumulator plants diagnosed.

References

1. Asri, F.Ö. and Sönmez, S. (2006). *Effects of Heavy Metal Toxicity on Plant Metabolism*. Antalya: Akdeniz University Faculty of Agriculture, Department of Soil Science.
2. Lam, Q.D., Schmalz, B., and Fohrer, N. (2010). Modelling point and diffuse source pollution of nitrate in a rural lowland catchment using the SWAT model. *Agricultural Water Management* 97 (2): 317–325.
3. Jacklin, D.M., Brink, I.C., and de Waal, J. (2019). Laboratory method design for investigating the phytoremediation of polluted water. *Water SA* 45: 4.
4. Oerke, E.C. and Dehne, H.W. (2004). Safeguarding production—losses in major crops and the role of crop protection. *Crop Protection* 23: 275–285.
5. Başaran, M.S. and Serim, A.T. (2010). Fragmentation of Herbicides in Soil, Selçuk University. *Selçuk Journal of Agriculture and Food Sciences* 24 (2): 54–61.
6. Gouma, S. (2009). Biodegradation of mixtures of pesticides by bacteria and white-rot fungi, Ph.D. thesis, School of Health Cranfield University. 458.
7. Surekha, R.M., Lakshmi, P.K.L., Suvarnalatha, D. et al. (2008). Isolation and characterization of a chlorpyrifos degrading bacterium from agricultural soil and its growth response. *African Journal of Microbiology Research* 2: 26–31.
8. Mohammed, M.S. (2009). Degradation of methomyl by the novel bacterial strain *Stenotrophomonas maltophilia* M1. *Electronic Journal of Biotechnology* 12: 1–6.
9. Coleman, J.O.D., Frova, C., Schröder, P. et al. (2002). Phytoremediation: Persistent herbicides exploiting plant metabolism for the phytoremediation of Persistent herbicides. *Environmental Science and Pollution Research* 9 (1): 18–28.
10. Gündüz, Ş., Kersting, U., and Kahramanoğlu, İ. (2006). *Weeds and Integrated Control Methods in Citrus Gardens*. Mersin: Mediterranean Exporters' Associations.

11 Pudelko, J.A., Wright, D.L., and Teare, I.D. (1993). A method for salvaging bird damaged pearl millet research. Fla. Agric. Exp. Stn. Res. Rep. No: N.F. 93–12:1–11.

12 Rice, P.J. (1996). The persistence, degradation, and mobility of metolachlor in soil and the fate of metolachlor and atrazine in surface water, surface water/sediment, and surface water/aquatic plant systems, Doctor of philosophy thesis, Iowa State University, Ames, Iowa.

13 Tiryaki, O., Canhilal, R., and Horuz, S. (2010). Usage and Risks of Pesticides. *Erciyes University Journal of Science Institute* 26 (2): 154–169.

14 Burken, J.G. (2003). Uptake and metabolism of organic compounds: Green-liver model. In: *Phytoremediation: Transformation and Control of Contaminants* (ed. S.C. McCutcheon and J.L. Schnoor). New York: Wiley.

15 Pilon-Smits, E. (2005). Phytoremediation. *Annual Review of Plant Biology* 56: 15–39.

16 Salt, D.E., Blaylock, M., Kumar, N.P. et al. (1995). Phytoremediation: A novel strategy for the removal of toxic metals from the environment using plants. *Bio/Technology* 13: 468–474.

17 Raskin, I., Smith, R.D., and Salt, D.E. (1997). Phytoremediation of metals: Using plants to remove pollutants from the environment. *Current Opinion in Biotechnology* 8: 221–226.

18 Caverzan, A., Piasecki, C., Chavarria, G. et al. (2019). Defenses against ROS in crops and weeds: The effects of interference and herbicides. *International Journal of Molecular Sciences* 20: 1086.

19 Gill, S.S. and Tuteja, N. (2010). Reactive oxygen species and antioxidant machinery in abiotic stresses tolerance in crop plants. *Plant Physiology and Biochemistry* 48: 909–930.

20 Varanasi, A., Prasad, P.V.V., and Jugulam, M. (2016). Impact of climate change factors on weeds and herbicide efficacy. *Advances in Agronomy* 135. ISSN 0065-2113.

21 Mander, L. and Liu, H.-W. (2010). *Comprehensive Natural Products II Chemistry and Biology*. Oxford: Elsevier.

22 Segura-Aguilar, J., Hakman, I., and Rydström, J. (1995). Studies on the mode of action of the herbicidal effect of 2,4,5-trichlorophenoxyacetic acid on germinating Norway spruce. *Environmental and Experimental Botany* 35 (3): 30–319.

23 Christie, W.W. (1990). *Gas Chromatography and Lipids*, 302. Glasgow: *The Oily Pres.*

24 Hara, A. and Radin, N.S. (1978). Lipid extraction of tissues with low-toxicity solvent. *Analytical Biochemistry* 90 (1): 420–426.

25 Ramalho, J.C., Campos, P.S., Teixeira, M. et al. (1998). Nitrogen dependent changes in antioxidant system and in fatty acid composition of chloroplast membranes from Coffea arabica L. plants submitted to high irradiance. *Plant Science* 135: 115–124.

26 Watanabe, K., Ohori, Y., Sato, Y. et al. (2001). Changes in fatty acid composition of neutral lipid in mung bean cotyledons by oxyfluorfen-induced peroxidation. *Pesticide Biochemistry and Physiology* 69: 166–173.

27 Canakci-Gülengül, S., Kirecci, O.A., and Karabulut, F. (2019). Changes based on oxidative stress in metolachlor and atrazine treated maize seedlings. *Pakistan Journal of Botany* 51 (2): 421–426.

28 EPA (Environmental Protection Agency), (2000). Introduction to Phytoremediation, EPA/600/R-99/107, National Risk Management Research Laboratory Office of Research and Development U.S. Environmental Protection Agency Cincinnati, Ohio 45268, USA.

29 Burken, J.G. and Schnoor, J.L. (1997). Uptake and metabolism of atrazine by poplar trees. *Environmental Science & Technology* 31: 1399–1406.

30 Belden, J.B., Phillips, T.A., and Coats, J.R. (2004). Effect of prairie grass on the dissipation, movement, and bioavailability of selected herbicides in prepared soil columns. *Environmental Toxicology and Chemistry* 23 (1): 125–132.

31 Piaia, B.R., Zago, M.L.C., Denardin, E.L.G. et al. (2017). Phytoremediation of 2,4-dichlorophenoxyacetic and propanil herbicides in soil cultivated by herb lemon balm (*Melissa officinalis*). *Ciência E Natura* Santa Maria 39 (1): 91–98.

32 Alves, C., Galon, L., Holz, C.M. et al. (2018). Physiological characteristics of winter cultivated plants with phytoremediation potential under the influence of the herbicides fomesafem and sulfentrazone. *Revista De Ciências Agroveterinárias* 18 (1).

33 Kruger, E.L., Anhalt, J.C., Sorenson, D. et al. (1997). Atrazine Degradation in Pesticide-Contaminated Soils: Phytoremediation Potential. In: *Phytoremediation of Soil and Water Contaminants* (ed. E. Kruger et al.). Washington, DC: ACS Symposium Series; American Chemical Society. 1997, Chapter 4.

34 Henderson, K.L., Belden, J.B., Zhao, S. et al. (2006). Phytoremediation of pesticide wastes in soil. *Zeitschrift Für Naturforschung C* 61 (3–4): 213–221.

35 Rice, P.J., Anderson, T.A., and Coats, J.R. (1997). Phytoremediation of Herbicide-Contaminated Surface Water with Aquatic Plants. In: *Phytoremediation of Soil and Water Contaminants* (ed. E. Kruger et al.). Washington, DC: ACS Symposium Series; American Chemical Society. Chapter 10.

36 Padmavathiamma, P.K. and Loretta, Y.L. (2007). Phytoremediation technology: Hyper-accumulation metals in plants. *Water Air Soil Pollut* 184: 105–126.

37 Baker, A.J.M., Revees, R.D., and Hajar, A.S.M. (1994). Heavy metal accumulation and tolerance in British populations of the metallophyte *Thlaspi caerulescens* J. & *C. Presl* (Brassicaceae). *New Phytologist* 127: 61–68.

38 Evangelou, M.W.H., Ebel, M., and Schaeffer, A. (2007). Chelate assisted phytoextraction of heavy metals from soil: Effect, mechanism, toxicity and fate of chelating agents. *Chemosphere* 68: 989–1003.

39 Blaylock, M.J. and Huang, J.W. (2000). Phytoextraction of metals. In: *Phytoremediation of Toxic Metals: Using Plants to Clean-up the Environment* (ed. I. Raskin and B.D. Ensley), 53–70. New York: Wiley

40 Rizzi, L., Petruzzelli, G., and Poggio, G. (2004). Soil physical changes and plant availability of Zn and Pb in a treatability test of phytostabilization. *Chemosphere* 57: 1039–1046.

41 Song, B., Xu, P., Chen, M. et al. (2019). Using nanomaterials to facilitate the phytoremediation of contaminated soil. *Critical Reviews in Environmental Science and Technology* 49 (9): 791–824.

42 Ghosh, M. and Singh, S.P. (2005). A review on phytoremediation of heavy metals and utilization of its by-products. *Applied Ecology and Environmental Research* 3: 1–18.

43 Pivetz, B.E. (2001). Phytoremediation of Contaminated Soil and Ground Water at hazardous Waste Sites United States Environmental Protection Agency EPA, 540/S-01/500, 36 p.

44 Gabor, T.S., North, A.K., Ross, L.C.M. et al. (2001). Beyond the pipe: The importance of wetlands and upland conservation practices in watershed management: Function and values for water quality and quantity. Ducks Unlimited Canada. 52.

45 Clemens, S., Palmgren, M.G., and Kramer, U. (2002). Along way ahead: Understanding and engineering plant metal accumulation. *Trends Plant* 7: 309–315.

46 Karenlampi, S., Schat, H., Vangronsveld, J. et al. (2000). Genetic engineering in the improvement of plants for phytoremediation of metal polluted soils. *Environmental Pollution* 107: 225–231.

47 Milner, M.J. and Kochian, L.V. (2008). Investigating heavy- metal hyperaccumulation using *Thlaspi caerulescens* as a model system. *Annals of Botany* 102: 3–13.

48 Chaney, R.L., Angle, J.S., McIntosh, M.S. et al. (2005). Using hyperaccumulator plants to phytoextract soil Ni and Cd. *Zeitschrift Naturforschung* 60: 190–198.

49 Meyers, B., Zaitsman, A., Lacroix, B. et al. (2010). Nuclear and plastid genetic engineering of plants: Comparison of opportunities and challenges. *Biotechnology Advances* 6: 747–756.

50 Memon, A.R., Aktopraklıgil, D., Özdemir, A. et al. (2000). *Heavy Metal Accumulation and Detoxification Mechanisms in Plants*. Kocaeli, Turkey: Tübitak MAM, Institute for Genetic Engineering and Biotechnology.

10

Microbial Prospects in Sediment Denitrification of Eutrophic Wetland Ecosystems

Rupak Kumar Sarma and Kamal Choudhury

Department of Botany, SBMS College, Sualkuchi, Assam 781103

10.1 Introduction

Nitrogen (N) has a significant impact on the proper functioning of ecological processes [1]. It is an essential element of many biomolecules required for the primary nutrients of life, which ultimately regulates productivity in different ecosystems. However, in the past few years, human-induced global alternation of the N cycle has led to general eutrophication and stoichiometric unbalances in wetland ecosystems [2]. The terrestrial mechanisms to release N back into the atmosphere to keep reactive N (mostly ammonium and nitrate) within the "safety" threshold has been subject to a thorough investigation [3]. However, the sediment denitrification processes in eutrophic wetland ecosystems need more investigations. The basic knowledge on sediment denitrification and nitrification–denitrification coupling in freshwater ecosystems is limited, including plant–microbe interaction in submerged aquatic vegetation [4]. Macrophytic populations are considered as major dominating plants in eutrophic systems [5]. The highly porous underground biomass of macrophytes is known to capture sediment carbon dioxide (CO_2) in carbon-rich eutrophic systems [6]. The carbon-rich rhizosphere of macrophytes may attract diverse groups of microbial populations, which may play a significant role in maintaining different biogeochemical cycles.

Moreover, concrete investigations are lacking in this aspect. Molecular characterization of nitrifiers and denitrifiers in macrophyte rhizosphere has been poor in urban eutrophic lakes [7]. Similarly, the abundance study of nitrite reducing biomarkers (*nirK* and *nirS*) and bacterial/archaeal ammonia-oxidizing genes (*amoA*-AOA and *amoA*-AOB) needs further improvements. Macrophyte rhizosphere ecosystems may significantly influence the diversity and abundance of *air* and *amoA* genes that enhance the process of denitrification and retard natural eutrophication. Overall, investigations are going on across the globe to understand hidden possibilities. This chapter gives a brief idea about the importance of macrophyte rhizosphere inhabiting nitrifiers and denitrifiers in coupled nitrification and denitrification in eutrophic water ecosystems.

Core Microbiome: Improving Crop Quality and Productivity, First Edition. Edited by Javid A. Parray, Nowsheen Shameem, Elsayed Fathi Abd-Allah, and Mohammad Yaseen Mir.

10.2 Macrophyte Rhizosphere for Habitat Structuring of Microorganisms

Plant rhizosphere is known as the hot spot of microbial colonization [8]. The persistent release of root surface organic compounds enhances active microbial colonization. Macrophytes and phytoplankton (microphytes) are significant plant populations in freshwater ecosystems. Macrophytes grow in water and the land–water transition zone. The dense underground rhizosphere biomass of macrophytes offers a continuous supply of organic carbon, nutrients, and other nutrients to root-inhibiting microbial populations and establishes an organized niche for microbial growth and development [9]. In return, microorganisms ensure mineral nutrients and defensive immunity to the host plant [10]. As the nutrient status has a significant role in distributing aquatic plants and microorganisms, the eutrophic condition is considered to be ideal for the proper growth and development of macrophytes and associated organisms. Similarly, a solid and fistulous stem empowers the root systems to release oxygen and other metabolites into the rhizosphere, which attracts microbial populations [11,12]. Again, high root oxygen-leaking (ROL) macrophyte rhizospheres are reported to be the best habitat for microbial assemblies [13].

The freshwater macrophyte rhizosphere harbors culturable bacterial groups that belong to actinobacteria, alpha-proteobacteria, beta-proteobacteria, gamma-proteobacteria, firmicutes, Bacteroides [14], and archaea [15,16]. However, minimal reports are published on the bacterial diversity of the eutrophic water ecosystem. Zhang et al. reported the dominance of Desulfobacteraceae, Helicobacteraceae, Vibrionaceae, and Woeseiaceae in the rhizosphere sediments of the eutrophic wetland ecosystem responsible for sulfur and nitrogen cycling, respectively [17]. Similarly, the dominance of methanotrophs (both types I and II) were evaluated in the rhizosphere system of eutrophic sites by Cui et al. [18]. Zhang et al. analyzed the sediment bacterial diversity in eutrophic Chinese lakes by metagenome analysis of the *16S rRNA* V4 region and observed the dominance of the *Proteobacteria*, *Acidobacteria*, *Chloroflexi*, and *Nitrospirae* groups [19].

10.3 Influence of Macrophyte Rhizosphere on the Removal of Excessive Nitrogen in the Aquatic Ecosystem

Removal of excess nitrogen from wetlands depends on the macrophyte populations and interchange of aerobic and anaerobic conditions to endorse nitrification and denitrification. In comparison with the water column, sediments perform a higher denitrification rate [20]. The benthic algal and macrophyte populations influence denitrification by releasing oxygen and carbon exudates [21–23]. The submerged macrophytes are regarded as unique aquatic vegetations for their physical structure to link water column and sediment. This arrangement plays a vital role in enriching the microbial populations of root rhizosphere sediment and simultaneously, the sediment nitrogen cycle. Moreover, some specific macrophytic communities, i.e., isoetids, significantly stimulate the sediment redox profile [24]. The high radial oxygen-leaking activity of isoetids is significant to produce sediment biogeochemistry.

Macrophytes play a direct or indirect role in the overall sediment denitrification process. They can provide sufficient surface area with heterogeneous oxygen conditions [25,26] that attract microbial populations required for coupled nitrification and denitrification. Alternatively, the microbial populations could be changed by changing the rhizosphere nutrient concentrations. The changes in nutrient concentrations are due to sequential progress in growth and development, changes in sediment chemistry, and availability of sediment organic carbon [27–29].

Coupled nitrification and denitrification (CND) are significantly crucial in mitigating eutrophication in wetlands. It has been suggested that an increase in microbial species composition was positively correlated with the denitrification rate [30]. Despite classical denitrification, coupled action of nitrifiers and denitrifiers is considered to be a practical pathway for the removal of excess nitrogen in eutrophic lake ecosystems. In an aquatic ecosystem, microbial rhizosphere-mediated CND is dependent upon their sediment counterparts [31]. Nitrification and denitrification are chemically defined as oxidation and reduction processes. Nitrification (conversion of ammonia to nitrite and nitrate) and denitrification (conversion of nitrate into gaseous nitrogen) occurs in aerobic and low-oxygen/anaerobic conditions, respectively. However, nitrite (N_2O) assimilation is highest at the interfaces of nitrification and denitrification. Nature solves this issue by CND. Finally, the nitrite (NO_2^-) or nitrate (NO_3^-) evolved during nitrification is rapidly utilized by the denitrifying bacteria and converted to atmospheric nitrogen.

Moreover, the actual mechanism of CND is still not adequately understood. First of all, nitrification is catalyzed by archaeal and bacterial ammonia oxidizers. Their abundance can be assessed by examining the presence of *amoA* gene. The *amoA* gene encrypts the catalytic subunit of ammonia monooxygenase. Accordingly, the complex denitrification processes are catalyzed by the enzyme toolbox constituting nitrate reductase, nitrite reductase, nitric oxide reductase, and nitrous oxide reductase. However, nitrite reductase (*nirK* and *nirS*) and nitrous oxide reductase (*nosZ*) are used as biomarkers to understand the abundance of denitrifiers. Overall, eutrophication promotes a nitrifying and denitrifying bacterial population with the proportional increase of functional biomarkers. Thus, different trophic levels of wetlands govern the richness/abundance of sediment nitrifiers and denitrifiers [19]. Nutritional status and substrate availability in the sediments play a crucial role in this aspect [32].

Moreover, the role of nitrifiers and denitrifiers in NO and N_2O emission and their impact on maintaining N concentration is not well understood in the eutrophic urban wetland ecosystem. Wang et al. made a preliminary investigation on the cumulative effect of macrophyte and integrated *Elodea nuttallii* immobilized nitrogen cycling bacteria technology on sediment denitrification of eutrophic water systems [33]. Zhao et al. made a comparative study on ammonia oxidizers in the rhizosphere and non-rhizosphere sediments of a eutrophic lake [34]. The investigators recorded the role of ammonia oxidizers (ammonia-oxidizing bacteria and ammonia-oxidizing archaea) in initiating the overall processes of coupled nitrification–denitrification. Damashek et al. studied the community composition of benthic ammonia oxidizers and their nitrification function in the Sacramento–San Joaquin riverine delta [35]. The comprehensive study showed the presence of a regional pattern of gene diversity and phylogenetically distinct ammonia-oxidizing microorganisms.

However, depth investigation on nitrifiers and denitrifiers in freshwater/urban eutrophic systems and their significance in CND is still lacking than in marine and soil environments [36]. Similarly, their distribution patterns are still not clear in lake ecosystems with different ecological types [37].

The knowledge on structural community and species richness of denitrifiers is considered the most important regulator to understand the overall denitrification processes [38]. Previous investigations reported the influence of macrophytes on enhancing the bacterial populations in their rhizospheres [39,40], suggesting the importance of the macrophyte rhizosphere niche in sustaining different biogeochemical cycles. Presently, an extensive study on *nirK* and *nirS* has been performed to understand the community structure of denitrifiers in freshwater and marine ecosystems [41–44]. However, investigating their proper quantification and role of aquatic plant vegetation in modulating denitrification is still much less explored at the molecular level [45,46].

10.3.1 Future Prospects and Conclusion

Fewer studies have been conducted on macrophyte–microbe interactions in sediment denitrification in eutrophic aquatic systems. Further results on exploration and functional characterization of microbial populations associated with macrophyte rhizosphere are required. The microbial diversity enumeration and their functional characterization would be helpful for a selection of efficient macrophytes having the propensity to host stable proportions of nitrifiers and denitrifiers. Efficient macrophytes could be tested for pilot-scale application to check their efficacy toward stable coupling of nitrification and denitrification in eutrophic systems. Further, improvement of the macrophyte rhizosphere system could be carried out through modern biotechnological intervention. Again, the overall findings would also be beneficial to protect and preserve the fragile and often endangered different aquatic plant subspecies.

Acknowledgments

This work was supported by a project sponsored by the Science and Engineering Research Board (Department of Science and Technology, New Delhi, Government of India, YSS/2015/000443).

References

1 Vitousek, P.M., Aber, J.D., Howarth, R.W., Likens, G.E., Matson, P.A., Schindler, D.W., Schlesinger, W.H., and Tilman, D.G. (1997). Human alteration of the global nitrogen cycle: Sources and consequences. *Ecological Applications* 7 (3): 737–750.

2 Mosier, A. and Francis, C. (2008). Relative abundance and diversity of ammonia-oxidizing archaea and bacteria in the San Francisco Bay estuary. *Environmental Microbiology* 10: 3002–3016.

3 McNeill, A. and Unkovich, M. (2007). The Nitrogen Cycle in Terrestrial Ecosystems. In: *Nutrient Cycling in Terrestrial Ecosystems. Soil Biology*, (eds. P. Marschner and Z. Rengel), vol. 10, Berlin, Heidelberg: Springer. https://doi.org/10.1007/978-3-540-68027-7_2.

4 De Vries, F.T. and Bardgett, R.D. (2012). Plant–microbial linkages and ecosystem nitrogen retention: Lessons for sustainable agriculture. *Frontiers in Ecology and Environment* 10: 425–432.

5 Bagousse-Pinguet, Y.L., Liancourt, P., Gross, N., and Straile, D. (2012). Indirect facilitation promotes macrophyte survival and growth in freshwater ecosystems threatened by eutrophication—. *Journal of Ecology* 100: 530–538.

6 Smolders, A.J.P., Lucassen, E.C.H.E.T., and Roelofs, J.G.M. (2002). The isoetid environment: Biogeochemistry and threats. *Aquatic Botany* 73: 325–350.

7 Hou, J., Cao, X., Song, C., and Zhou, Y. (2013). The predominance of ammonia-oxidizing archaea and *nirK*-gene-bearing denitrifiers among ammonia-oxidizing and denitrifying populations in sediments of a large urban eutrophic lake (Lake Donghu). *Canadian Journal of Microbiology* 59: 456–464.

8 Bowen, G.D. and Rovira, A.D. (1991). The rhizosphere, the hidden half of the hidden half. In: *Plant Roots, the Hidden Half*, (eds. Y. Waisel, A. Eshel and U. Kafkafi), 641–629. New York, USA: Marcel Dekker, Inc.

9 Stottmeister, U., Wießner, A., Kuschk, P., Kappelmeyer, U., Ka¨stner, M., Bederski, O., Mu¨ ller, R.A., and Moormann, H. (2003). Effects of plants and microorganisms in constructed wetlands for wastewater treatment. *Biotechnology Advances* 22: 93–117.

10 Srivastava, J.K., Chandra, H., Kalra, S.J.S., Mishra, P., Khan, H., and Yadav, P. (2016). Plant-microbe interaction in aquatic system and their role in the management of water quality: A review. *Applied Water Science* DOI: 10.1007/s13201-016-0415-2.

11 Herrmann, M., Saunders, A.M., and Schramm, A. (2009). Effect of lake trophic status and rooted macrophytes on community composition and abundance of ammonia-oxidizing prokaryotes in freshwater sediments. *Applied and Environmental Microbiology* 75: 3127–3136.

12 Toyama, T., Furukawa, T., Maeda, N., Inoue, D., Sei, K., Mori, K., Kikuchi, S., and Ike, M. (2011). Accelerated biodegradation of pyrene and benzo[a]pyrene in the *Phragmites australis* rhizosphere by bacteria–root exudate interactions. *Water Research* 45 (4): 1629–1638.

13 Hernández-Del Amo, E., Dolinová, I., la Ramis-Jorba, G., Gich, F., and Bañeras, L. (2020). Limited effect of radial oxygen loss on ammonia oxidizers in *Typha angustifolia* root hairs. *Scientific Reports 24* 10 (1): 15694. doi: 10.1038/s41598-020-72653-9.

14 Calheiros, C.S.C., Duque, A.E., Moura, A., Henriques, I.S., Correia, A., Rangel, A.O.S.S., and Castro, P.M.L. (2009). Changes in the bacterial community structure in two-stage constructed wetlands with different plants for industrial wastewater treatment. *Bioresource Technology* 100 (13): 3228–3235.

15 Wang, Y., Inamori, R., Kong, H., Xu, K., Inamori, Y., Kondo, T., and Zhang, J. (2008). Influence of plant species and wastewater strength on constructed wetland methane emissions and associated microbial populations. *Ecological Engineering* 32: 22–29.

16 Wei, B., Yu, X., Zhang, S., and Gu, L. (2011). Comparison of the community structures of ammonia-oxidizing bacteria and archaea in rhizoplanes of floating aquatic macrophytes. *Microbiological Research* 166: 468–474.

17 Zhang, X., Zhao, C., Yu, S., Jiang, Z., Liu, S., Wu, Y., and Huang, X. (2020). Rhizosphere Microbial Community Structure Is Selected by Habitat but Not Plant Species in Two Tropical Seagrass Beds. *Frontiers in Microbiology* 11: 161. DOI: 10.3389/fmicb.2020.00161.

18 Cui, J., Zhao, J., Wang, Z., Cao, W., Zhang, S., Liu, J., and Bao, Z. (2020). Diversity of active root-associated methanotrophs of three emergent plants in a eutrophic wetland in northern China. *AMB Express* 10: 48. https://doi.org/10.1186/s13568-020-00984-x.

19 Zhang, Y., Song, C., Zhou, Z., Cao, X., and Zhou, Y. (2019). Coupling between Nitrification and Denitrification as well as Its Effect on Phosphorus Release in Sediments of Chinese Shallow Lakes. *Water* 11 (9): 1809.

20 Seitzinger, S.P. (1988). Denitrification in freshwater and coastal marine ecosystems-ecological and geochemical significance. *Limnology and Oceanography* 33: 702–724.

21 Veraart, A.J., de Bruijne, W.J., de Klein, J.J., Peeters, E.T., and Scheffer, M. (2011). Effects of aquatic vegetation type on denitrification. *Biogeochemistry* 104: 267–274.

22 Nizzoli, D., Welsh, D.T., Longhi, D., and Viaroli, P. (2014). Influence of *Potamogeton pectinatus* and microphytobenthos on benthic metabolism, nutrient fluxes and denitrification in a freshwater littoral sediment in an agricultural landscape: N assimilation versus N removal. *Hydrobiologia* 737: 183–200.

23 Vila-Costa, M., Pulido, C., Chappuis, E., Calviño, A., Casamayor, E.O., and Gacia, E. (2016). Macrophyte landscape modulates lake ecosystem-level nitrogen losses through tightly coupled plant-microbe interactions. *Limnology and Oceanography* 61: 78–88.

24 Gacia, E., Chappuis, E., Lumbreras, A., Riera, J.L., and Ballesteros, E. (2009). Functional diversity of macrophyte communities within and between Pyrenean lakes. *Journal of Limnology* 68: 25–36.

25 Eriksson, P.G. and Weisner, S.E.B. (1996). Functional differences in epiphytic microbial communities in nutrient-rich freshwater ecosystems: An assay of denitrifying capacity. *Freshwater Biology* 36: 555–562.

26 Eriksson, P.G. and Weisner, S.E.B. (1999). An experimental study on effects of submersed macrophytes on nitrification and denitrification in ammonium-rich aquatic systems. *Limnology and Oceanography* 44: 1993–1999.

27 Carpenter, S.R. and Lodge, D.M. (1986). Effects of submersed macrophytes on ecosystem processes. *Aquatic Botany* 26: 341–370.

28 Weisner, S.E.B., Eriksson, P.G., Graneli, W., and Leonardson, L. (1994). Influence of macrophytes on nitrate removal in wetlands. *Ambio* 23: 363–366.

29 Körner, S. (1999). Nitrifying and denitrifying bacteria in epiphytic communities of submerged macrophytes in a treated sewage channel. *Acta Hydrochimica Et Hydrobiological* 27: 27–31.

30 Groffman, P.M. (2012). Terrestrial denitrification: Challenges and opportunities. *Ecological Processes* 1: 11. https://doi.org/10.1186/2192-1709-1-11.

31 Thamdrup, B. and Dalsgaard, T. (2008). Nitrogen cycling in sediments. In: *Microbial Ecology of the Oceans*, (ed. D.L. Kirchman), 2nd ed., 527–568. John Wiley and Sons.

32 Lisa, J.A., Song, B., Tobias, C.R., and Hines, D.E. (2015). Genetic and biogeochemical investigation of sedimentary nitrogen cycling communities responding to tidal and seasonal dynamics in Cape Fear River Estuary. *Estuarine Coastal and Shelf Science* 167: A313–A323.

33 Wang, Y., Li, Z., Li, Z., Feng, L., Fan, N., and Shen, J. (2013). Effects of macrophyte associated nitrogen cycling bacteria on denitrification in the sediments of the eutrophic Gonghu Bay, Taihu Lake. *Hydrobiologia* 700: 329–341.

34 Zhao, D., Luo, J., Zeng, J., Wang, M., Yan, W., Huang, R., and Wu, Q.L. (2014). Effects of submerged macrophytes on the abundance and community composition of ammonia-oxidizing prokaryotes in a eutrophic lake. *Environmental Science and Pollution Research* 21: 389–398.

35 Damashek, J., Smith, J.M., Mosier, A.C., and Francis, C.A. (2015). Benthic ammonia oxidizers differ in community structure and biogeochemical potential across a riverine delta. *Frontiers in Microbiology* 5: 743.

36 Bollmann, A., Bullerjahn, G.S., and McKay, R.M. (2014). Abundance and Diversity of Ammonia Oxidizing Archaea and Bacteria in Sediments of Trophic End Members of the Laurentian Great Lakes, Erie and Superior. *PLoS ONE* 9 (5): e97068. DOI: 10.1371/journal.pone.0097068.

37 Sun, S., Bao, Z., and Sun, D. (2014). Study on emission characteristics and reduction strategy of nitrous oxide during wastewater treatment by different processes. *Environmental Science and Pollution Research* DOI: 10.1007/s11356-014-3654-5.

38 Peralta, A.L., Matthews, J.W., and Kent, A.D. (2010). Microbial community structure and denitrification in a wetland mitigation bank. *Applied and Environmental Microbiology* 76: 4207–4215.

39 Munch, C., Kuschk, P., and Roske, I. (2005). Root stimulated nitrogen removal: Only a local effect or important for water treatment? *Water Science and Technology* 51 (9): 185–192.

40 Gagnon, V., Chazarenc, F., Comeau, Y., and Brisson, J. (2007). Influence of macrophyte species on microbial density and activity in constructed wetlands. *Water Science and Technology* 56: 249–254.

41 Yan, T., Fields, M.W., Wu, L., Zu, Y., Tiedje, J.M., and Zhou, J. (2003). Molecular diversity and characterization of nitrite reductase gene fragments (*nirK* and *nirS*) from nitrate- and uranium-contaminated groundwater. *Environmental Microbiology* 5: 13–24.

42 Santoro, A.E., Boehm, A.B., and Francis, C.A. (2006). Denitrifier community composition along a nitrate and salinity gradient in a coastal aquifer. *Applied and Environmental Microbiology* 72: 2102–2109.

43 Oakley, B.B., Francis, C.A., Robert, K.J., Fuchman, C.A., Srinivasan, S., and Staley, J.T. (2007). Analysis of nitrite reductase (*nir*K and *nir*S) genes and cultivation reveal depauperate community of denitrifying bacteria in the Black Sea suboxic zone. *Environmental Microbiology* 9 (1): 118–130.

44 Kim, Y.M., Cho, H.U., Lee, D.S., Park, D., and Park, J.M. (2011). Influence of operational parameters on nitrogen removal efficiency and microbial communities in a full-scale activated sludge process. *Water Research* 45: 5785e5795.

45 Mosier, A.C. and Francis, C.A. (2010). Denitrifier abundance and activity across the San Francisco Bay estuary. *Environmental Microbiology Reports* 2: 667–676.

46 Rissanen, A.J., Tilrola, M., and Ojala, A. (2011). Spatial and temporal variation in denitrification and the denitrifier community in a boreal lake. *Aquatic Microbial Ecology* 64 (1): 27–40.

11

Role of Plant Microbiome in Carbon Sequestration for Sustainable Agriculture

Ranjith Sellappan[1], Aswini Krishnan[2], and Kalaiselvi Thangavel[3]

[1]*Department of Agricultural Microbiology, T.N.A.U., Coimbatore*
[2]*Division of Microbiology, ICAR-IARI, New Delhi*
[3]*Department of Agricultural Microbiology, T.N.A.U., Coimbatore*

11.1 Introduction

Carbon dioxide (CO_2) is a predominant greenhouse gas in Earth's atmosphere, seeking much attention worldwide. In preindustrial times, the concentration of atmospheric CO_2 was 280 ppm; presently, it is 412 ppm and expected to reach 600–800 ppm by the end of the century [2,3]. It is essential to reduce the carbon concentration in the atmosphere by controlling carbon emissions and sequestrate carbon for sustainable living on Earth.

Carbon is the abundant element present in Earth's crust, and it is also the primary component of all biomolecules, viz proteins, lipids, nucleic acids, and carbohydrates. Besides being preponderated in the living environment, carbon also prevails in the non-living environment in many forms such as CO_2, natural gas, coal, petroleum, carbonate rocks, the biomass of plants, dissolved in seawater, and soil carbon [4]. By using microorganisms in carbon sequestration, many industries produce carbon-based commercial products that are highly profitable and eco-friendly [5].

Soil organic carbon comes from atmospheric CO_2, trapped by plant photosynthesis and the amount of atmospheric CO_2 capture during photosynthesis converted into biomass is about 120 pentagrams of carbon per year (Pg yr^{-1}) [6]. Plants allocate 40% of photosynthetically fixed carbon to soil through rhizodeposition and root exudation. Root exudates are utilized by soil microbiota, which is attracted to the rhizosphere. Assimilation of rhizodeposition by soil and microbial biomass increases the carbon pool in soil. The routing for the assimilated carbon efflux to the atmosphere is through soil, root, and microbial respiration. Because of soil and microbial respiration, 60 Pg yr^{-1} is returned to the atmosphere [7]. In carbon sequestration, there is nearly a 17–24-fold increase in carbon capture [8]. Autotrophic microbes capture the atmospheric CO_2 for biomass production.

Plant-associated microbes play a pivotal role in carbon capture. Modern agricultural practices change the microbial community and composition; these will simultaneously affect the carbon capturing capacity of microbes [9]. Understanding plant–microbe interaction and carbon sequestration is necessary to reduce carbon concentration in the atmosphere. Huge researchers explain soil carbon sequestrations, but the plant microbes in carbon sequestration are not clearly understood. The objective of this chapter is to

Core Microbiome: Improving Crop Quality and Productivity, First Edition. Edited by
Javid A. Parray, Nowsheen Shameem, Elsayed Fathi Abd-Allah, and Mohammad Yaseen Mir.

elucidate (i) the necessity of carbon sequestration; (ii) role of plant microbes in carbon capture; (iii) tripartite link between agricultural practices, microbes, and soil carbon; (iv) behavior of plant-associated microbes under high CO_2 condition.

11.2 Importance of Carbon Sequestration

Carbon cycle is a very crucial process for maintaining life on Earth, which involves the interconnected continuous exchange of carbon between the atmosphere, organisms, and the planet [10,11]. It has been assessed that the global organic carbon stocks of around 680 billion tons of carbon are found in the first 30 cm of soil. Moreover, this quantity is proportionate to approximately three times the amount of atmospheric carbon [12]. In the terrestrial carbon cycle, about 8% of the carbon pool is balanced between primary production via photosynthesis (by plants as well as photo- and chemo-autotrophic microbes) and heterotrophic respiration (mainly by microbes) [13]. However, the total atmospheric carbon stock would be devoured within 12 years if the soil refrains microbial respiration [14]. On the other hand, the carbon which is left in the atmosphere in greenhouse gases contributes to global warming. Thus, capturing and storing carbon through abiotic and biotic processes is inherent when making them inaccessible to contribute to global heating [15].

The process of trapping the carbon and hoarding them for short or long term is known as *carbon sequestration*, which can decrease the level of CO_2 in the atmosphere [16,17]. The central portion of the sequestered carbon is stored in soils, vegetation (especially forests), oceans, and other geological formations. Moreover, around 75% of the total organic carbon is preserved as organic carbon in soil [18]. The soil organic matter is composed of two different fractions, viz soil organic carbon and soil inorganic carbon. Soil organic carbon comprises a heterogeneous mixture of carbon-based compounds [19], such as plant and animal residues, microbial cells, hummus, charcoal, graphite, and coal.

In contrast, soil inorganic carbon constitutes elemental carbon and carbonate minerals [20]. The retention of carbon varies in the soil with different vegetation, soil, climatic circumstances and drainage. Microorganisms have a central role in sequestering carbon through biomass production and converting soil organic carbon and soil inorganic carbon, which can be exploited by other living organisms [12]. Thus, it is concluded that necessary actions should be taken to protect the natural carbon-rich sources, including soil organic matter that inherently circumvent emission into the atmosphere.

11.3 Prokaryotic Microbes in Carbon Sequestration

Prokaryotic microorganisms existing in the soil, especially bacteria, have their inherent role in soil carbon capture and soil carbon release [21]. The process of bio-sequestration aids the natural sequestration of CO_2 by soil microbes and other photosynthetic organisms [22]. There are two primary functional groups of soil bacteria, copiotrophs and oligotrophs, which are ecologically classified based on their trophic state [23]. Copiotrophs include Bacteroidetes, α and γ Proteobacteria that utilize more labile carbon [24] in nutrient-rich soil, which increases the rate of growth. In contrast, oligotrophs encompass Actinobacteria, Acidobacteria, and δ Proteobacteria, which exploit the recalcitrant forms of carbon and decrease the rate of growth in soils with lesser carbon turnover rates, resulting in low CO_2 emission with more soil carbon sequestration [25]. In agricultural soils, a higher amount of organic carbon in the soil would be retained by changes in certain land management practices, altering the soil microorganism diversity to reduce soil carbon loss [26]. Soil carbon sequestration is augmented by several microbes facilitating the aggregate formation and stabilization [27,28]. Besides, some bacteria can reduce CO_2 and diminish the atmospheric CO_2 levels, thereby sequestering carbon [29,30]. Many studies have reported the ingenuity of bacteria like *Bacillus cereus* [31], *Bacillus pumilus* [80], *Bacillus pasteurii*, and *Bacillus mucilaginosus* [32] in sedimenting carbonates and cosubsidized carbon storage. These reports suggest the use of such bacteria along with other soil microorganisms in potential carbon sequestration. Otherwise, microbes should be genetically engineered to overpower the efficacy of carbon fixation and sequestration. For instance, *Synechococcus elongatus* PCC 7942 was genetically altered to combine CO_2 fixation by the biological system with the chemical transformation of CO_2 [33]. Several archaea and bacteria also participate in global carbon sequestration utilizing CO_2 as a source of carbon and H_2 as a source of energy, thus designated as CO_2-type hydrogenotrophic methanogens. Some of them include *Methanothermobacter thermautotrophicus*, *Methanobrevibacter aboriphilus*, *Methanosarcina marburgensis*, and *Methanothermobacter barkeri* [34].

Microorganisms employ different metabolic pathways for capturing carbon. Autotrophic archaea, such as *Metallosphaera* spp., *Sulfolobus* spp., *Archaeoglobus* spp., and *Cenarchaeum* spp., involve a 3-hydroxypropionate-4-hydroxybutyrate cycle to sequester CO_2 [35]. *Clostridium autoethanogenum* fixes carbon dioxide and carbon monoxide using core metabolite acetyl-CoA by the Wood–Ljungdahl pathway [36]. α-Proteobacteria, such as *Xanthobacter flavus*, *Oligotropha carboxidovorans*, *Rhodobacter capsulatus*, and *Rhodobacter sphaeroides* [37–39]; some β-proteobacteria, such as *Herbaspirillum autotrophicum* and *Ralstonia eutropha*, and some γ-proteobacteria, such as *Hydrogenovibrio marinus*, *Acidithiobacillus thiooxidans*, and *Acidithiobacillus ferrooxidans* [40–43], utilize line Calvin cycle to sequester CO_2.

Like bio-inoculants that are being extensively applied to promote plant development and production, many bacteria and other microorganisms capable of capturing carbon by natural means or by overexpression of their genes can also be employed to sequester higher amounts of carbon, thereby significantly decreasing the effects of global warming.

11.4 Eukaryotic Microbes in Carbon Sequestration

Eukaryotes are unicellular or more complex multicellular organisms, which play a significant role in carbon sequestration. Most of the soil fungi are associated with plants for their nutrient requirement. Fungi uptake plant photosynthetically derived carbon source for hyphal growth. The fungi cell wall consists of chitin and melanin molecules that are very recalcitrant and persist in soil for a prolonged period.

Recently, fewer studies explain the importance of plant fungi in the carbon cycle, and they store 70% of organic carbon in soil (Figure 11.1) [44]. Growth of microbes on leaf litter

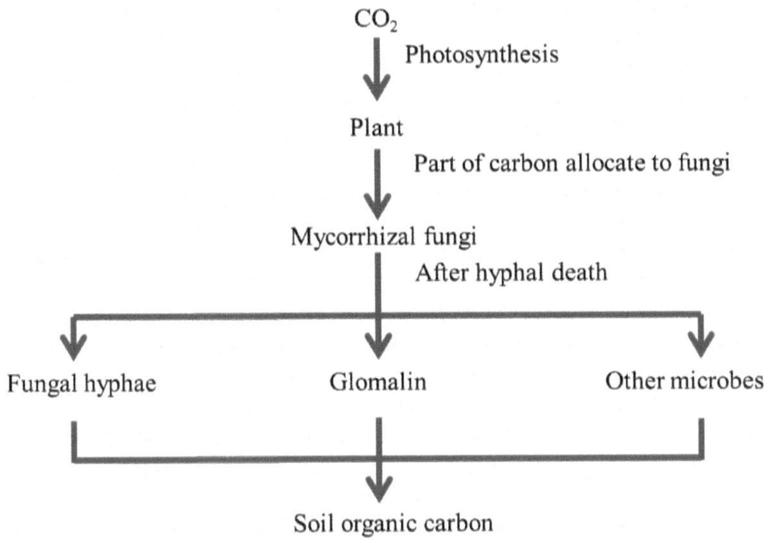

Figure 11.1 Schematic diagram of how CO_2 is transferred from atmosphere to plant, plant to mycorrhiza, and toward mycorrhizal carbon storage.

was evaluated. The result showed carbon storage by fungi was 26 times higher than bacterial cells [45]. Mycorrhiza fungi are one of the symbiotic eukaryotes associated with most plant roots, helping plants with nutrient uptake from soil. Mycorrhiza have a significant role in soil carbon storage.

Arbuscular mycorrhizal fungi (AMF) received carbon stored in the soil as fungal cell wall chitin and glomalin-related soil proteins (GRSP). Compared with plant-originated carbon microbes, accumulated carbon persists longer in soil [46]. Residual hyphal biomass of AMF contributes to carbon sequestration; approximately 54–900 kg ha^{-1} of soil carbon is stored. Further, AMF hyphae interact with soil particles from water-soluble soil aggregates and afford a physical barrier for soil carbon [47].

AMF produce hyphal exudates, namely glomalin (glycoprotein, GRSP), which contains carbohydrate chains of asparagine, organic matter, and aliphatic compounds, viz methylene and methines [48]. GRSP improves soil fertility, aggregation, and metal-binding character [49]. In soil, GRSP has a higher residence period of 6–42 years [50]. In sandy soil and sun, *Glomus claroideum* and *Rhizophagus irregularis*, formerly known as *Glomus intraradices*, biomass contribute approximately 30% of microbial biomass [51]. *Gigaspora margarita* mycelium contributes nearly 16–92% of root biomass [52]. *Acaulospora* spp. and *Glomus* spp. and GRSP contain 1.04–12.24 and 2.85–6.45 mg g^{-1} of two sample areas [53]. Ectomycorrhiza fungi play a pivotal role in carbon capture. Some ectomycorrhizal fungi are proliferating to elevated atmospheric concentration and reduce after nitrogen increase [54]. Tall fescue (grass) often infected with some fungal endophytes, viz *Neotyphodium coenophialum*. These fungal endophytes produce alkaloids that affect herbivore attack, production, and litter decomposition. Aboveground fungal endophytes infection of tall fescue grasses increases the soil carbon and total nitrogen pools, on average by 5–6%. Carbon mineralization rates were higher in the endophyte-infected area (29 mg C^{-1}soil h^{-1}) over the control (26 mg C^{-1}soil h^{-1}). These fungal endophytes improved the soil physical and biological factors. The mutualistic association of these fungal endophytes increases the carbon sequestration potential [55].

Plant photosynthetic rate and soil respiration rate affect soil carbon storage. However, the effect of AMF in plant photosynthesis and soil respiration remains unclear. Wang et al.'s [56] experiment showed the development of photosynthetic rate and soil respiration on AMF-inoculated coalfields. Treatments include inoculation of 100 g AMF per seedling (*Prunus discadenia, Prunus dictyoneura, Xanthoceras sorbifolium, Armeniaca sibirica*) and maintain control. AMF increased photosynthetic rate by 15.3–33.1% and organic carbon storage in soil by 17.2% compared with the control. Easily extractable glomalin and total glomalin were significantly enhanced in AMF-treated soil.

In forest ecosystems, boreal forest soil plays an essential proportion for global carbon. Clemmensen et al. [57] reported that cord-forming ectomycorrhizal and ericoid mycorrhiza play a crucial role in carbon storage. Mycorrhiza possessed an ideal position in carbon and nitrogen storage in the boreal forest ecosystem. Tropical rain forest soil glomalin content was 1.45 Mg C ha^{-1}, contributing to 3.2% of total soil carbon and 5% of total soil nitrogen [58].

Photosynthetic algae capture CO_2 for their biomass production. Algae can use CO_2 through the Calvin Benson Bassham cycle with the key enzyme ribulose 1, 5-bisphosphate carboxylase/oxygenase (RuBisCo). Algal concentration mechanisms of carbon and carbonic anhydrases (CAs) enzymes effectively capture atmospheric CO_2 for their growth (Figure 11.2).

Algae grow commercially for carbon sequestration because they are autotrophic. Algal cultivation reduces atmospheric carbon and has industrial importance, including nutritional supplements, bioplastics, biodiesel, and biochemicals. Algae are also used to treat pollution involving flue gas emissions from wastewater treatment [59]. Albio fertilizers and algal-based products, such as pesticides and herbicides, play a vital role in maintaining agriculture production. Wang et al. [60] evaluated the carbon sequestration potential of nitrogen-fixing albio fertilizers in plastic-covered tunnels. Algal powder (21.0 ton hm^{-2}) fixed carbon 39.4 ton hm^{-2} near factory areas and 1182.6 ton hm^{-2} from rice fields. The approximate reduction of direct emission was 1222 ton hm^{-2}, and the net capacity of CO_2

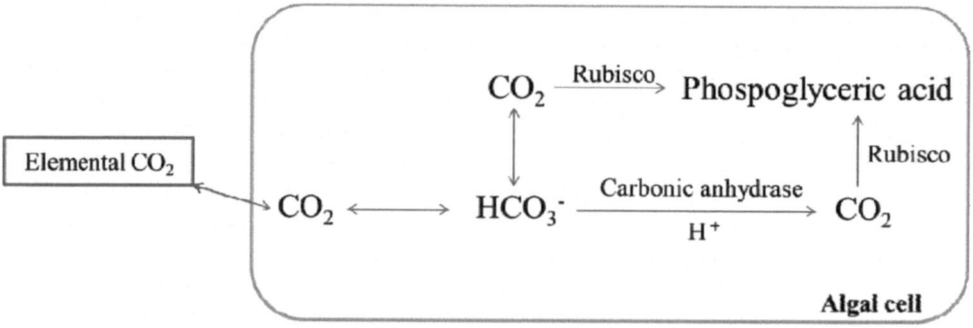

Figure 11.2 The figure depicts the process of carbon dioxide and bicarbonate conversion into phospoglyceric acid with the use of carbonic anhydrase and RuBisCo enzymes and hydrogen ions in an algal cell.

sequestration was 1172.6 ton hm^{-2}. Eukaryotic microbes play the main role in soil carbon capture. Mycorrhiza present ubiquitously in soil, play an important role in soil carbon pool, and improve plant growth and health. The autotrophic nature of algae can capture CO_2 for their metabolism. Soil algae also play a role in carbon capture, and currently, algal carbon capture and bioproduction are some of the trending, commercially promising technologies.

11.5 Agricultural Practices and Carbon Sequestration

Improper agricultural practices negatively affect soil carbon sequestration. The carbon content of agricultural soil is one-third compared with non-disturbed native lands as forests or grasslands. Agricultural practices disturb the biotic and abiotic components of soil. The severe negative causes are loss of organic matter in soil, increase in soil erosion, and effect on soil quality [61], which reduce the capability of soil carbon sequestration. Agricultural land management, comprising cropping system and tillage practices, may affect the physical, chemical, and biological properties of soil.

Alternative agricultural practices, such as minimum tillage, improve soil quality, precisely required fertilizer use, and proper irrigation [62]. In Brazil, the government has been recommending the practice of the zero tillage method in 8 million ha of arable land, and nearly 8 Tg C yr^{-1} has been stored [63]. Nitrogen management in cropland influences soil carbon sequestration, and intensive nitrofertilizer application achieves higher grain yield and carbon capture potential [64]. A cover crop is also a good management practice to enhance carbon sequestration. Planting cover crop as a promising technique to capture carbon in agricultural land has helped capture massive carbon up to 5.3 t C ha^{-1} yr^{-1} [64]. Maize cultivated with cover crop showed increased AMF hyphal density and stored carbon 10–35 mg g^{-1} of soil [65,66]. Amendment of organic fertilizer also increased the carbon sequestration rate [67]. Improper agricultural land use leads to loss of soil carbon because it affects the soil physical, chemical, and biological components.

11.6 Impact of Agricultural Practices on Soil Microbes and Carbon Sequestration

Healthy soil contains huge and diverse microbiota. Microorganisms in soil play a prominent role in the decomposition of carbon-based compounds and intense agricultural practices affect the microbial load and community. Fungi and bacteria highly contribute to the transformation and prolong storage of soil carbon. Any minor changes in soil bacteria and fungi affect the storage and fluxes of carbon. Tillage reduces soil microbiota and enzymatic activity. For example, it severely damages fungal hyphae. No-tillage practices enhance the biomass of fungi quantitatively and also improve soil organic matter [68]. Intensive agricultural practices reduce the organic weight and microbial community change and biomass reduction in soil. An experiment was carried out to analyze microbial community composition in barren land and cultivated land [69].

Long-term application of fertilizers on agricultural land affects microbes both directly and indirectly. Zhong and Cai [70] reported that the continuous application of nitrogen and phosphorus fertilizers have an adverse impact on arable soil microbes. The result showed that microbial biomass carbon ranged from 30 mg g^{-1} in degraded plots to 85 mg g^{-1} in control plots, and 263 mg g^{-1} in NPK applied plots. Following continuous use of inorganic fertilizers for 13 years, the cultivated microorganisms and microbial biomass were significantly increased. The application of inorganic fertilizers does not directly affect the microbial population, but it indirectly affects the microbial load of the soil. Nakhro and Dkhar [71] reported the total microbial biomass carbon present in organic and inorganic fertilizers applied to soil. The result showed that application of organic fertilizer, namely farmyard manure (928 kg $plot^{-1}$), rock phosphate (262 kg $plot^{-1}$), and neem cake (350 kg $plot^{-1}$) resulted in higher microbial load than inorganic fertilizer application, viz urea (250 kg $plot^{-1}$), single superphosphate (525 kg $plot^{-1}$), and muriate of potassium (100 kg $plot^{-1}$). Reduction of microbial community increases the carbon turnover rates; a change in the microbial community affects carbon sequestration [72]. Application of inorganic fertilizers and other modern agricultural practices may affect soil carbon storage. Minimum disturbance in arable land could improve carbon storage, indigenous microbial population, and crop growth.

11.7 Carbon Capture by Plant Microbiome

Plant-associated microbes play a pivotal role in their health and productivity. Beneficial plant microbiome increases plant growth through nutrient acquisition and various biotic and abiotic stress tolerances. Plant microbiome not only provide benefits to plants but also capture carbon (elemental CO_2) for their metabolism. Therefore, researchers have shown interest in plant microbes in carbon capture and storage, and it is an underexplored area of research.

Plants utilize CO_2 for photosynthesis. Plant carbons are transferred to the rhizosphere through rhizodeposition. Organic carbons present in the soil are wholly derived from plants through plant death, root exudation, and root respiration. Plant root exudates contain a wide variety of carbon compounds, such as amino acids, including leucine, tryptophan, proline, cysteine, and glycine; organic acids such as citric, oxalic, fumaric, and acetic acids; sugars such as glucose, galactose, fructose, and maltose; vitamins, namely biotin, thiamine, and niacin; nucleosides; enzymes, such as invertase and amylase; and inorganic ions/gaseous molecules [73]. Carbon transfer from atmosphere to soil depends on gradient maintained as constant removal of carbon exudates by abiotic (soil uptake) and biotic (microbes) components. About 20–30% of total assimilated carbon of cereals is transferred into the soil. In this carbon, half of the carbon is found in the roots and one-third is utilized through root respiration and microbes. In pasture lands, plants translocate 30–50% of carbon to the soil. Plants transfer 5–21% of their photosynthetic carbon into the soil. Maize plants transfer 0.5–10% of fixed carbon into the soil [74]. Microbes utilize hizodeposition and root exudation for their metabolic cell signaling [75].

Plant growth-promoting bacteria (PGPB) colonization improves the health and growth of the plant. Few research reports indicate that some PGPB reduce CO_2 and contribute to carbon sequestration [76]. *Pseudomonas fluorescens* increased plant productivity under high CO_2 condition [29,77]. *Bacillus mucilaginosus* produces CA, capturing atmospheric CO_2 for biomass build-up [78]. Xiao et al. [79] reported that the CA genome from *Bacillus mucilaginosus* has successfully transcribed and translated in *Escherichia coli*. Bacterial species, such as *Bacillus pumilus*, *Bacillus mucilaginous*, *Bacillus cereus*, and *Bacillus pasteurii*, have been reported to capture atmospheric CO_2 [80]. Ahmed et al. reported that PGPB inoculation enhances the plant C : N under elevated CO_2 condition [81]. It is hypothesized that increasing CO_2 also improves rhizodeposition in plants, and higher C : N reduces decomposition (Figure 11.3) [82].

Belowground microbial carbon sequestration is well studied, but the role of aboveground plant microbes in CO_2 capture is still an unexplored area. Microbes associated with aboveground plant parts are present in the phyllosphere, and they provide considerable benefits to plants (stress tolerance, nutrients); autotrophic phyllosphere microbes utilize CO_2, and some microbes use plant carbon (e.g. methanol by methylotrophs) for energy-building [83]. So, phyllosphere microbes also play a role in carbon sequestration, but it is still an underexplored area of research. Research on phyllosphere microbes provides new findings in the microbial ecology field and clarifies the problem associated with carbon sequestration, climate change, etc. [84].

Endophytes are symbiotic microbes that dwell inside the plant tissue. They improve the health and growth of the host plants. Endophytic microorganisms depend on the host plant for their nutrients. Aboveground endophytes may change litter chemistry and root exudate

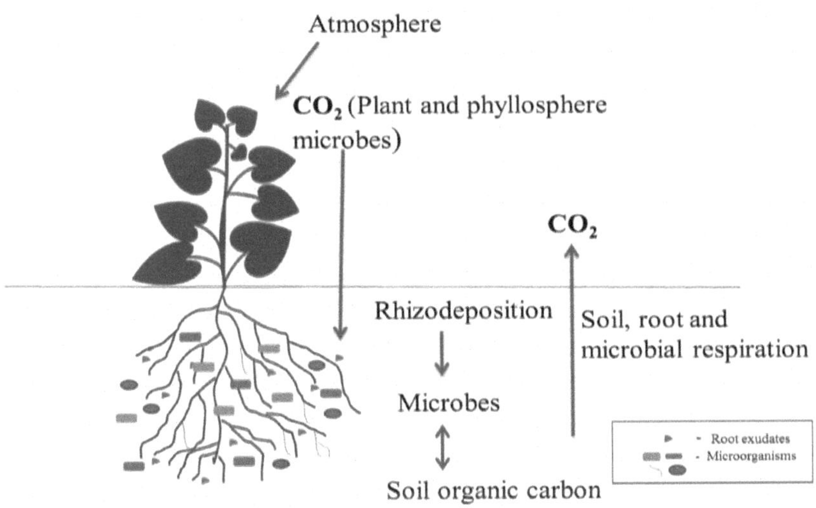

Figure 11.3 Plant and phyllosphere (autotrophic) microbes capture CO_2 and part of the carbon transfer through root from rhizodeposition and through root exudates. Microbes are attracted to rhizodeposition and use the carbon source for their metabolism; after root death, carbon is stored as organic carbon in the soil. The stored carbon returns to environmental CO_2 by soil, root, and microbial respiration.

pattern of their host, which impacts the soil microbial load and community, which, in turn, sequestrates soil carbon [85,86]. *Pseudomonas putida, Burkholderia vietnamiensis, Acinetobacter* sp., *Rhanella* sp., *Sphingomonas* spp., and *Herbaspirillum* sp. endophytic consortium were inoculated in famous willow and sweet corn plants, which increased growth and CO_2 assimilation [87]. Plant halobiont contains a substantial microbial population with diverse characters, and autotrophic microbes can capture atmospheric carbon for their energy. Heterotrophic microbes derive carbon from organic sources for their energy. Plant microbes thereby play an indispensable role in carbon storage.

11.8 Plant–Microbe Interaction in an Elevated CO2 Ecosystem

Day by day, the CO_2 level in the atmosphere has been increasing. In the future, plants must adapt to high CO_2 condition. Uptake of carbon by roots and microbial biomass increases the carbon pool in soil. If any change happens in this, it directly influences atmospheric carbon. Elevated CO_2 condition alters the rhizodeposition of plants, which affects the microbial load and diversity of the rhizosphere. Therefore, it is also necessary to investigate the plant–microbe interaction during high CO_2. *Rhizobium leguminosarum* inoculated in elevated CO_2 (600/350 ppm for 3 years) grew white clover (*Trifolium repens*). The population increased twofold, and nodulation efficiency (17%) also enhanced [88]. *Pseudomonas* spp. inoculated in perennial ryegrass (*Lolium perenne*) grown in elevated CO_2 (60/36 Pa) stimulated siderophore production [89].

Soil microbial community used in *Populus deltoides* tree grown under elevated CO_2 (800 and 1200/400 ppm) condition recorded increased beta-proteobacteria and fungal biomass [90]. There was no significant difference in the colonization ion of *Glomus* sp. in *Plantago lanceolate* under elevated CO_2 condition [91]. A high CO_2 level affects the root, causing infection and hyphal growth of AMF and an increase in hyphal growth (21%) [92]. Some mycorrhizal species show no effect under elevated CO_2 condition. Gavito et al. [93] reported no effect of elevated CO_2 (700 ppm) on AMF colonization in *Pisum sativum*. The effect of colonization of AMF in elevated CO_2 varies based on plant types; for example, the C_3 plant shows no effect, but the C_4 plant shows improved colonization. *Glomus mosseae, Glomus geosporum, Glomus versiforme*, and *Scutellospora* spp. inoculated in C_4 barnyard grass under elevated CO_2 (700/400 ppm) condition showed increased colonization of mycorrhiza in both single (16.8%) and mixed culture (36.9%), increased nutrient content, and productivity [94]. Experiments have been conducted on the carbon budget of serpentine and sandstone grassland ecosystems under elevated CO_2 condition for three years. Under raised CO_2 (200 ppm) conditions, disease incidence also reduced; the ratio of C : N increased by 20%, phenolic, by 15%, and tannins, by 14%, and stomatal conductance decreased by 21–36% in the red maple tree.

Elevated CO_2 increases the carbon uptake, while also rapidly increasing carbon partitioning into various components of the soil. From this study, it can be concluded that the formal CO_2 ecosystem has increased carbon storage [6]. Elevated CO_2 condition improves the microbial community colonization compared with the typical ecosystem, and it also increases the residence time of soil carbon, thus helping in soil carbon storage [7].

11.9 Conclusion

With regard to carbon sequestration, biological sequestration is a possible method to alleviate climate change. Soil and plant microbiome play an imperative role in terrestrial carbon sequestration. Microbial inoculants are used in agricultural practices to obtain the desirable result in soil and plant. The introduction of microbial inoculants into the agroecosystem for carbon sequestration is an effective and eco-friendly method. Microbial composition and community changes that happen during various stresses (agricultural practices, environmental stress, etc.) highly affect the carbon sequestration rate. Many reports indicate that PGPB contribute to climate-change mitigation through carbon capture. Bacteria that promote plant growth can uptake plant-derived and atmospheric carbon for biomass production. Plant-associated soil fungi also can sequester carbon-like bacteria. Fungi sequester more carbon through their hyphae, produce glomalin-like protein, and can form soil aggregates. In the terrestrial carbon sequestration system, soil organic carbon (through plants, soil, and microbes) plays a vital role. Numerous studies on plant, soil, and microbial carbon sequestration have been conducted. Carbon capture through plant microbes is still in the exploratory phase. Understanding the role of the plant microbiome in sequestering carbon will pave the way for soil carbon storage.

References

1 Mann, M.E., Steinman, B.A., and Miller, S.K. (2014). On forced temperature changes, internal variability, and the AMO. *Geophysical Research Letters* 41 (9): 3211–3219.

2 Knohl, A. and Veldkamp, E. (2011). Indirect feedbacks to rising CO_2. *Nature* 475 (7355): 177–178.

3 Buis, A. (2019). The Atmosphere: Getting a Handle on Carbon Dioxide. *N.A.S.A. Global Climate Change*.

4 Shelake, R.M., Waghunde, R.R., Verma, P.P. et al. (2019). Carbon sequestration for soil fertility management: Microbiological perspective. In: *Soil Fertility Management for Sustainable Development* (ed. D. Panpatte and Y. Jhala), 25–42. Singapore: Springer

5 Jajesniak, P., Ali, H., and Wong, T.S. (2014). Carbon dioxide capture utilization using biological systems: Opportunities and challenges. *Journal of Bioprocess Biotech* 4 (3): 1–15.

6 Grover, M., Maheswari, M., Desai, S. et al. (2015). Elevated CO_2: Plant associated microorganisms and carbon sequestration. *Applied Soil Ecology* 95: 73–85.

7 Drigo, B., Kowalchuk, G.A., Knapp, B.A. et al. (2013). Impacts of 3 years of elevated atmospheric CO_2 on rhizosphere carbon flow and microbial community dynamics. *Global Change Biology* 19 (2): 621–636.

8 Kheirfam, H. (2020). Increasing soil potential for carbon sequestration using microbes from biological soil crusts. *Journal of Arid Environments* 172: 104022.

9 Merino, C., Nannipieri, P., and Matus, F. (2015). Soil carbon controlled by plant, microorganism and mineralogy interactions. *Journal of Soil Science and Plant Nutrition* 15 (2): 321–332.

10 Lu, Y. and Conrad, R. (2005). In situ stable isotope probing of methanogenic archaea in the rice rhizosphere. *Science* 309 (5737): 1088–1090.

11 Trumbore, S. (2006). Carbon respired by terrestrial ecosystems–recent progress and challenges. *Global Change Biology* 12 (2): 141–153.

12 Gougoulias, C., Clark, J.M., and Shaw, L.J. (2014). The role of soil microbes in the global carbon cycle: Tracking the below-ground microbial processing of plant-derived carbon for manipulating carbon dynamics in agricultural systems. *Journal of the Science of Food and Agriculture* 94 (12): 2362–2371.

13 Prentice, I.C., Farquhar, G.D. Fasham. et al. (2001). *The Carbon Cycle and Atmospheric Carbon Dioxide.* Cambridge University Press. The Scientific Basis. *Contribution of Working Group I to the Third Assessment Report of the Intergovernmental Panel on Climate Change,* ed. by Houghton JT. Cambridge University Press, Cambridge, pp. 183–238.

14 Jones, D.L., Nguyen, C., and Finlay, R.D. (2009). Carbon flow in the rhizosphere: Carbon trading at the soil–root interface. *Plant and Soil* 321 (1–2): 5–33.

15 Trivedi, P., Anderson, I.C., and Singh, B.K. (2013). Microbial modulators of soil carbon storage: Integrating genomic and metabolic knowledge for global prediction. *Trends in Microbiology* 21 (12): 641–651.

16 Lal, R. (2008). Carbon sequestration. Philosophical Transactions of the Royal Society. *Biological Sciences* 363 (1492): 815–830.

17 Olson, K.R. and Al-Kaisi, M.M. (2015). The importance of soil sampling depth for accurate account of soil organic carbon sequestration, storage, retention and loss. *Catena* 125: 33–37.

18 Mondal, A., Khare, D., Kundu, S. et al. (2017). Spatial soil organic carbon (SOC) prediction by regression kriging using remote sensing data. *The Egyptian Journal of Remote Sensing and Space Science* 20 (1): 61–70.

19 Stout, B., Lal, R., and Monger, C. (2016). Carbon capture and sequestration: The roles of agriculture and soils. *International Journal of Agricultural and Biological Engineering* 9 (1): 1–8.

20 Lal, R. (2004). Agricultural activities and the global carbon cycle. *Nutrient Cycling in Agroecosystems* 70 (2): 103–116.

21 Fang, H., Cheng, S., Wang, Y. et al. (2014). Changes in soil heterotrophic respiration, carbon availability, and microbial function in seven forests along a climate gradient. *Ecological Research* 29 (6): 1077–1086.

22 Macreadie, P.I., Baird, M.E., Trevathan-Tackett, S.M. et al. (2014). Quantifying and modelling the carbon sequestration capacity of seagrass meadows–a critical assessment. *Marine Pollution Bulletin* 83 (2): 430–439.

23 Fierer, N., Bradford, M.A., and Jackson, R.B. (2007). Toward an ecological classification of soil bacteria. *Ecology* 88 (6): 1354–1364.

24 Fierer, N., Lauber, C.L., Ramirez, K.S. et al. (2012). Comparative metagenomic, phylogenetic and physiological analyses of soil microbial communities across nitrogen gradients. *The ISME Journal* 6 (5): 1007–1017.

25 Singh, B.K., Bardgett, R.D., Smith, P. et al. (2010). Microorganisms and climate change: Terrestrial feedbacks and mitigation options. *Nature Reviews Microbiology* 8 (11): 779–790.

26 Finn, D., Kopittke, P.M., Dennis, P.G. et al. (2017). Microbial energy and matter transformation in agricultural soils. *Soil Biology and Biochemistry* 111: 176–192.

27 Woodward, F.I., Bardgett, R.D., Raven, J.A. et al. (2009). Biological approaches to global environment change mitigation and remediation. *Current Biology* 19 (14): 615–623.

28 Lennon, J.T., Aanderud, Z.T., Lehmkuhl, B.K. et al. (2012). Mapping the niche space of soil microorganisms using taxonomy and traits. *Ecology* 93 (8): 1867–1879.

29 Nie, M., Bell, C., Wallenstein, M.D. et al. (2015). Increased plant productivity and decreased microbial respiratory C loss by plant growth-promoting rhizobacteria under elevated CO_2. *Scientific Reports* 5: 9212.

30 Peng, Y.L., Zhao, H.Z., Yang, Q.Z. et al. (2010). Progress in microbial and enzimmobilization of carbon dioxide. *Chemistry & Bioengineering* 27 (7): 10–13.

31 Han, J., Lian, B., and Ling, H. (2013). Induction of calcium carbonate by. *Bacillus Cereus. Geomicrobiology Journal* 30 (8): 682–689.

32 Zhang, Z., Lian, B., Hou, W. et al. (2011). *Bacillus mucilaginosus* can capture atmospheric CO_2 by carbonic anhydrase. *African Journal of Microbiology Research* 5 (2): 106–112.

33 Chen, P.H., Liu, H., Chen, Y.J. et al. (2012). Enhancing CO_2 bio-mitigation by genetic engineering of cyanobacteria. *Energy & Environmental Science* 5 (8): 8318–8327.

34 Rittmann, S., Seifert, A., and Herwig, C. (2015). Essential prerequisites for successful bioprocess development of biological CH_4 production from CO_2 and H_2. *Critical Reviews in Biotechnology* 35: 141–151.

35 Berg, I.A., Kockelkorn, D., Bucke, L.W. et al. (2007). A 3-hydroxypropionate/4-hydroxybutyrate autotrophic carbon dioxide assimilation pathway in Archaea. *Science* 318: 1782–1786.

36 Liew, F., Henstra, A.M., Winzer, K. et al. (2016). Insights into CO_2 fixation pathway of *Clostridium autoethanogenum* by targeted mutagenesis. *mBio* 7: e00427–16.

37 Meijer, W.G. (1994). The Calvin cycle enzyme phosphoglycerate kinase of *Xanthobacter favus* required for autotrophic CO_2 fixation is not encoded by the CBB operon. *Journal of Bacteriology* 176: 6120–6126.

38 Paoli, G.C. and Tabita, F.R. (1998). Aerobic chemolithoautotrophic growth and RubisCO function in *Rhodobacter capsulatus* and a spontaneous gain of function mutant of. *Rhodobacter Sphaeroides. Archives Microbiology* 170: 8–17.

39 Ding, L. and Yokota, A. (2004). Proposals of *Curvibacter gracilis* gen. nov., sp. nov. and *Herbaspirillum putei* sp. nov. for bacterial strains isolated from well water and reclassification of *Pseudomonas huttiensis, Pseudomonas lanceolata, Aquaspirillum delicate* and *Aquaspirillum autotrophicum* as *Herbaspirillum huttiense* comb. nov., *Curvibacter lanceolatuscomb.* nov., *Curvibacter delicatus* comb. nov. and *Herbaspirillum autotrophicum* comb. *International Journal of Systematic Evolutionary Microbiology* 54: 2223–2230.

40 Hugler, M., Huber, H., Stetter, K.O. et al. (2003). Autotrophic CO_2 fixation pathways in archaea (Crenarchaeota). *Archives Microbiology* 179: 160–173.

41 Yoshizawa, Y., Toyoda, K., Arai, H. et al. (2004). CO_2-responsive expression and gorganization of three Ribulose-1, 5-Bisphosphate Carboxylase/ Oxygenase enzymes and carboxysomes in *Hydrogenovibrio marinus* Strain MH-110. *Journal of Bacteriology* 186: 5685–5691.

42 Suzuki, I. and Werkman, C.H. (1958). Chemoautotrophic carbon dioxide fixation by extracts of *Thiobacillus thiooxidans* II formation of phosphoglyceric acid. *Archives of Biochemistry and Biophysics* 77: 112–123.

43 Wang, W., Xiao, S., Chao, J. et al. (2008). Regulation of CO_2 fixation gene expression in *Acidithiobacillus ferrooxidans* ATCC 23270 by Lix984n Shock. *Journal of Microbiology and Biotechnology* 18: 1747–1754.

44 Averill, C. and Hawkes, C.V. (2016). Ectomycorrhizal fungi slow soil carbon cycling. *Ecology Letters* 19 (8): 937–947.

45 Mehar, S.K. and Sundaramoorthy, S. (2018). Carbon Sequestration and the Significance of Soil Fungi in the Process. In: *Fungi and Their Role in Sustainable Development: Current Perspectives* (ed. Praveen Gehlot and Joginder Singh), 467–482. Singapore: Springer

46 Prescott, C.E. (2010). Litter decomposition: What controls it and how can we alter it to sequester more carbon in forest soils. *Biogeochemistry* 101 (1–3): 133–149.

47 Zhu, Y.G. and Miller, R.M. (2003). Carbon cycling by arbuscular mycorrhizal fungi in soil-plant systems. *Trends in Plant Science* 8 (9): 407–409.

48 Rillig, M.C. (2004). Arbuscular mycorrhizae, glomalin, and soil aggregation. *Canadian Journal of Soil Science* 84 (4): 355–363.

49 Schindler, F.V., Mercer, E.J., and Rice, J.A. (2007). Chemical characteristics of glomalin-related soil protein (GRSP) extracted from soils of varying organic matter content. *Soil Biology and Biochemistry* 39 (1): 320–329.

50 Parihar, M., Rakshit, A., Meena, V.S. et al. (2020). The potential of arbuscular mycorrhizal fungi in C cycling: A review. *Archives of Microbiology*, 1–16.

51 Olsson, P.A. and Wilhelmsson, P. (2000). The growth of external AM fungal mycelium in sand dunes and in experimental systems. *Plant and Soil* 226 (2): 161.

52 Fujiyoshi, M., Nakatsubo, T., Ogura, S. et al. (2000). Estimation of mycelial biomass of arbuscular mycorrhizal fungi associated with the annual legume *Kummerowia striata* by ergosterol analysis. *Ecological Research* 15 (2): 121–131.

53 Violi, H.A., Barrientos-Priego, A.F., Wright, S.F. et al. (2008). Disturbance changes arbuscular mycorrhizal fungal phenology and soil glomalin concentrations but not fungal spore composition in montane rainforests in Veracruz and Chiapas, Mexico. *Forest Ecology and Management* 254 (2): 276–290.

54 Farley, K.A., Bremer, L.L., Harden, C.P. et al. (2013). Changes in carbon storage under alternative land uses in biodiverse Andean grasslands: Implications for payment for ecosystem services. *Conservation Letters* 6 (1): 2127.

55 Iqbal, J., Siegrist, J.A., Nelson, J.A. et al. (2012). Fungal endophyte infection increases carbon sequestration potential of southeastern U.S.A. tall fescue stands. *Soil Biology and Biochemistry* 44 (1): 81–92.

56 Wang, F., Liu, X., Shi, Z. et al. (2016). Arbuscular mycorrhizae alleviate negative effects of zinc oxide nanoparticle and zinc accumulation in maize plants – a soil microcosm experiment. *Chemosphere* 147: 88–97.

57 Clemmensen, K.E., Finlay, R.D., Dahlberg, A. et al. (2015). Carbon sequestration is related to mycorrhizal fungal community shifts during long-term succession in boreal forests. *New Phytologist* 205 (4): 1525–1536.

58 Lovelock, C.E., Wright, S.F., Clark, D.A. et al. (2004). Soil stocks of glomalin produced by arbuscular mycorrhizal fungi across a tropical rain forest landscape. *Journal of Ecology* 92(2): 278–287.

59 Bai, X., Huang, Y., Ren, W. et al. (2019). Responses of soil carbon sequestration to climate-smart agriculture practices: A meta-analysis. *Global Change Biology* 25 (8): 2591–2606.

60 Wang, M., Joel, A.S., Ramshaw, C. et al. (2015). Process intensification for post-combustion CO_2 capture with chemical absorption: A critical review. *Applied Energy* 158: 275–291.

61 Bronick, C.J. and Lai, R. (2005). Soil structure and management: A review. *Geoderma* 124 (1–2): 3–22.

62 Ventorino, V., Pascale, A., Fagnano, M. et al. (2019). Soil tillage and compost amendment promote bioremediation and biofertility of polluted area. *Journal of Cleaner Production* 239: 118087.

63 Corbeels, M., Marchao, R.L., Neto, M. et al. (2016). Evidence of limited carbon sequestration in soils under no-tillage systems in the Cerrado of Brazil. *Scientific Reports* 6: 21450.

64 Dolan, M.S., Clapp, C.E., and Allmaras, R.R. (2006). Soil organic carbon and nitrogen in a Minnesota soil as related to tillage, residue and nitrogen management. *Soil and Tillage Research* 89 (2): 221–231.

65 Vicente-Vicente, J.L., García-Ruiz, R., Francaviglia, R. et al. (2016). Soil carbon sequestration rates under Mediterranean woody crops using recommended management practices: A meta-analysis. *Agriculture, Ecosystems & Environment* 235: 204–214.

66 Kabir, Z. and Koide, R.T. (2002). Effect of autumn and winter mycorrhizal cover crops on soil properties, nutrient uptake and yield of sweet corn in Pennsylvania, U.S.A. *Plant and Soil* 238 (2): 205–215.

67 Farooqi, Z.U.R., Sabir, M., Zeeshan, N. et al. (2018). Enhancing carbon sequestration using organic amendments and agricultural practices. Carbon Capture, Utilization and Sequestration.

68 Six, J. and Paustian, K. (2014). Aggregate-associated soil organic matter as an ecosystem property and a measurement tool. *Soil Biology and Biochemistry* 68: A4–A9.

69 Bastida, F., Moreno, J.L., Hernandez, T. et al. (2006). Microbiological degradation index of soils in a semiarid climate. *Soil Biology and Biochemistry* 38 (12): 3463–3473.

70 Zhong, W.H. and Cai, Z.C. (2007). Long-term effects of inorganic fertilizers on microbial biomass and community functional diversity in a paddy soil derived from quaternary red clay. *Applied Soil Ecology* 36 (2–3): 84–91.

71 Nakhro, N. and Dkhar, M.S. (2010). Populations and biomass carbon in paddy field soil. *Agronomy Journal* 9: 102–110.

72 Vesterdal, L., Elberling, B., Christianse, J.R. et al. (2012). Soil respiration and rates of soil carbon turnover differ among six common European tree species. *Forest Ecology and Management* 264: 185–196.

73 Kumar, R., Pandey, S., and Pandey, A. (2006). Plant roots and carbon sequestration. *Current Science* 885–890.

74 Kuzyakov, Y. and Domanski, G. (2000). Carbon input by plants into the soil- Review. *Journal of Plant Nutrition and Soil Science* 163 (4): 421–431.

75 Wu, H., Lin, L., Zhang, J. et al. (2016). Purification ability and carbon dioxide flux from surface flow constructed wetlands treating sewage treatment plant effluent. *Bioresource Technology* 219: 768–772.

76 Li, J., Roche, B., Bull, J.M. et al. (2020). Passive acoustic monitoring of a natural CO_2 seep site–Implications for carbon capture and storage. *International Journal of Greenhouse Gas Control* 93: 102899.

77 Nie, M. and Pendall, E. (2016). Do rhizosphere priming effects enhance plant nitrogen uptake under elevated CO_2? *Agriculture, Ecosystems & Environment* 224: 50–55.
78 Zhang, F., Pant, D., and Logan, B.E. (2011). Long-term performance of activated carbon air cathodes with different diffusion layer porosities in microbial fuel cells. *Biosensors and Bioelectronics* 30 (1): 49–55.
79 Xiao, L., Lian, B., Hao, J., Liu, C., and Wang, S. (2015). Effect of carbonic anhydrase on silicate weathering and carbonate formation at present day CO_2 concentrations compared to primordial values. *Scientific reports* 5(1): 1–10.
80 Komala, T. and Khun, T.C. (2014). Biological carbon dioxide sequestration potential of. *Bacillus Pumilus. Sains Malaysiana* 43 (8): 1149–1156.
81 Ahmed, I.U., Mengistie, H.K., Godbold, D.L. et al. (2019). Soil moisture Integrates the influence of land use and season on soil microbial community composition in the Ethiopian highlands. *Applied Soil Ecology* 135: 85–90.
82 Wu, C.H., Bernard, S.M., Andersen, G.L. et al. (2009). Developing microbe–plant interactions for applications in plant-growth promotion and disease control, production of useful compounds, remediation and carbon sequestration. *Microbial Biotechnology* 2 (4): 428–440.
83 Vacher, C., Hampe, A., Porte, A.J. et al. (2016). The phyllosphere: Microbial jungle at the plant–climate interface. *Annual Review of Ecology, Evolution, and Systematics* 47: 1–24.
84 Thapa, S., Prasanna, R., Ranjan, K. et al. (2017). Nutrients and host attributes modulate the abundance and functional traits of phyllosphere microbiome in rice. *Microbiological Research* 204: 55–64.
85 Franzluebbers, A.J. and Stuedemann, J.A. (2002). Particulate and non-particulate fractions of soil organic carbon under pastures in the Southern Piedmont U.S.A. *Environmental Pollution* 116: S53–S62.
86 Sahoo, U.K., Singh, S.L., Gogoi, A. et al. (2019). Active and passive soil organic carbon pools as affected by different land-use types in Mizoram, Northeast. *India. PloS One* 14 (7): e0219969.
87 Knoth, J.L., Soo – Hyungkim, Gregoryj, E.T.T.L. et al. (2012). Effects of cross host species inoculation of nitrogen-fixing endophytes on growth and leaf physiology of maize. *Bioenergy* 75 (2): 407–418.
88 Montealegre-Gracia, A.L., Lamelas-Gracia, M.T., Garcia-Martin, A. et al. (2017). Using low-density discrete Airborne Laser Scanning data to assess the potential carbon dioxide emission in case of a fire event in a Mediterranean pine forest. *G.I. Science & Remote Sensing* 54 (5).
89 Tarnawski, S. and Aragno, M. (2006). The influence of elevated [CO_2] on diversity, activity and biogeochemical functions of rhizosphere and soil bacterial communities. In: *Managed Ecosystems and CO_2* (ed. J. Nosberger, S.P. Long, R.J. Norby, M. Stitt, G.R. Hendrey, H. Blum), 393–412. Berlin: Springer.
90 Lipson, D.A., Blair, M., Barron-Gafford, G. et al. (2006). Relationships between microbial community structure and soil processes under elevated atmospheric carbon dioxide. *Microbial Ecology* 51 (3): 302–314.
91 Staddon, P.L., Gregersen, R., and Jakobsen, I. (2004). The response of two *Glomus mycorrhizal* fungi and a fine endophyte to elevated atmospheric CO_2, soil warming and drought. *Global Change Biology* 10 (11): 1909–1921.

92 Alberton, O., Kuyper, T.W., and Gorissen, A. (2005). Taking mycocentrism seriously: Mycorrhizal fungal and plant responses to elevated CO_2. *New Phytologist* 167 (3): 859–868.

93 Gavito, M. E., Curtis, P. S., Mikkelsen, T. N., & Jakobsen, I. (2000). Atmospheric CO2 and mycorrhiza effects on biomass allocation and nutrient uptake of nodulated pea (Pisum sativum L.) plants. *Journal of Experimental Botany* 51(352): 1931–1938.

94 Tang, J., Xu, L., Chen, X. et al. (2009). Interaction between C_4 barnyard grass and C_3 upland rice under elevated CO_2: Impact of mycorrhizae. *Acta Oecologia* 35 (2): 227–235.

12

Functions and Emerging Trends of the Microbial Community in Heavy Metals Bioremediation

A Review

Nida Idrees, Robeena Sarah, and Baby Tabassum

Toxicology Laboratory, Department of Zoology, Government Raza P.G. College Rampur, 244901, Uttar Pradesh, India

12.1 Introduction

The distribution of microorganisms is widely organized in the atmosphere because of their significant metabolic activity. They can survive easily under unfavorable environmental conditions. Microorganisms have specific nutritional versatility, which can also be exploited for the biodegradation of toxicants. The procedure is defined as bioremediation. The process may be executed by the capability of various microorganisms that transform, remodel, and utilize toxic contaminants to obtain energy and generate biomass [1]. Bioremediation could be a microbiological and well-performed strategic mechanism enforced to reconstruct or break down pollutants into low-toxic or nontoxic substances despite merely accumulating pollutants and procuring them for further applications. Biological agents are applicable for the process of bioremediation to clean up contaminated areas. The examples of typical prime bioremediation involve microorganisms, including bacteria, archaea, algae, and fungi [2].

Bioremediation is a biotechnological process that involves particular microorganisms to solve and clear specific threats created by toxicants from the environment via biodegradation. Both approaches, i.e. biodegradation and bioremediation, are related and interchangeable in terms of working mechanisms. Microorganisms act as specific toxicant-remediation tools in water, sediments, and soil because of their significance with regard to alternative remedy procedural protocols. Microorganisms are helpful because they restore or transform polluted natural surroundings and are helpful in mitigating various types of pollution [3]. This review aims to enunciate the present trends, methods, strategies, and applications of microorganisms within bioremediation and devote it to the relevant background to fill the associated gaps in this specific field. It is currently a trending research field because microorganisms are eco-friendly and have a promising and valuable genetic material to mitigate various environmental threats rising from complex forms of pollution.

Core Microbiome: Improving Crop Quality and Productivity, First Edition. Edited by
Javid A. Parray, Nowsheen Shameem, Elsayed Fathi Abd-Allah, and Mohammad Yaseen Mir.

12.2 Heavy Metals Toxicity: A Threat to the Biosphere

The adulteration of toxic metals in water and soil causes a threat to the biosphere. It adversely affects not only water quality and soil fertility but also agricultural production directly or indirectly, which impacts human health. Heavy metal contamination is taken into account as a consequence of expeditious industrialization, urbanization, and intensive agricultural farming, in addition to mining, manufacturing, municipal wastes, excessive irrigation, and uncontrolled use of herbicides and fungicides.

Hence, the reports from the U.S. Environmental Protection Agency have alerted that around forty million loads of waste products have been created annually from within the United States alone. Moreover, the Agency for Toxic Substance and Disease Register (ATSDR) issued a list of 275 compounds referred to as priority substances. ATSDR listed arsenic as number one followed by lead, cadmium listed at number seven, cobalt at fifty-one, nickel at fifty-seven, and zinc at seventy-five. These elements cause serious health hazards to humans and are generally found in hazardous waste [4].

Historically, toxic metals and metalloids, mainly As, Pb, Cd, Co, Zn, and Hg, have significant adverse impacts on human health because of their widespread poisoning [5]. Many case studies and investigations related to these toxic metal pollutions are reported in almost every nation due to their severe impact, e.g. rice is a staple food in Bangladesh and India, reportedly countries with primary arsenic contamination sources. Due to arsenic contamination, people of these countries (around 100 million) are at a very high risk of metal toxicity [6]. A report indicated that cadmium accumulation was found within the kidneys of livestock in New Zealand [5]. Similarly, India also reported the presence of these toxic metals beyond the threshold levels in grazing animals, particularly in muscle and other tissues used as meat, with an ultimate outcome of intense economic losses and severe human toxicity [7].

12.3 Factors Affecting Microbial Bioremediation

The mechanism of bioremediation by microorganisms works against pollutants only if they are compatible with specific materials and some compounds that can assist them in providing energy and nutrition for many new cells. The process relies upon many factors, such as the concentration and chemical structure of a particular pollutant, the psycho-chemical nature of the surroundings, and their availability to microorganisms for the new and additional procedures [8]. The universal fact that commonly influences the speed of degradation is that microorganisms and pollutants do not contact one another; additionally, pollutants and microbes do not seem to be evenly unfolding within the surroundings.

The optimization and regulation of the bioremediation procedure could involve a complicated system of many factors, such as temperature, pH, moisture, the structure of the soil, in water, available nutrients, reaction potential and content of oxygen, chemical element bioavailability, the inadequacy of skilled human resources in the related field of specialization and physico-chemical bioavailability of various pollutants, such as solubility, concentration, solubility type, chemical toxicity, and structure, which determine the degradation mechanics [9]. The wide range of pH typically influences biodegradation; therefore, a pH in the range of 6.5–8.5 is principally optimal for further procedure in different terrestrial and aquatic systems. Moreover, moisture content also affects the pollutant metabolism rate due to its kind, form, and quantity of soluble materials, which are accessible and also the pH and osmotic pressure required in aquatic and terrestrial systems [10].

12.4 Different Microorganisms and Related Pollutants

Heavy metals can neither be degraded nor be destroyed biologically; although changes might occur within the nuclear element structure, they can be transformed from one oxidation state into another. Moreover, bacteria are considerably adequate substitutes in weighty, heavy metals bioremediation. Microorganisms are capable of safeguarding themselves from the toxicity of heavy metals by different mechanisms, such as bioaccumulation, uptake, adsorption, methylation, oxidization, and reduction [11].

Microorganisms may uptake these toxic metals actively (bioaccumulation) and/or passively (adsorption). The alternate technique referred to as microbial methylation plays a vital role in remediating heavy metals; however, the method is based on the volatile nature of methylated compounds. For example, mercury/Hg (II) is often biomethylated by many different species of bacteria (Table 12.1). Some groups or species of microorganisms also take part in oil bioremediation (Table 12.2)

Table 12.1 Bioremediation of heavy metals by specific microorganisms.

Microorganisms	Heavy metals	References
Aerococcus sp., *Rhodopseudomonas palustris*	Cd, Cr, Pb	[12,13]
Aeromonas sp., *Pseudomonas aeruginosa*	Cr, Cu, Ni, U	[14]
Aspergillus fumigatus, A. versicolor, Cladosporium sp., *Microsporum* sp., *Paecilomyces* sp., *Trichoderma* sp.	Cd	[15]
Bacillus safensis (JX126862) strain *(RSA-4* and *PB-5)*	Cd	[16]
Cunninghamella elegans	Heavy metals	[17]
Geobacter spp.	Fe (III), U (VI)	[18]
Lysinibacillus sphaericus CBAM5	Co, Cr, Cu, Pb	[19]
Microbacterium profundi strain *Shh49T*	Fe	[20]
Pseudomonas aeruginosa and *Pseudomonas fluorescens*	Cu, Fe^{2+}, Mn^{2+}, Pb^{2+}, Zn^{2+}	[21]
Saccharomyces cerevisiae	Hg, Ni, Pb, heavy metals	[22–24]

Table 12.2 Microorganisms as oil remediators.

Microorganisms	Compound	References
Alcaligenes odorans, Pseudomonas aeruginosa, Bacillus subtilis, Corynebacterium propinquum	Oil	[25]
Aspergillus niger, Candida krusei, Candida glabrata, and *Saccharomyces cerevisiae*	Crude oil	[26]
B. licheniformis, B. brevis, P. aeruginosa KH6, and *B. sphaericus*	Crude oil	[27]
Bacillus cereus A	Diesel oil	[28]
Fusarium sp.	Oil	[29]
Pseudomonas aeruginosa, Arthobacter, P. putida sp., and *Bacillus* sp.	Diesel oil	[30]
Pseudomonas cepacia, Bacillus cereus, Bacillus coagulans, Serratia ficaria, and *Citrobacter koseri*	Diesel oil, crude oil	[31]

12.5 Conclusion

Biodegradation is an effective option to repair, clean, recover, and mitigate environmental pollution with the help of the microbial community. The intensity of unwanted venturesome waste substances and degradation is considered to be in competition with alternative biological agents. It is inadequate to provide essential nutrients and a few uncomfortable external abiotic conditions, such as pH, moisture, aeration, temperature, and low bioavailability of a waste product. The above factors suggested that biodegradation in the natural state is a smaller amount of triple crown that has been accepted as the less favorable method. Therefore, bioremediation is often effectively used wherever surroundings might allow microbial growth and regulate metabolic activity. Bioremediation has been elaboratively used in recent years in several sites globally, with varying degrees of success. Moreover, the benefits from the technique performed are larger than the disadvantages evident from the number of elite sites that favor the use of this technology; it is also continuously gaining popularity with time. Different species explored from various sites are generally effective in their activity-control mechanisms.

Nowadays, it is a perpetual challenge for researchers and scientists to get rid of these toxic metals and scale back them to below the threshold limits, beneath the favorable conditions with the help of specific microbes, plants, and alternative biological processes. The increased innovative technologies supporting microbes and associated plants might give a lot of advanced strategies to cut back on environmental and human health risks directly related to severe heavy metal contamination.

References

1 Tang, C.Y., Criddle, C.S., and Leckie, J.O. (2007). Effect of flux (transmembrane pressure) and membranes properties on fouling and rejection of reverse osmosis and nanofiltration membranes treating perfluorooctane sulfonate containing wastewater. *Environmental Science & Technology* 41: 2008–2014.

2 Strong, P.J. and Burgess, J.E. (2008). Treatment methods for wine-related ad distillery wastewaters: A review. *Bioremediation Journal* 12: 70–87.

3 Demnerova, K., Mackova, M., Spevakova, V., Beranova, K., Kochankova, L. et al. (2005). Two approaches to biological decontamination of groundwater and soil polluted by aromatics characterization of microbial populations. *International Microbiology* 8: 205–211.

4 Glick, B.R. (2015). *Beneficial Plant–bacterial Interactions*. Switzerland: *Springer* International Publishing.

5 Bolan, N., Kunhikrishnan, A., Thangarajan, R., Kumpiene, J., Park, J., Makino, T. et al. (2014). Remediation of heavy metal(loid)s contaminated soils—To mobilize or to immobilize? *Journal of Hazardous Materials* 266 (4): 141–166.

6 Bhattacharya, S., Gupta, K., Debnath, S., Ghosh, U.C., Chattopadhyay, D., and Mukhopadhyay, A. (2012). Arsenic bioaccumulation in rice and edible plants and subsequent transmission through food chain in Bengal basin: A review of the perspectives for environmental health. *Toxicological and Environmental Chemistry* 94 (3): 1–13.

7 Rajaganapathy, V., Xavier, F., Sreekumar, D., and Mandal, P.K. (2011). Heavy metal contamination in soil, water and fodder and their presence in livestock and products: A review. *Journal of Environmental Science and Technology* 4 (3): 234–249.

8 El Fantroussi, S. and Agathos, S.N. (2005). Is bioaugmentation a feasible strategy for pollutant removal and site remediation? *Current Opinion in Microbiology* 8: 268–275.

9 Madhavi, G.N. and Mohini, D.D. (2012). Review paper on – Parameters affecting bioremediation. *International Journal of Life Science and Pharma Research* 2: 77–80.

10 Cases, I. and de Lorenzo, V. (2005). Genetically modified organisms for the environment: Stories of success and failure and what we have learned from them. *International Microbiology* 8: 213–222.

11 Abatenh, E., Gizaw, B., Tsegaye, Z., and Wassie, M. (2017). The role of microorganisms in bioremediation- A review. *Open Journal of Environmental Biology* 2 (1): 38–46.

12 Sinha, S.N. and Biswas, K. (2014). Bioremediation of lead from river water through lead-resistant purple-nonsulfur bacteria. *Global Journal of Microbiology and Biotechnology* 2: 11–18.

13 Sinha, S.N. and Paul, D. (2014). Heavy metal tolerance and accumulation by bacterial strains isolated from wastewater. *Journal of Chemical, Biological and Physical Sciences* 4: 812–817.

14 Sinha, S.N., Biswas, M., Paul, D., and Rahaman, S. (2011). Biodegradation potential of bacterial isolates from tannery effluent with special reference to hexavalent chromium. *Biotechnology Bioinformatics and Bioengineering* 1: 381–386.

15 Soleimani, N., Fazli, M.M., Mehrasbi, M., Darabian, S., Mohammadi, J. et al. (2015). Highly cadmium tolerant fungi: Their tolerance and removal potential. *Journal of Environmental Health Science and Engineering* 13: 1–9.

16 Priyalaxmi, R., Murugan, A., Raja, P., and Raj, K.D. (2014). Bioremediation of cadmium by *Bacillus safensis* (JX126862), a marine bacterium isolated from mangrove sediments. *International Journal of Current Microbiology and Applied Sciences* 3: 326–335.

17 Tigini, V., Prigione, V., Giansanti, P., Mangiavillano, A., Pannocchia, A. et al. (2010). Fungal biosorption, an innovative treatment for the decolourization and detoxification of textile effluents. *Water* 2: 550–565.

18 Mirlahiji, S.G. and Eisazadeh, K. (2014). Bioremediation of Uranium by *Geobacter* spp. *Journal of Research and Development* 1: 52–58.

19 Pena-Montenegro, T.D., Lozano, L., and Dussán, J. (2015). Genome sequence and description of the mosquitocidal and heavy metal tolerant strain *Lysinibacillus sphaericus CBAM5. Standards in Genomic Sciences* 10: 1–10.

20 Wu, Y.H., Zhou, P., Cheng, H., Wang, C.S., and Wu, M. (2015). Draft genome sequence *of Microbacterium profundi Shh49T*, an Actinobacterium isolated from deep-sea sediment of a polymetallic nodule environment. *Genome Announcements* 3: 1–2.

21 Paranthaman, S.R. and Karthikeyan, B. (2015). Bioremediation of heavy metal in paper mill effluent using *Pseudomonas* spp. *International Journal of Microbiology* 1: 1–5.

22 Chen, C. and Wang, J.L. (2007). Characteristics of Zn2+ biosorption by *Saccharomyces cerevisiae. Biomedical and Environmental Sciences* 20: 478–482.

23 Talos, K., Pager, C., Tonk, S., Majdik, C., Kocsis, B. et al. (2009). Cadmium biosorption on native *Saccharomyces cerevisiae* cells in aqueous suspension. *Acta Universitatis Sapientiae, Agriculture and Environment* 1: 20–30.

24 Infante, J.C., De Arco, R.D., and Angulo, M.E. (2014). Removal of lead, mercury and nickel using the yeast. *Saccharomyces Cerevisiae. Revista MVZ Córdoba* 19: 4141–4149.

25 Singh, A., Kumar, V., and Srivastava, J.N. (2013). Assessment of bioremediation of oil and phenol contents in refinery waste water via bacterial consortium. *Journal of Petroleum & Environmental Biotechnology* 4: 1–4.

26 Burghal, A.A., Najwa, M.J.A., and Al-Tamimi, W.H. (2016). Mycodegradation of crude oil by fungal species isolated from petroleum contaminated soil. *International Journal of Innovative Research in Science, Engineering and Technology* 5: 1517–1524.

27 El-Borai, A.M. Eltayeb, K.M., Mostafa, A.R., and El-Assar, S.A. (2016). Biodegradation of industrial oil-polluted wastewater in Egypt by bacterial consortium immobilized in different types of carriers. *Polish Journal of Environmental Studies* 25: 1901–1909.

28 Maliji, D., Olama, Z., and Holail, H. (2013). Environmental studies on the microbial degradation of oil hydrocarbons and its application in Lebanese oil-polluted coastal and marine ecosystem. *International Journal of Current Microbiology and Applied Sciences* 2: 1–18.

29 Hidayat, A. and Tachibana, S. (2012). Biodegradation of aliphatic hydrocarbon in three types of crude oil by Fusarium sp. F092 under stress with artificial seawater. *Journal of Environmental Science and Technology* 5: 64–73.

30 Sukumar, S. and Nirmala, P. (2016). Screening of diesel oil-degrading bacteria from petroleum hydrocarbon contaminated soil. *International Journal of Advanced Research in Biological Sciences* 3: 18–22.

31 Kehinde, F.O. and Isaac, S.A. (2016). Effectiveness of augmented consortia of *Bacillus coagulans, Citrobacter koseri* and *Serratia ficaria* in the degradation of diesel polluted soil supplemented with pig dung. *African Journal of Microbiology Research* 10: 1637–1644.

13

Microbiomics and Sustainable Agriculture

New Frontiers

Shabeer Ahmad Dar[1], Mohammad Yaseen Mir[1], Azra N. Kamili[1], Irshad Ahmad Nawchoo[2], and Shabir Ahmad Bhat[1]

[1]*Centre of Research for Development, University of Kashmir, Srinagar, India, Jammu and Kashmir 190006*
[2]*Department of Botany, University of Kashmir, Srinagar, India, Jammu and Kashmir 190006*

13.1 Introduction

Microbes started to inhabit privileged organisms during evolution, forming symbiotic associations ranging from mutualistic to antagonistic ones [1]. In humans, the symbiotic association between the intestinal microbiota and the host has been broadly studied. Intestinal microbiota perform various metabolic activities, including compound assimilation, production, and breakdown, and are linked to host health in several cases [1]. Plant bodies can be thought of as a dynamic interplay of environmental niches that host a wide variety of microorganisms in their rhizosphere, surface-tissues (rhizoplane and phylloplane), and internal tissues (endosphere) with which they create a wide range of beneficial, neutral, and harmful interactions [2]. Microorganisms have evolved many mechanisms to invade and proliferate within eukaryotic cells and tissues over 500 million years ago, with the growth and diversification of plants and animals [3]. The microbiota's high genetic and metabolic diversity (primarily bacteria) can be due to their high adaptability to all forms of ecological niches [3]. Symbionts offer a range of advantages to the host in the environments where they live. In the case of plants, the microbiota plays a role in the acquirement of soil nutrients, abiotic stress tolerance, and controls of disease, both of which are linked to the endorsement of plant development and, as a result, an improvement in ecological health in the natural world, or the production of food, fiber, and energy.

Given the ability of microorganisms to boost plant production performance, several studies have been conducted to determine which variables influence the modulation of symbiotic interactions [4]. So far, it has been discovered that the evolution of bacterial symbiosis will take many different paths. The majority of the time, this symbiosis develops from environmental relatives, bacteria with beneficial traits for the host, or by a horizontal gene transfer of symbiotic features [4]. According to Sachs et al. [4], bacterial symbiosis can also be innate from parasitic infections through vertical genetic transmission. The "microbiome," which is characterized as the "set of genes found in association with the organisms that colonize a given environment" [5], is the native microbial community of plants. The microbiome dynamically interacts with the plant host to form synergistic relationships that

Core Microbiome: Improving Crop Quality and Productivity, First Edition. Edited by Javid A. Parray, Nowsheen Shameem, Elsayed Fathi Abd-Allah, and Mohammad Yaseen Mir.

affect the host's functioning [6]. Several studies are underway to determine how the microbiome is constructed and what factors influence its dynamics to shape plant success in the ecosystem.

The microbiome's potential has sparked research into the biotic factors and abiotic factors that influence the taxonomic and functional arrangement of plants. These variables are linked to plant genes, the dealings of microbial species, particular agricultural methods, and specific ecological conditions. However, little interest has been paid to the transactions between "native" plant microorganisms and the "microbial inoculants" [7], which are products based on microorganism formulations that have been introduced. Most technical bio-products for plant development promotion and disease control have emerged because of the dynamics of indigenous microbial communities and their relations with the environment; the use of these individual microbial strains has mixed results [8] as bio-inoculant formulations containing a single cultivable microorganism [8]. Thus, the influence of bio-inoculants on the stability of the non-cultivable microbial plant community is yet to be determined. Still, it can be due to one of three factors: (1) direct impact on plant metabolism; (2) effect on microbiome structure displacement; or (3) a combination of the two. Most researches, however, ignore the consequence of bio-inoculants on the structure and dynamics of the bacterial microbiome. In-plant microbiome science, together with new sequencing creation technologies and bioinformatics methods have exposed the taxonomic structure and relative profusion of microorganisms [8]. It is possible to understand how the microbiome is organized in diverse habitats and study their function in these environments using these independent-cultivation techniques [7]. Although beneficial microbial communities have been underutilized in modern agriculture, attempts to introduce individual microbes into crop-growing can be traced back to the 1800s, when the U.S. Department of Agriculture proposed inoculations for legume crops after experiments showed that *Rhizobium* bacteria inhabit nodules and fix nitrogen for their plant hosts [9]. Individual microbes have been inoculated onto crop plants to encourage plant development, nitrogen and phosphorus uptake [10], and resistance against diseases [11] over the past few decades. However, these efforts have mainly concentrated on single microbial strains [12] as we argue that for plant benefit, the emphasis should be on understanding and managing the spectrum from single strains to synthetic consortia to entire microbiomes, and have varying degrees of effectiveness, which is usually due to the heterogeneity of microbial communities in field settings and their interactions with the environment. The creation and implementation of protocol standards is a critical step in organizing the analysis of whole plant microbiomes in an agricultural context. The American Academy of Microbiology, the American Society for Microbiology [13], the American Society for Microbiology, and the National Institute of Standards and Technology in the United States have been guiding this endeavor. Many of the same programs with identified knowledge gaps and research priorities in plant microbiome research have also defined knowledge gaps and research priorities; earlier research focused on taxonomic diversity discovery and explanation, while more recent efforts have focused on elucidating the functions of microbiomes and this chapter aims not to evaluate these efforts; instead, we want to communicate a collection of research goals that will help speed up the incorporation of plant-associated microbiomes into sustainable agriculture. Recent efforts have focused on elucidating the functions of microbiomes. These priorities reveal essential and valuable goals that can be achieved in the next 5–10 years. Plant "health" will undoubtedly benefit from advancements in this sector. In addition to isolating the "central microbiome" or "specific microbiome" of cultures for the formulation of novel bio-inoculants, it is expected that influencing the bacterial population of plants to encourage their paramount growth is achievable [14].

This chapter aims to identify the site reinforcement factors that affect the plant microbiome's organization. Finally, a model will be considered that transforms this information into biotechnological agricultural applications.

13.2 Concepts and Development of Studies on Plant Microbiomes

The modern microbiome originated about 3.7 billion years ago, 2 billion years prior to the eukaryotic species [15]. The concept of "microbiome," on the other hand, has a shorter history. Lederberg and McCray [16] coined the phrase "the ecological population of commensal, symbiont, or pathogenic microorganisms, which occupy space in our body," referring to microorganisms linked with humans. The concept of "microorganisms associated with humans" was simplified in 2002 by "The Human Microbiome Project Consortium" [17]. Currently, the word microbiome refers to any combination of microorganisms. However, there is an issue with its definition, which can be read as "microbiome," which is known as the "collection of resident microorganisms that inhabit a given host/environment," or "microbiome," which is defined as the "set of genes found in connection with the organisms that colonize a given environment." Both meanings are appropriate in this situation. For several years, the study of particular microorganisms, such as arbuscular mycorrhizal fungi and rhizobia, has dominated the study of the microbial communities of the plant kingdom [1]. Thus, a holistic observation of microbial groups connected with plants is a new viewpoint [18]. The high number of publications on this subject demonstrates the importance of plant microbiome research. Many studies (i) examine various plant compartments from the soil to plant range; (ii) show changes in the microbiome modulated by ecological factors; and (iii) show potential agricultural benefits [8]. Numerous study groups in Brazil have contributed to significant microbiome findings. Brazilian research groups are responsible for about 2–3% of scientific publications in the area of plant microbiomes [8]. Brazilian contributions to research articles on plant endosphere, rhizosphere, and phyllosphere account for 4%, 2%, and 1% of the literature, respectively [8]. These studies investigated the microbiome composition in various plant compartments, factors that affect microbiome assemblage, and the use of molecular tools to reach microbial communities, as well as the possible benefits for sustainable agriculture. The "Brazilian Microbiome Project (BMP)" (http://www.brmicrobiome.org) is a recent Brazilian initiative that aims to organize data from various Brazilian microbiomes [15]. The findings of the BMP are incorporated into a more comprehensive project called the "Earth Microbiome Project (EMP)" (http://www.earthmicrobiome.org), which provides a broader study of Brazilian microbiomes [15]. From an international viewpoint to the plant microbiome theme, it is prominent that the latest initiatives have attempted to incorporate it into agricultural invention. The first was initiated by the White House. The microbiome was founded on May 14, 2016, with $450 million to develop microbiome research and commercialization.

The European Union Commission launched a second initiative on October 13, 2016, when it launched the International Bioeconomy Forum, utilizing microbiomes for foodstuff and nutritional protection as the core theme [14]. The illumination of taxonomic and functional trends of microorganism relations and plant–microorganism interactions is a path for upcoming studies based on the works presented. It is worth noting that, in tandem with the need to unlock the "code" of microbiomes, there has been an increase in demand for new methodologies for studying them.

13.3 Valuable Microorganisms for Crop Plants

13.3.1 Plant Growth-Promoting Bacteria

Plant growth-promoting bacteria (PGPBs) make up a significant percentage of plant microbes and can stimulate plant growth in various ways under diverse ecological conditions. Inplant roots, PGPBs, including beneficial plant-symbiotic bacteria such as rhizobia and *Frankia* spp. form a symbiotic arrangement [19–21]. Leguminous and non-leguminous plants (e.g., actinorhizal plants) are the hosts of *rhizobia* and *Frankia* spp., respectively [20]. Rhizobia's symbiosis with legumes is a host-specific relationship [22], and each genus's rhizobia are well known for nodulating a specific host legume to produce nodules that fix atmospheric nitrogen (N_2) [22]. *Rhizobium* spp. are commonly found to nodulate chickpea (*Cicer arietinum*) and common bean plants, while *Bradyrhizobium* spp. are mostly found to nodulate cowpea (*Vigna unguiculata*) and soybean (*Glycine max*) plants [23]. *B. japonicum*, for example, fixes $N_2$to promote plant growth, improving soybean production and lowering fertilizer requirements in fields [24]. Due to the effectiveness of rhizobia's N_2 fixation, the Rhizobium–legume relationship is utilized as a biofertilizer in many countries [25]. It is identified that rhizobia inoculation helps preserve N for crop production by a decline in N fertilizers, which in turn saves the atmosphere [21,26,27]. Wu et al. found that inoculating rice with one species of *rhizobia*, *Sinorhizobium meliloti*, aided rice growth by the expression of genes that promote cell expansion and speed up cell division [28]. Another study found that combining arbuscular mycorrhizal (*Glomus mosseae*) and rhizobial (*R. leguminosarum*) symbioses reduced root hemiparasiticPedicularis species' damage to clover (*Trifolium repens*)[29]. The actinobacteria genus Frankia forms actinorhizal nodules with actinorhizal plants. It promotes the development and environmental resistance of many woody plants, including *Alnus rubra*, *Hippophae tibetana*, *Elaeagnus angustifolia*, and *Casuarina glauca* [29–31]. The formation of root nodules in plants by *Frankia* spp. is a host plant-specific process. It can be divided into three groups, each of which includes some known species: (1) *Frankia alni*, *F. casuarinae*, and *F. torreyi*; (2) *Frankia coriariae* and *F. datiscae*; and (3) *Frankia elaeagni* and *F. irregularis*, which represent the associated Alnus, Dryas, and Elaeagnus hosts, respectively [31–33]. *Frankia* spp. in Cluster 1 can also encourage the progress of *Casuarina glauca* by symbiosis, according to Ngom et al. [34].

13.3.2 Plant Growth-Promoting Fungi

Plant growth-promoting fungi (PGPFs) have received a lot of attention as biofertilizers because they can sustain plant quality and quantity while still being environmentally friendly [35]. Sage (*Salvia miltiorrhiza*), grapevine (*Vitis vinifera*), and lettuce (*Lactuca sativa*) have all made strides in the production of PGPFs for crop cultivation [35–45]. *Trichoderma* spp., *Ganoderma* spp., and yeasts (*Saccharomyces* spp.) have all been used as PGPFs in studies under both natural and environmental stress conditions. For example, Jaroszuk-Scisel et al. recorded a *Trichoderma* strain isolated from healthy rye (*Secale cereale*) rhizosphere and produced auxin, gibberellins, and ACC deaminase in vitro as well as inhibit *Fusarium* spp. growth, which could colonize the roots of wheat (*Triticum aestivum*) seedlings and enhance stem growth [46]. Another study found that *T. asperellum* could induce peroxidase and polyphenol oxidase in lettuce plants, as well as the cell wall degrading enzymes chitinase and 1,3-glucanase, to protect infected plants from leaf spot disease [47–49]. Several isolated killer yeast *S. cerevisiae* strains have been shown to function as PGPFs in grape plants to control *Colletotrichum gloeosporioides* [50]. By inducing the activities of defense-associated enzymes, including 1,3-glucanase and chitinase, as well as upregulating the expression of pathogenesis-related genes in plants, a cell wall isolated from yeast (*Rhodosporidiumpaludigenum*) could induce disease resistance against *Penicillium expansum* in pear (*Pyrus pyrifolia*) fruits [51]. By acting as symbionts to their hosts, arbuscular mycorrhizal fungi (AMF), also known as "Glomeromycota" at the fungal phylum level, can support plants as PGPFs during nutrient absorption and stress resistance [52,53] (figure 1a). Arbuscular mycorrhizal fungi help liverworts (Marchantiophyta), hornworts (*Ceratophyllum demersum*), and lycophytes adapt to land during the early years by becoming a community of obligate endomycorrhizas existent for nearly 400 million years [54]. According to Spagnoletti et al., AMF encourage soybean production by assisting soybean plants in resisting arsenic (As)-polluted soils [53]. Cely et al. reported that inoculation of *Rhizophagusclarus* increased soybean yield under field conditions [55]. Furthermore, AMF increased rice yield by lowering total As levels and improving the nitrogen/carbon (N: C) ratio in rice grains [56,57]. Furthermore, one study found that selenium (Se) AMF interactions improved the quality of soybean and forage grass (*Urochloadecumbens*) used as animal feed by raising Se absorption [58]. One model species of AMF, *Rhizoglomusintraradices*, aids rice, soybean, ginseng (Panax ginseng), and other crops to resist drought and cold stress and promote nutrient absorption [59,60]. *R. intraradices* also helped rice plants avoid blast disease (caused by *Magnaporthe oryzae* spores), and transcriptome data showed that in response to *M. oryzae*, *R. intraradices*-inoculated rice plants had significantly higher expression levels of genes involved in the auxin and salicylic acid signaling pathways than non-inoculated rice plants [61].

13.3.3 Biocontrol Agents

Several scientists have studied biocontrol agents (BCAs) in the past 30 years [61–63], including different valuable AMFs, PGPBs, and PGPFs. *Pseudomonas, Saccharomyces,Streptomyces*, and *Trichoderma* organisms, for example, are the main essential BCAs for assisting plants in acquiring nutrients and controlling plant pathogens [48,64,65]. These organisms have a variety of antagonistic mechanisms, as well as direct parasitism, competition for nutrients and space with pathogenic fungi or bacteria, development of antifungal and bacterial compounds to resist pathogenic fungi and bacteria, and direct parasitism [48,64,65]. Furthermore, these microbes can boost plant resistance by reinforcing the cell wall and improving the physiological and biochemical responses to stress [66].

13.4 Influences of the Community Compositions of Rhizosphere, Phyllosphere, and Endosphere Microbiota on Growth and Performance of Crop Plants

The variety and population structure of microbes linked with crop plants and their association with changes in plant metabolites and gene expression has been demonstrated in a few studies [20,67,132]. Plant health and nutrient content are indicated by the community compositions of the rhizosphere, phyllosphere, or endosphere microbiota connected with them [68–74]. In contrast, the presence of highly enriched pathogenic fungi in the rhizosphere indicates stress-like conditions that could damage plant growth [75,76]. Beneficial rhizospheric microbes, on the other hand, increase plant development, absorption of nutrients through various mechanisms, such as macrobiotic matter mineralization, disease control against soil-borne pathogens, nitrogen fixation, solubilization of potassium (K) and phosphate, and the production of IAA and ACC deaminases [71,72]. As a result, the soil/plant/microbial association must be managed appropriately [71,72]. Beneficial mutualists improve plant growth and resistance to environmental stresses; commensals use the leaf habitat or antagonistic pathogens for their growth, development, and reproduction [77–79]. Phyllospheric microbes, primarily bacteria and fungi, may function as: (i) beneficial mutualists that increase plant development and resistance to ecological stresses; (ii) commensals that use the leaf habitat for their growth, development, and reproduction; or (iii) antagonists. Bacillus and Stenotrophomonas genera were found in great abundance in the phyllosphere microbial communities of 14 phylogenetically diverse plant species grown under regulated greenhouse conditions, showing antagonistic potential against *Botrytis cinerea* [72], a foliar pathogen that causes gray mold disease in more than 200 dicot crop plants [80]. Endophytic bacteria and fungi live within plant tissues and play essential roles in plant growth, fitness, development, and defense without harming host plants [81,82]. Endophytic bacteria and fungi live for some time on the intercellular spaces of plants, obtaining carbohydrates, amino acids, and inorganic nutrients from their hosts [83]. IAA, soluble phosphate, and siderophores are all produced by endophytic bacteria and fungi. They can help host plants grow [84]. Borah et al., for example, investigated the endophytic microbes of cultivated rice (*O. Sativa*) and wild rice (*O. rufipogon*) plants and found a wide range of functional characteristics linked to multiple traits correlated with plant growth and development [85]. Murphy et al. screened fungal endophytes from wild barley (*Hordeum murinum*) for beneficial and promoting traits in cultivated barley (*Hordeum vulgare*) and discovered that certain fungi could promote developed barley growth [86].

Furthermore, a few studies have found that artificially modifying plant growth environments and conditions and plant domestication impact the community structure of rhizosphere and endosphere microbiomes [87–90]. For their survival against stress tolerance, wild plant species retain more diversity linked with microbiota than domesticated crops. According to a recent analysis, standard wild rice (*O. rufipogon*) grown in low-nutrient wildland revealed high rhizobacterial diversity than the cultured rice grown under cultivated field conditions with higher soil nutrients [91]. Martín-Robles et al. investigated the effects of domestication on the arbuscular mycorrhizal symbioses of 14 crop species under various phosphorus-availability conditions, finding that when phosphorus availability was increased, domestication reduced AMF colonization in domesticated plants relative to their wild progenitors [92]. More research into the impact of plant domestication on microbiome compositions is required. The rhizosphere microbiota of wild and cultivated species grown in field conditions is essential for studying the relationship between microbiota and host plants.

A plant may release various chemical compounds into the atmosphere that have an allelochemical effect on another plant, either directly or indirectly [93]. Furthermore, studies have shown that the development of allelochemicals by plants alters the structure of associated microbiomes significantly [94–96]. The impact of microbiota on allelochemical development in plants, on the other hand, is unknown. Furthermore, the relationships between specific allelochemicals and specific microbiota are unknown. *Sanqi ginseng* (*P. notoginseng*) replanting is not recommended, according to Li et al. [97], owing to the accumulation of soil-borne pathogens and allelochemicals (e.g., ginsenosides). Li et al. also demonstrated that a reductive soil disinfestation method combined with allelochemical degradation and pathogen suppression could effectively mitigate Sanqior Pseudoginseng replanting failure [97]. The explanation for the persistent cropping obstacle on microbiome population compositions, on the other hand, will be the subject of future study.

13.5 Applications of Individual Microbes for Improvement of Crop Performance and Soil Ameliorations

Many beneficial microorganisms have been commercialized [98], but their effectiveness in terms of advantages for crop plants in fields has not always been consistent. However, it has been demonstrated that combining beneficial microbes for application to crop plants improves plant production very effectively (Table 13.1). Over the past 30 years, PGPBs and PGPFs have been used in agricultural practices. For example, AMFs have been shown to increase cotton (*Gossypium hirsutum*) and soybean yields and soybean resistance to the *Macrophominaphaseolina* pathogen and stress caused by As [53,55]. According to Tian et al.'s analysis, in carrot (*Daucus carota*) [99], the authors observed AMFs enhance nitrogen transfer from soil to plants. *R. intraradices*-mediated nitrogen transfer also induced the expression of nitrogen metabolism-related genes in the intraradical and extraradical mycelia as well as in host plants [99]. The interaction of the AMF*R. irregularis*and the PGPB*P. putida* increased the growth, development, and pathogen resistance of wheat host plants [100]. Drought has been one of the main limiting factors of crop improvement due to global warming [67]. AMFs improved the plant physiological index of rice and wheat, allowing them to withstand drought better [101,102]. Under local water conditions, the alkaloid content of two medicinal plants,

Table 13.1 Some representative applications of microbes in crop plants.

Research aspects	Microorganism	Beneficial Traits	References
Rhizobia/Franki in colonization with host plants	*Rhizobia/Frankia* spp.	*Rhizobium* spp. promoted the growth of rice seedlings and solubilized silicate. *Bradyrhizobium japonicum* helped soybean in resisting arsenic contamination in soil.	[103,104]
Positive effects ofmutualistic symbioses	AMFs	*Rhizoglomusintraradices* reduced the relative abundance of pathogenic fungi in plant soil.	[75]
PGPB/PGPF applications in improving plant growth, productivity, and resistance	PGPBs, PGPFs	Production of IAA, ACC in improving plant deaminases, JA and ABA. Trichoderma spp. and yeasts help plants in resisting diseases.	[107,10847,46]
Allelochemical effects/soil ameliorations	Microbiomes in soil	Reductive soil disinfestations can reduce the allelochemical effect in soil. Soil amendments associated with changes in the compositions of soil microbiomes.	[97, 109,88]

ACC, 1-aminocyclopropane-1-carboxylic acid; AMFs, arbuscular mycorrhizal fungi; IAA, indole-3-acetic acid; JA, jasmonic acid; ABA, abscisic acid; PGPBs, plant growth-promoting bacteria; PGPFs, plant growth-promoting fungi.

W. somnifera and *H. niger*, was improved by using PGPBs, such as *Azospirillum*, *Azotobacter*, *Pseudomonas*, and *Bacillus* species [38,39]. In terms of endosymbiotic PGPBs, Chandrakala et al. screened and isolated one *Rhizobium* species from the rice rhizosphere that could solubilize silicate and boost rice plant growth potential [103]. *B. japonicum* has been shown to help soybeans withstand high levels of As in the soil [104]. In contrast, *S. fredii* has been shown to effectively modulate various legumes, thus promoting host plant growth [105]. Certain bacterial species can produce JA and ABA, according to Forchetti et al. [106]. However, it is unknown if the contents of these hormones in plants can be directly affected by the related microbes. As a result, studies on the mutual effects of microbes and host plants on hormone metabolism in both plants and microbes in the sense of plant and microbiota relationships have been novel, and fascinating, and this subject merits further research attention.

13.6 Root and Shoot Microbiome

A leaf's surface area can support an average of 105–107 cm^2 bacteria and fungi with 101–108 CFU g^{-1}, which is significantly less than the rhizosphere variety [110]. Even though the rhizosphere is home to diverse microbiota, some have evolved mechanisms for physically communicating with plants. Bacteria and fungi bind to the epidermal-cell wall root surface through chemotaxis processes followed by microbial adhesion mechanisms. These epiphytic interactions in the rhizoplane can entail anchoring structures and adhesive proteins, resulting in the firm adhesion of microorganisms to the plant cell wall, with the formation of aggregates and biofilms as a consequence [111]. Systemic spreading through the xylem vessel elements caused by rising water flux in the plant's transpiratory processes or direct access of microorganisms to the surface of the stem and leaves may trigger epiphytic interactions in the phylloplane [1]. Endophytes may be categorized as obligatory or optional, depending on their life strategy. Mandatory endophytes rely on the host plant for growth and survival, and they are passed down through the generations through seeds or vectors [112,113].

The ability of microorganisms to colonize and distribute plant resources will influence the structure of microbial communities in the various compartments that make up plant microbiomes [114]. Root endophytes often bind to the surface of the roots and multiply for efficient colonization. Then they infiltrate the intercellular atmosphere via natural openings, which are primarily caused by the emission of hydrogen sulfide from lateral roots [111]. Some endophytes will travel to the plant's shoot after initial colonization [111]. The distribution pattern of endophytic communities on tissues is also influenced by resource allocation in different plant compartments [114]. The tendency of the microbiota to compete with other microorganisms and the effects of adverse environmental conditions also play a role in its persistence in plant compartments [115]. The microbiota of the phyllosphere is found in different leaf-niche microenvironments. It must adapt to a more hostile environment characterized by average temperature, light, UV radiation levels, and water and nutrient availability [115]. These are the factors that change the phyllosphere's composition and microbial diversity. Bacteria predominate in this region, and the species are diverse. *Firmicutes, Acidobacteria, Actinobacteria*, and *Cyanobacteria* are all common bacteria found in the environment, including *Cryptococcus, Sporobolomyces*, and *Rhodotorula* yeasts and filamentous fungi from the genera *Cladosporium, Alternaria, Penicillium, Acremonium*, and *Mucor*. The root and its microbiome have identical compositions. The domination of bacteria has been observed in the rhizosphere. *Proteobacteria, Actinobacteria*, and *Bacteroidetes* are three species of bacteria. The bacterial diversity in the rhizosphere, on the other hand, is higher in the compartment and lower in the root's interior [116]. The bacteria *Actinophytocola, Streptomyces, Pseudonocardia*, and the fungi *Canalisporium, Aspergillus, Minimelanolocus*, and *Macrophomina*were found in the endophytic microbial population associated with olive roots (*Olea europaea*) [117]. The root-epiphytic population is likely to be selectively influenced by the plant's physical and biochemical barriers. Because of the selective pressure to obtain access to plant tissue, the endophytic microbiota is less diverse. Its composition appears to stay constant within the plant while varying significantly in the rhizosphere [118]. Recognizing the plant body as a network of related niches that house the microbiota in the rhizosphere, rhizoplane/phylloplane, and endosphere domains is a step toward figuring out what factors influence plant microbiome composition.

References

1 Hunter, P. (2016). Plant microbiomes and sustainable agriculture. *EMBO Reports* 1–4.
2 Turner, T.R., James, E.K., and Poole, P.S. (2013). The plant microbiome. *Genome Biology* 14: 209. https://doi.org/10.1186/gb-2013-14-6-209.
3 Toft, C. and Andersson, S.G.E. (2010). Evolutionary microbial genomics: Insights into bacterial host adaptation. *Nature Reviews. Genetics* 11: 465–475. https://doi.org/10.1038/nrg2798.
4 Sachs, J.L., Skophammer, R.G., and Regus, J.U. (2011). Colloquium Paper: Evolutionary transitions in bacterial symbiosis. *Proceedings of the National Academy of Sciences* 108 (Suppl): 10800–10807. https://doi.org/10.1073/pnas.1100304108.
5 Boon, E., Meehan, C.J., Whidden, C., Wong, D.H.J., Langille, M.G.I., and Beiko, R.G. (2014). Interactions in the microbiome: Communities of organisms and communities of genes. *FEMSMicrobiol. Rev* 38: 90–118. https://doi.org/10.1111/1574-6976.12035.
6 Foo, J.L., Ling, H., Lee, Y.S., and Chang, M.W. (2017). Microbiome engineering: Current applications and its future. *Biotechnology Journal* 1600099. https://doi.org/10.1002/biot.201600099.
7 Andreote, F.D., Gumiere, T., and Durrer, A. (2014). Exploring interactions of plant microbiomes. *Scientia Agricola* 71: 528–539. https://doi.org/10.1590/0103-9016-2014-0195.
8 Busby, P.E., Soman, C., Wagner, M.R., Friesen, M.L., Kremer, J., Bennett, A., Morsy, M., Eisen, J.A., Leach, J.E., and Dangl, J.L. (2017). Research priorities for harnessing plant microbiomes in sustainable agriculture. *PLoS Biology* 1–14. https://doi.org/10.1371/journal.pbio.2001793.
9 Schneider, A. (1892). Observations on Some American *Rhizobia*. *Bulletin of the Torrey Botanical Club* 19 (7): 203–218.
10 Afzal, A. and Bano, A. (2008). *Rhizobium* and Phosphate Solubilizing Bacteria Improve the Yield and Phosphorus Uptake in Wheat (*Triticum aestivum*). *International Journal of Agriculture & Biology* 10: 85–88.
11 Harman, G.E., Howell, C.R., Viterbo, A., Chet, I., and Lorito, M. (2004). *Trichoderma* Species—Opportunistic, Avirulent Plant Symbionts. *Nature Reviews. Microbiology* 2 (1): 43–56. https://doi.org/10.1038/nrmicro797PMID:15035008.
12 Kandula, D.R.W., Jones, E.E., Stewart, A., McLean, K.L., and Hampton, J.G. (2015). *Trichoderma* Species for Biocontrol of Soil-Borne Plant Pathogens of Pasture Species. *Biocontrol Science and Technology* 25(9): 1052–1069. PLOS Biology | https://doi.org/10.1371/journal.pbio.2001793 March 28, 2017 11/1433. Reid A, Greene SE. How Microbes Can Help Feed the World. Issues. 2013(105):33–37.
13 Alivisatos, A.P., Blaser, M.J., Brodie, E.L., Chun, M., Dangl, J.L., Donohue, T.J. et al. (2015). A Unified Initiative to Harness Earth's Microbiomes. *Science* 350 (6260): 507–508. https://doi.org/10.1126/science.aac8480PMID:26511287.
14 Johns, N.I., Blazejewski, T., Gomes, A.L.C., and Wang, H.H. (2016). Principles for designing synthetic microbial communities. *Current Opinion in Microbiology* 31: 146–153. https://doi.org/10.1016/j.mib.2016.03.010.
15 Torsvik, V., Overseas, L., and Thingstad, T.F. (2002). Prokaryotic diversity - magnitude, dynamics, and controlling factors. *Science* 296: 1064–1066. https://doi.org/10.1126/science.1071698.

16 Lederberg, B.J. and McCray, A.T. (2001). 'Ome sweet' omics – a genealogical treasury of words. *Science* 15: 8.
17 Consortium, H.M.P. (2012). Structure, function and diversity of the healthy human microbiome. *Nature* 486: 207–214. https://doi.org/10.1038/nature11234.
18 Singh, B.K. and Trivedi, P. (2017). Microbiome and the future for food and nutrient security. *MicrobBiotechnol* 10: 50–53. https://doi.org/10.1111/1751-7915.12592.
19 Pylro, V.S., Roesch, L.F.W., Ortega, J.M., do Amaral, A.M., Tortola, M.R., Hirsch, P.R., Rosado, A.S., Goes-Neto, A. et al. (2014). Brazilian microbiome project: Revealing the unexplored microbial diversity-challenges and prospects. *Microbial Ecology* 67: 237–241. https://doi.org/10.1007/s00248-013-0302-4.
20 Cissoko, M., Hocher, V., Gerbi, H., Gully, D., Carré-Mlouka, A., Sane, S., Pignoly, S., Champion, A., Ngom, M., Pujic, P. et al. (2018). Actinorhizal signaling molecules: Frankia root hair deforming factor shares properties with NIN inducing factor. *Frontiers in Plant Science* 9: 1494.
21 Poole, P., Ramachandran, V., and Terpolilli, J. (2018). Rhizobia: From saprophytes to endosymbionts. *Nature Reviews. Microbiology* 16: 291–303.
22 Batista, L., Irisarri, P., Rebuffo, M., Cuitiño, M.J., Sanjuán, J., and Monza, J. (2015). Nodulation competitiveness as a requisite for improved rhizobial inoculants of *Trifolium pratense. Biology and Fertility of Soils* 51: 11–20.
23 Koskey, G., Mburu, S.W., Kim, J.M., Ombori, O., Maingi, J.M., and Njeru, E.M. (2018). Genetic characterization and diversity of Rhizobium isolated from root nodules of mid-altitude climbing bean (*Phaseolus vulgaris L.*) varieties. *Frontiers in Microbiology* 9: 968.
24 Ronner, E., Franke, A.C., Vanlauwe, B., Dianda, M., Edeh, E., Ukem, B., Bala, A., Van Heerwaarden, J., and Giller, K.E. (2016). Understanding variability in soybean yield and response to P-fertilizer and *rhizobium* inoculants on farmers' fields in northern Nigeria. *Field Crops Research* 186: 133–145.
25 Mia, M.B. and Shamsuddin, Z. (2010). Rhizobium as a crop enhancer and biofertilizer for increased cereal production. *African Journal of. Biotechnology* 9: 6001–6009.
26 Crespo-Rivas, J.C., Navarro-Gomez, P., Alias-Villegas, C., Shi, J., Zhen, T., Niu, Y., Cuellar, V., Moreno, J., Cubo, T., Vinardell, J.M. et al. (2019). *Sinorhizobiumfredii*HH103RirA is required for oxidative stress resistance and efficient symbiosis with soybean. *International Journal of Molecular Sciences* 20: 787.
27 Ozturk, A., Caglar, O., and Sahin, F. (2003). Yield response of wheat and barley to inoculation of plant growth-promoting rhizobacteria at various levels of nitrogen fertilization. *Journal of Soil Science and Plant Nutrition* 166: 262–266.
28 Wu, Q., Peng, X., Yang, M., Zhang, W., Dazzo, F.B., Uphoff, N., Jin, Y., and Shen, S. (2018). *Rhizobia* promote the growth of rice shoots by targeting cell signaling, division and expansion. *Plant Molecular Biology* 97: 507–523.
29 Sui, X.L., Zhang, T., Tian, Y.Q., Xue, R.J., and Li, A.R. (2019). A neglected alliance in battles against parasitic plants: Arbuscular mycorrhizal and rhizobial symbioses alleviate damage to a legume host by root hemiparasiticPedicularis species. *The New Phytologist* 221: 470–481.
30 Van Nguyen, T. and Pawlowski, K. (2017). *Frankia* and *Actinorhizal* Plants: Symbiotic Nitrogen Fixation. In: *Rhizotrophs: Plant Growth Promotion to Bioremediation* (ed. S. Mehnaz), 237–261. Singapore: Springer.

31 Samant, S., Dawson, J.O., and Hahn, D. (2016). Growth responses of introduced *Frankia* strains to edaphic factors. *Plant and Soil* 400: 123–132.

32 Wolters, D.J., Van Dijk, C., Zoetendal, E.G., and Akkermans, A.D.L. (1997). Phylogenetic characterization of ineffective Frankia in *Alnus glutinosa* (L.) Gaertn. nodules from wetland soil inoculants. *Molecular Ecology* 6: 971–981.

33 Noxious, I., Cortés-albayay, C., Carro, L., Castro, J.F., Atari, M., Ghodhbane-Gtari, F., Klenk, H.P., Tisa, L.S., Sangal, V., and Goodfellow, M. (2019). Genomic insights into plant-growth-promoting potentialities of the genus Frankia. *Frontiers in Microbiology* 10: 1457. doi: 10.3389/fmicb.2019.01457.

34 Ngom, M., Diagne, N., Laplace, L., Champion, A., and Sy, M.O. (2016). Symbiotic ability of diverse Frankia strains on Casuarina glauca plants in hydroponic conditions. *Symbiosis* 70: 79–86.

35 Santoyo, G. (2012). Orozco-Mosqueda, M.d.C.; Govindappa, M. Mechanisms of biocontrol and plant growth-promoting activity in soil bacterial species of Bacillus and Pseudomonas: A review. *Biocontrol Science and Technology*22: 855–872.

36 Jiang, Y., Ran, C., Chen, L., Yin, W., Liu, Y., Chen, C., and Gao, J. (2019). Purification and characterization of a novel antifungal flagellin protein from endophyte *Bacillus methylotrophicus*NJ13 against *Ilyonectriarobusta*. *Microorganisms* 7: 605.

37 Maiyappan, S., Amalraj, E., Santhosh, A., and Peter, A. (2010). Isolation, evaluation and formulation of selected microbial consortia for sustainable agriculture. *Journal of Biopesticides (Jbiopest)* 2: 2.

38 Rajasekar, S. and Elango, R. (2011). Effect of microbial consortium on plant growth and improvement of alkaloid content in *Withaniasomnifera* (Ashwagandha). *Current Botany* 2: 27–30.

39 Ghorbanpour, M., Hatami, M., and Khavazi, K. (2013). Role of plant growth-promoting rhizobacteria on antioxidant enzyme activities and tropane alkaloid production of *Hyoscyamus niger* under water deficit stress. *Turkish Journal of Biology* 37: 350–360.

40 Maggini, V., Mengoni, A., Bogani, P., Firenzuoli, F., and Fani, R. (2020). Promoting model systems of microbiota–medicinal plant interactions. *Trends in Plant Science* 25(3): 223-225.

41 Mishra, J., Fatima, T., and Arora, N.K. (2018). Role of Secondary Metabolites from Plant Growth-Promoting Rhizobacteria in Combating Salinity Stress. In: *In-Plant Microbiome: Stress Response*, (ed. D. Egamberdieva and P. Ahmad), 127–163. Singapore: Springer.

42 Kannojia, P., Choudhary, K.K., Srivastava, A.K., and Singh, A.K. (2019). Chapter Four—PGPRBioelicitors: Induced Systemic Resistance (I.S.R.) and Proteomic Perspective on Biocontrol. In: *PGPR Amelioration in Sustainable Agriculture*, (ed. A.K. Singh, A. Kumar and P.K. Singh), 67–84. Cambridge, UK: Woodhead Publishing.

43 Narendra Babu, A., Jogaiah, S., Ito, S.I., Kester Nagaraj, A., and Tran, L.S.P. (2015). Improvement of growth, fruit weight and early blight disease protection of tomato plants by rhizosphere bacteria is correlated with their beneficial traits and induced biosynthesis of antioxidant peroxidase and polyphenol oxidase. *Plant Science* 231: 62–73.

44 Berger, B., Brock, A.K., and Ruppel, S. (2013). Nitrogen supply influences plant growth and transcriptional responses induced by *Enterobacter radicincitans* in *Solanum lycopersicum*. *Plant and Soil* 370: 641–652.

45 Zhou, L., Tang, K., and Guo, S. (2018). The plant growth-promoting fungus (PGPF) Alternaria sp. A13 markedly enhances *Salvia miltiorrhiza* root growth and active ingredient accumulation under greenhouse and field conditions. *International Journal of Molecular Sciences* 19: 270.

46 Jaroszuk-Scisel, J., Tyskiewicz, R., Nowak, A., Ozimek, E., Majewska, M., Hanaka, A., Tyskiewicz, K., Pawlik, A., and Janusz, G. (2019). Phytohormones (auxin, gibberellin) and A.C.C. deaminase in vitro synthesized by the mycoparasiticTrichoderma DEMTkZ3A0 strain and changes in the level of auxin and plant resistance markers in wheat seedlings inoculated with this strain conidia. *International Journal of Molecular Sciences* 20: 4923.

47 Bailee, B., Ito, S.I., and Sunpapao, A. (2019). *Trichoderma asperellum*T1 mediated antifungal activity and induced defense response against leaf spot fungi in lettuce (*Lactuca sativa*L.). *Physiological and Molecular Plant Pathology* 106: 96–101.

48 El-Sharkawy, H.H.A., Abo-El-Wafa, T.S.A., and Ibrahim, S.A.A. (2018). Biological control agents improve the productivity and induce the resistance against downy mildew of grapevine. *Journal of Plant Pathology* 100: 33–42.

49 El-Sharkawy, H.H.A., Rashad, Y.M., and Ibrahim, S.A. (2018). Biocontrol of stem rust disease of wheat using arbuscular mycorrhizal fungi and Trichoderma spp. *Physiological and Molecular Plant Pathology* 103: 84–91.

50 Liu, Z., Du, S., Ren, Y., and Liu, Y. (2018). Biocontrol ability of killer yeasts (*Saccharomyces cerevisiae*) isolated from wine against *Colletotrichum gloeosporioides* on grape. *Journal of Basic Microbiology* 58: 60–67.

51 Sun, C., Lin, M., Fu, D., Yang, J., Huang, Y., Zheng, X., and Yu, T. (2018). Yeast cell wall induces disease resistance against *Penicillium expansum* in pear fruit and the possible mechanisms involved. *Food Chemistry* 241: 301–307.

52 Stürmer, S.L., Bever, J.D., and Morton, J.B. (2018). Biogeography of arbuscular mycorrhizal fungi (Glomeromycota): A phylogenetic perspective on species distribution patterns. *Mycorrhiza* 28: 587–603.

53 Spagnoletti, F., Carmona, M., Gómez, N.E.T., Chiocchio, V., and Lavado, R.S. (2017). Arbuscular mycorrhiza reduces the negative effects of *M. phaseolina* on soybean plants in arsenic-contaminated soils. *Applied Soil Ecology* 121: 41–47.

54 MacLean, A.M., Bravo, A., and Harrison, M.J. (2017). Plant signaling and metabolic pathways enabling arbuscular mycorrhizal symbiosis. *The Plant Cell* 29: 2319–2335.

55 Cely, M.V.T., de Oliveira, A.G., de Freitas, V.F., de Luca, M.B., Barazetti, A.R., Dos Santos, I.M.O., Gionco, B., Garcia, G.V., Prete, C.E.C., and Andrade, G. (2016). Inoculant of arbuscular mycorrhizal fungi (*Rhizophagusclarus*) increase yield of soybean and cotton under field conditions. *Frontiers in Microbiology* 7: 720.

56 Li, H., Chen, X.W., and Wong, M.H. (2016). Arbuscular mycorrhizal fungi reduced the ratios of inorganic/organic arsenic in rice grains. *Chemosphere* 145: 224–230.

57 Zhang, X., Wang, L., Ma, F., Yang, J., and Su, M. (2017). Effects of arbuscular mycorrhizal fungi inoculation on carbon and nitrogen distribution and grain yield and nutritional quality in rice (*Oryza sativa*L.). *Journal of the Science of Food and Agriculture* 97: 2919–2925.

58 Bamberg, S.M., Ramos, S.J., Carneiro, M.A.C., and Siqueira, J.O. (2019). Effects of selenium (Se) application and arbuscular mycorrhizal (A.M.F.) inoculation on soybean ('*Glycine max*') and forage grass ('*Urochloadecumbens*') development in oxisol. *Australian Journal of Crop Science* 13: 380.

59 Liu, Z., Li, Y., Ma, L., Wei, H., Zhang, J., He, X., and Tian, C. (2014). Coordinated regulation of arbuscular mycorrhizal fungi and soybean MAPK pathway genes improved mycorrhizal soybean drought tolerance. *Molecular Plant-Microbe Interactions* 28: 408–419.

60 Jafari, M., Yari, M., Ghabooli, M., Sepehri, M., Ghasemi, E., and Jonker, A. (2018). Inoculation and co-inoculation of alfalfa seedlings with root growth-promoting microorganisms (*Piriformospora indica*, *Glomus intraradices* and *Sinorhizobiummeliloti*) affect molecular structures, nutrient profiles and availability of hay for ruminants. *Animal Nutrition* 4: 90–99.

61 Tian, L., Chang, C., Ma, L., Nasir, F., Zhang, J., Li, W., Tran, L.S.P., and Tian, C. (2019). Comparative study of the mycorrhizal root transcriptomes of wild and cultivated rice in response to the pathogen *Magnaportheoryzae*. *Rice* 12: 35.

62 Gramsci, B.R., Lutz, M.C., Lopes, C.A., and Sangorrín, M.P. (2018). Enhancing the efficacy of yeast biocontrol agents against postharvest pathogens through nutrient profiling and the use of other additives. *Biological Control* 121: 151–158.

63 Chandra, S. and Singh, B.K. (2016). Trichoderma spp.: As potential biocontrol agents (B.C.A.s) against fungal plant pathogens. *Indian Journal of Life Sciences* 5: 105. 92. Elsen, A.; Gervacio, D.; Swennen, R.; De Waele, D. AMF-induced biocontrol against plant-parasitic nematodes in Musa sp.: A systemic effect. Mycorrhiza 2008, 18, 251–256.

64 Chandra, S. and Singh, B.K. (2016). Trichoderma spp.: As potential biocontrol agents (B.C.A.s) against fungal plant pathogens. *Indian Journal of Life Sciences* 5: 105.

65 Cabot, C., Bosch, R., Martos, S., Poschenrieder, C., and Perelló, A. (2018). Salinity is a prevailing factor for amelioration of wheat blast by biocontrol agents. *Biological Control* 125: 81–89.

66 Siahmoshteh, F., Siciliano, I., Banani, H., Hamidi-Esfahani, Z., Razzaghi-Abyaneh, M., Gullino, M.L., and Spadaro, D. (2017). Efficacy of *Bacillus subtilis* and *Bacillus amyloliquefaciens* in the control of *Aspergillus parasiticus* growth and aflatoxins production on pistachio. *International Journal of Food Microbiology* 254: 47–53.

67 Senapati, N., Stratonovitch, P., Paul, M.J., and Semenov, M.A. (2018). Drought tolerance during reproductive development is important for increasing wheat yield potential under climate change in Europe. *Journal of Experimental Botany* 70: 2549–2560.

68 Shi, S., Tian, L., Nasir, F., Li, X., Li, W., Tran, L.S.P., and Tian, C. (2018). Impact of domestication on the evolution of rhizomicrobiome of rice in response to the presence of *Magnaportheoryzae. Plant Physiology And Biochemistry: Ppb / Societefrancaise de physiologievegetale* 132: 156–165.

69 Tian, L., Shi, S., Ma, L., Nasir, F., Li, X., Tran, L.S.P., and Tian, C. (2018). Co-evolutionary associations between root-associated microbiomes and root transcriptomes in wild and cultivated rice varieties. *Plant Physiology And Biochemistry* 128: 134–141.

70 Musilova, L., Ridl, J., Polivkova, M., Macek, T., and Uhlik, O. (2016). Effects of secondary plant metabolites on microbial populations: Changes in community structure and metabolic activity in contaminated environments. *International Journal of Molecular Sciences* 17: 1205.

71 Meena, V.S., Meena, S.K., Verma, J.P., Kumar, A., Aeron, A., Mishra, P.K., Bisht, J.K., Pattanayak, A., Naveed, M., and Dotaniya, M.L. (2017). Plant beneficial rhizospheric microorganism (PBRM) strategies to improve nutrients use efficiency: A review. *Ecological Engineering* 107: 8–32.

72 Zhang, X., Zhang, R., Gao, J., Wang, X., Fan, F., Ma, X., Yin, H., Zhang, C., Feng, K., and Deng, Y. (2017). Thirty-one years of rice-rice-green manure rotations shape the rhizosphere microbial community and enrich beneficial bacteria. *Soil Biology & Biochemistry* 104: 208–217.

73 Hardin, P.R., van Overbeek, L.S., and Elsas, J.D. (2008). Properties of bacterial endophytes and their proposed role in plant growth. *Trends in Microbiology* 16: 463–471.

74 Ortega, R.A., Mahnert, A., Berg, C., Müller, H., and Berg, G. (2016). The plant is crucial: Specific composition and function of the phyllosphere microbiome of indoor ornamentals. *FEMS Microbiology Ecology* 92: 173.

75 Tian, L., Shi, S., Ma, L., Zhou, X., Luo, S., Zhang, J., Lu, B., and Tian, C. (2019). The effect of *Glomus intraradices* on the physiological properties of *Panax ginseng* and on rhizospheric microbial diversity. *Journal of Ginseng Research* 43: 77–85.

76 Liu, D., Sun, H., and Ma, H. (2019). Deciphering microbiome related to rusty roots of *Panax ginseng* and evaluation of antagonists against pathogenic Ilyonectria. *Frontiers in Microbiology* 10: 1350.

77 Stone, B.W., Weingarten, E.A., and Jackson, C.R. (2018). The role of the phyllosphere microbiome in plant health and function. *Annual Plant Review Online* 6: 533–556.

78 Vacher, C., Hampe, A., Porté, A.J., Sauer, U., Company, S., and Morris, C.E. (2016). The phyllosphere: Microbial jungle at the plant–climate interface. *Annual Review of Ecology, Evolution, and Systematics* 47: 1–24.

79 Bringel, F. and Couée, I. (2015). Pivotal roles of phyllosphere microorganisms at the interface between plant functioning and atmospheric trace gas dynamics. *Frontiers in Microbiology* 6: 486.

80 AbuQamar, S., Moustafa, K., and Tran, L.S. (2017). Mechanisms and strategies of plant defense against *Botrytis cinerea*. *Critical Reviews in Biotechnology* 37: 262–274.

81 Farrar, K., Bryant, D., and Cope-Selby, N. (2014). Understanding and engineering beneficial plant-microbe interactions: Plant growth promotion in energy crops. *Plant Biotechnology Journal* 12: 1193–1206.

82 Truyens, S., Weyens, N., Cuypers, A., and Vangronsveld, J. (2015). Bacterial seed endophytes: Genera, vertical transmission and interaction with plants. *Environmental Microbiology Reports* 7: 40–50.

83 Bacon, C.W. and Hinton, D.M. (2006). Bacterial endophytes: The Endophytic Niche, Its Occupants, and Its Utility. In: *Plant-Associated Bacteria*, (ed. S.S. Gnanamanickam) Springer: Dordrecht, The Netherlands. 155–194. 143. Zhao, Y.; Xiong, Z.; Wu, G.; Bai, W.; Zhu, Z.; Gao, Y.; Parmar, S.; Sharma, V.K.; Li, H. Fungal endophytic communities of two wild rosa varieties with different powdery mildew susceptibilities. Front. Microbiol. 2018, 9, 2462).

84 Zhao, Y., Xiong, Z., Wu, G., Bai, W., Zhu, Z., Gao, Y., Parmar, S., Sharma, V.K., and Li, H. (2018). Fungal endophytic communities of two wild rosa varieties with different powdery mildew susceptibilities. *Frontiers in Microbiology* 9: 2462.

85 Borah, M., Das, S., Baruah, H., Boro, R.C., and Barooah, M. Diversity of culturable endophytic bacteria from wild and cultivated rice showed potential plant growth-promoting activities. bioRxiv 2018.

86 Murphy, B.R., Hodkinson, T.R., and Doohan, F.M. (2018). Endophytic Cladosporium strains from a crop wild relative increase grain yield in barley. *Environmental Biology* 118: 147–156.

87 Shenton, M., Iwamoto, C., Kurata, N., and Ikeo, K. (2016). Effect of wild and cultivated rice genotypes on rhizosphere bacterial community composition. *Rice* 9: 42–53.

88 Shi, S., Tian, L., Nasir, F., Bahadur, A., Batool, A., Luo, S., Yang, F., Wang, Z., and Tian, C. (2019). Response of microbial communities and enzyme activities to amendments in saline-alkaline soils. *Applied Soil Ecology* 135: 16–24.

89 Sawers, R.J.H., Ramirez-Flores, M.R., Olalde-Portugal, V., and Paszkowski, U. (2018). The impact of domestication and crop improvement on arbuscular mycorrhizal symbiosis in cereals: Insights from genetics and genomics. *The New Phytologist* 220: 1135–1140.

90 Majeed, A., Muhammad, Z., and Ahmad, H. (2018). Plant growth-promoting bacteria: Role in soil improvement, abiotic and biotic stress management of crops. *Plant Cell Reports* 37: 1599–1609.

91 Tian, L., Zhou, X., Ma, L., Xu, S., Nasir, F., and Tian, C. (2017). root-associated bacterial diversities of *Oryza rufipogon* and *Oryza sativa* and their influencing environmental factors. *Archives of Microbiology* 199: 563–571.

92 Martín-Robles, N., Lehmann, A., Seco, E., Aroca, R., Rillig, M.C., and Milla, R. (2018). Impacts of domestication on the arbuscular mycorrhizal symbiosis of 27 crop species. *The New Phytologist* 218: 322–334.

93 Chiapusio, G., Sánchez, A.M., Reigosa, M.J., González, L., and Pellissier, F. (1997). Do germination indices adequately reflect allelochemical effects on the germination process? *Journal of Chemical Ecology* 23: 2445–2453.

94 Vivanco, J.M., Bais, H.P., Stermitz, F.R., Thelen, G.C., and Callaway, R.M. (2004). Biogeographical variation in community response to root allelochemistry: Novel weapons and exotic invasion. *Ecology Letters* 7: 285–292.

95 Niro, E., Marzaioli, R., De Crescenzo, S., D'Abrosca, B., Castaldi, S., Esposito, A., Fiorentino, A., and Rutigliano, F.A. (2016). Effects of the allelochemical coumarin on plants and soil microbial community. *Soil Biology & Biochemistry* 95: 30–39.

96 Scavo, A., Abate, C., and Mauromicale, G. (2019). Plant allelochemicals: Agronomic, nutritional and ecological relevance in the soil system. *Plant and Soil* 442: 23–48.

97 Li, Y., Wang, B., Chang, Y., Yang, Y., Yao, C., Huang, X., Zhang, J., Cai, Z., and Zhao, J. (2019). Reductive soil disinfestation effectively alleviates the replant failure of *Sanqi ginseng* through allelochemical degradation and pathogen suppression. *Applied Microbiology and Biotechnology* 103: 3581–3595.

98 Berg, G. (2009). Plant-microbe interactions promoting plant growth and health: Perspectives for controlled use of microorganisms in agriculture. *Applied Microbiology and Biotechnology* 84: 11–18.

99 Tian, C., Kasiborski, B., Koul, R., Lammers, P.J., Bucking, H., and Shachar-Hill, Y. (2010). Regulation of the nitrogen transfer pathway in the arbuscular mycorrhizal symbiosis: Gene characterization and the coordination of expression with nitrogen flux. *Plant Physiology* 153: 1175–1187.

100 Pérez-de-Luque, A., Tille, S., Johnson, I., Pascual-Pardo, D., Ton, J., and Cameron, D.D. The interactive effects of arbuscular mycorrhiza and plant growth-promoting *rhizobacteria* synergistically enhance host plant defences against pathogens. Sci. Rep. UK 2017, 7, 16409].

101 Mathur, S., Tomar, R.S., and Jajoo, A. (2019). Arbuscular Mycorrhizal Fungi (A.M.F.) Protects Photosynthetic Apparatus of Wheat under Drought Stress. *Photosynth. Res.* 139: 227–238.

102 Ruiz-Sánchez, M., Aroca, R., Muñoz, Y., Polón, R., and Ruiz-Lozano, J.M. (2010). The arbuscular mycorrhizal symbiosis enhances the photosynthetic efficiency and the antioxidative response of rice plants subjected to drought stress. *Journal of Plant Physiology* 167: 862–869.

103 Chandrakala, C., Voleti, S.R., Bandeppa, S., Sunil Kumar, N., and Latha, P.C. (2019). Silicate solubilization and plant growth-promoting potential of Rhizobium sp. isolated from rice rhizosphere. *Silicon* 11: 2895–2906.

104 Armendariz, A.L., Talano, M.A., Olmos Nicotra, M.F., Escudero, L., Breuer, M.L., Porporatto, C., and Agostini, E. (2019). Impact of double inoculation with *Bradyrhizobium japonicum*E109 and *Azospirillumbrasilense*Az39 on soybean plants grown under arsenic stress. *Plant Physiology And Biochemistry: Ppb / Societefrancaise de physiologievegetale* 138: 26–35.

105 Lopez-Baena, F.J., Ruiz-Sainz, J.E., Rodriguez-Carvajal, M.A., and Vinardell, J.M. (2016). *Bacterial molecular signals in the Sinorhizobiumfredii*–soybean symbiosis. *International Journal of Molecular Sciences* 17: 755.

106 Forchetti, G., Masciarelli, O., Alemano, S., Alvarez, D., and Abdala, G. (2007). Endophytic bacteria in sunflower (*Helianthus annuus* L.): Isolation, characterization, and production of jasmonates and abscisic acid in culture medium. *Applied Microbiology and Biotechnology* 76: 1145–1152.

107 Misra, S., Dixit, V.K., Khan, M.H., Kumar Mishra, S., Dwivedi, G., Yadav, S., Lehri, A., and Singh Chauhan, P. (2017). Exploitation of agro-climatic environment for selection of 1-aminocyclopropane-1-carboxylic acid (A.C.C.) deaminase producing salt-tolerant indigenous plant growth-promoting *rhizobacteria*. *Microbiological Research* 205: 25–34.

108 Josiah, S., Kurjogi, M., Govind, S.R., Huntrike, S.S., Basappa, V.A., and Tran, L.S.P. (2016). Isolation and evaluation of proteolytic actinomycete isolates as novel inducers of pearl millet downy mildew disease protection. *Scientific Reports UK* 6: 30789.

109 Luo, S., Tian, L., Chang, C., Wang, S., Zhang, J., Zhou, X., Li, X., Tran, L.S.P., and Tian, C. (2018). Grass and maize vegetation systems restore saline-sodic soils in the SonnenPlain of northeast China. *Land Degradation & Development* 29: 1107–1119.

110 Berg, G., Rybakova, D., Grube, M., and Koberl, M. (2015). The plant microbiome explored: Implications for experimental botany. *Journal of Experimental Botany* 67: 995–1002. http://doi.org/10.1093/jxb/erv466.

111 Ambrosini, A., de Souza, R., and Passaglia, L.M.P. (2016). Ecological role of bacterial inoculants and their potential impact on soil microbial diversity. *Plant and Soil* 400: 193–207. https://doi.org/10.1007/s11104-015-2727-7.

112 Mitter, B., Pfaffenbichler, N., Flavell, R., Company, S., Antonelli, L., Petric, A., Berninger, T., Naveed, M., Sheibani-Tezerji, R., von Maltzahn, G., and Sessitsch, A. (2017). A new approach to modify plant microbiomes and traits by introducing beneficial bacteria at flowering into progeny seeds, Front. *Microbiol* 8: 11. http://doi.org/10.3389/fmicb.2017.00011.

113 Bergottini, V.M., Herv, V., Sosa, D., Otegui, M.B., Zapata, P.D., and Junier, P. (2017). Exploring the diversity of the root-associated microbiome of *Ilex paraguariensis*St. Hil. *(Yerba Mate), Agriculture, Ecosystems & Environment Applied Soil Ecology* 109: 23–31. https://doi.org/10.1016/j.apsoil.2016.09.013.

114 GaieroCa, J.R.McCall, Ka.,Thompson, N.J.,Day, A.S.,Best, K.E. (2013). Inside the root microbiome: Bacterial root endophytes and plant growth promotion. *American Journal of Botany* 100: 1738–1750. https://doi.org/10.3732/ajb.1200572.

115 Sivakumar, N., Sathishkumar, R., Selvakumar, G., Shyamkumar, R., and Arjunekumar, K. (2020). Phyllospheric microbiomes: Diversity, ecological significance, and biotechnological applications. *Plant Microbiomes for Sustainable Agriculture* 25: 113–172. https://doi.org/10.1007/978-3-030-38453-1_5.

116 Hacquard, S., Garrido-Other, R., González, A., Spaepen, S., Ackermann, G., Lebeis, S., McHardy, A.C., Dangl, J.L., Knight, R., Ley, R., and Schulze-Lefert, P. (2015). Microbiota and host nutrition across plant and animal kingdoms. *Cell Host & Microbe* 17: 603–616. https://doi.org/10.1016/j.chom.2015.04.009.

117 Fern′ Andez-González, A.J., Villadas, P.J., ′ Cabanas, C.G.-L., ValverdeCorredor, A., Belaj, A., Mercado-Blanco, J., and Fernandez-l ′ Opez, M. (2019). Defining the root endosphere and rhizosphere microbiomes from the World Olive Germplasm Collection. *Scientific Reports* 9: 1–13. https://doi.org/10.1038/s41598-019-56977-9.

118 Zhang, Y., Zheng, L., Zheng, Y., Xue, S., Zhang, J., Huang, P., Zhao, Y., Hao, X., He, Z., Hu, Z., Zhou, C., Chen, Q., Liu, J., Wang, G., Sang, M., Sun, X., Wang, X., Xiao, X., and Li, C. (2020). Insight into the assembly of root-associated microbiome in the medicinal plant *Polygonum cuspidatum. Industrial Crops and Products* 145: 112163. https://doi.org/10.1016/j.indcrop.2020.112163.

14

Role of Nanotechnology in Soil Microbiome and Agricultural Development

Bisma Farooq[1], Shahnaz Anjum[2], Madiha Farooq[2], Gulzar Ahmed Rather[2], Asma Nazir[2], Bijaya Kumar Nayak, and Anima Nanda**

[1]Department of Biomedical Engineering, Sathyabama Institute of Science and Technology, deemed to be university, Rajiv Gandhi Salai, Chennai – 600119, India.
[2]KM Centre Post Graduate Studies, Pondicherry
*Corresponding authors: bijuknayak@gmail.com and animananda72@gmail.com

14.1 Introduction

In the word "nanotechnology," the prefix "nano" describes one-billionth or 10^{-9} of something. At first, in a famous lecture entitled "There is a plenty room at the bottom," a physics Nobel laureate Richard P Feynman introduced "nanotechnology" [1]. Ever since, many revolutionary developments have been made and demonstrated, which explains

the idea of Feynman in the fields of physics, chemistry, and biology. In 1974, a professor of science at the Tokyo University, namely Norio Taniguchi, invented "nanotechnology" and described it as ultra-fine measurements and extra high precision dimensions. After ten years, "Bottom-Up approach" was introduced by a scientist, namely K Eric Drexler, in which he described and explained how everything in the universe is made of atoms and molecules and thus indeed predicting the future of nanotechnology [2].

Some of the definitions of nanotechnology are as follows:

- It is the study or use of structures between 1 and 1000 nanometers in size having novel properties and functions due to the arrangements of their atoms on the 1–100-nm scale.
- It is the study of fine-tuning of materials at atomic, molecular, and macromolecular scales, where properties differ significantly from those at a larger scale.

These definitions offer a broader and exciting perspective of the role of nanotechnology in the world.

Nowadays, nanotechnology is considered a breakthrough technology for the manipulation and fabrication of materials at the nanometer scale. The technology is thought to have immense potential benefits in various fields like pharmaceuticals, information, and other communication technologies. It has brought enormous benefits in the areas, such as water decontamination and production of lighter and stronger materials. However, it was not until 1981, with the introduction of the scanning tunneling microscope, one could "see" the discrete atoms that modern technology started. It is difficult to imagine just how

Core Microbiome: Improving Crop Quality and Productivity, First Edition. Edited by Javid A. Parray, Nowsheen Shameem, Elsayed Fathi Abd-Allah, and Mohammad Yaseen Mir.

minute nanotechnology is; a billionth of a meter is equal to one nanometer. Here are a few illustrative examples:

- An inch comprises 25,400,000 nm.
- The thickness of a newspaper is around 100,000 nm.
- On a relative scale, supposing a marble as a nanometer, then one meter would approximately be the size of the earth.

Although nanotechnology is considered a modern disciplinary science in physics, chemistry, and biology, it has been there since the pre-Christian Era, for example, the famous Lycurgus cup, which displays different shades of colors depending upon the illumination that is being subjected externally, either green or red, which contains nanoparticles of gold and silver. Another example is when Romans introduced metals with nanometric dimensions. Later on, the colloidal synthesis of gold and its interaction with light was explained [3]. While nanotechnology was explained by every field scientist, in the science of magnetic field, nanoparticles were explained by Blakemore in 1975 when he published pioneer work where the biochemically precipitated magnetite was seen in the tissues of mammals, birds, and microorganisms. Many of these organisms use biogenic magnetite to evaluate the earth's magnetic field for locomotion and navigation. When concentrated to nanoscale by different nanotechnologies, regular and straightforward materials underwent strange changes and results in unpredictable characteristics, including extraordinary strength, chemical reactions, and electrical conductivity superparamagnetic behavior, which the same material does not possess on a macro scale.

Many other types of research and development are under process, which leads to the fabrication of larger molecules into smaller units with better and tremendous functionalities. More appropriately, nanoparticles are defined as the aggregation of atoms and molecules ranging from 1 to 100 nm. They exhibit various size-dependent features and can be made up of two or more species (atoms and molecules). Nanoparticles are classified based on their dimensionality, composition, morphology, aggregation, and similarity [4, 5]. Nanoparticles with one dimension are typically the outer surface coating or thin films. These films have been used for several years in electronic and optic systems. However, these thin films can get dumped by different processes that could have been advanced at the atomic level [6]. Two-dimensional (2D) nanoparticles are commonly used, but their properties are less advanced than those of fibers, fibrils, etc. In contrast, three-dimensional (3D) nanoparticles are very significant and possess great challenges for production and understanding their properties. Examples include zinc oxide and copper oxide.

As concerns the composition of nanoparticles, they are of three types: the first one is mixed; the second one has core-shell nanoparticles, and the third one comprises layered nanoparticles.

- Mixed nanoparticles can either be ordered or random. For example, ordered nanoparticles can be a mixture of two atoms or molecules, while the random one consists of the solutions of solids.
- Core-shell nanoparticles are those individual nanoparticles that consist of one type of atom, which surrounds the core of a different kind of atom. This type of pattern of nanoparticles is common for large classes.

- Layered nanoparticles, commonly referred to as "dumbbells like" in the literature by Janus. These types of nanoparticles consist of different kinds of atoms and molecules linked together by a standard bridge. The help of such nanoparticles minimizes the bond between the two other elements, thus ultimately facilitating phase separation.

Because of the increasing need for multifunctional nanoparticles, they take on the shape of a dumbbell.

14.2 Ancient Agricultural Concepts and Techniques

Human beings started out as hunter–gatherers with a distinct way of life and began to wander along with the seasons, searching for food and shelter. This lifestyle lasted for thousands of years, but then the changes were experienced along with the growth areas and life patterns of plants in response to the retreat of glaciers. Ancient men were capable of differentiating crops for exploitation on a different basis. A semi-nomadic lifestyle eventually followed the nomadic lifestyle. For many years, people set up camps and used them to obstruct wild crops with the amplest resources [7]. It is assumed that the semi-nomads had a herd of cattle and livestock, which they used to travel with wherever they migrated, protecting them, and providing them shelter to use them in many ways, such as feeding on their meat, wearing their fur, and using other resources. The urge to protect food for the next season sprung from the fear that they might not find food. Archaeologists found that people started farming about 10,000 years, at some place in the Indus Valley, which later began as an isolated development community alongside the Yangtze River in China. It has been found that firstly crops and later the livestock were domesticated in large numbers by humans to exploit them in many ways in future [7]. The survival of the human species directly connects with agriculture, as it has intertwined its position in society with its integration of sustainability. In agricultural science and technology, initial civilization could be well thought of as an affluent time. Much of the growing techniques were adopted by the Greeks and Romans from other civilizations with whom they were in contact – best markedly that of the ancient Eastern areas, including Mesopotamia. During the confinement of modern civilization, the actual comprehensive plant and animal agricultural practices were seen [8]. The new technologies made agriculture effective enough to sustain life in contemporary cities; otherwise, technology might not have significantly progressed. Around the eleventh century, there were few changes before the upsurge of the kingdoms in Europe. Through feudalism, the systems instigated, and the secular lords and other church authorities pursued expanding and improvising the yields with the increasing populace. Indeed, momentous technological developments were also present in that period. In that period, there was enormous selective crossbreeding, predominantly in livestock and other organizations. In many places of the world, the fragments of the agricultural system can still be seen. For example, in some European areas, it can be seen as the medieval ridge and furrow-strip farming and some advances called "marginal landscapes," which are the areas lacking in quantitative crop production. Nowadays, agricultural engineering has caught an unprecedented scale that includes drainage of the floodplains, alteration of the low fertility areas for proper use; for example, on the salty soil, barley grows well and thus can be ideally grown on floodplains, and conversion of bracken into pastures and wild woodlands to plains. The final development in modern agricultural practices was seen in the sixteenth century, when the concept of crop rotation was introduced by farmers, which helped them increase crop production by swapping land uses not to deplete the soil. In the beginning, the land was used to grow crops but later on in the next year, the same field was used to domesticate livestock, and in the following third year, the land was kept fallow to protect it from soil depletion. This type of land-use pattern was found to be beneficial to farmers as it resulted in increased economy and crop management proved to be economical as well.

By the twenty-first century, the definition of agriculture has been changed as it primarily deals with the production of staple crops, medicinal plants, poultry products, fisheries, and other cash crops important for shelter, textiles, and horticulture as compared with ancient times where it only revolved around the production of food grain [9]. Agriculture holds the power to solve various societal issues, and innovative, eco-friendly agricultural practices that promote social harmony can be explored. However, ecological imbalances, such as water pollution, soil pollution, air pollution, climate change, pest infestation, and various crop diseases, continue to be significant worldwide. Advances in agricultural practices pave the way to bumper harvesting of food grain, which need suitable processing and packaging for lasting shelf life. In almost all developing countries as India, pests, weeds, insects, and diseases cause significant damage to the agricultural production levels. According to research, pest attack results in enormous losses in rice, wheat, pulses, oilseeds, and cotton [10]. Pests also cause massive damage to the storage go-downs and packages, damaging the quality of the stored food grain. One of the most famous examples of pest management was toxins produced from the bacterium *Bacillus thuringiensis*, which were widely applied against lepidopteran and dipteran species. Incessant practice of using chemical fertilizers, irrigating devices, or Agri-machinery or growing two or more crops in a year result in higher yields, while causing degradation of the environment. As farmers incurred extraordinary production expenses and financial losses, they began to look for less input and maintainable agricultural systems.

Further, with urbanization and modern lifestyles, the food habits of people have changed drastically, to where people have become accustomed to and prefer packaged and ready-to-use food materials. Along with these changes in food habits, alarming rates of food poisoning have been reported worldwide. Therefore, for better production and avoiding any hazardous side effects on human health or the ecosystem, the implementation of various recent innovative farming techniques can be explored.

14.3 Synthesis of Nanoparticles

Production methods involve the processing or formation of nanoparticles. These nanoparticles could be obtained from considerably larger molecules or synthesized by the "bottom-up" approach, which, for instance, nucleates and cultivates particles from fine molecular distributions in both liquid and vapor phase. Synthesis of nanoparticles may also comprise functionalization via conjugation to bioactive molecules. Nanoparticles can be manufactured by adopting many methods, such as co-precipitation, ion sputtering scattering, inert gas condensation, hydrothermal synthesis, spark discharge, pulse laser ablation, microwave, microemulsion, template synthesis, and biological synthesis. Metallic nanocomposites are synthesized by spray pyrolysis, liquid infiltration, high energy ball mailing, physical vapor deposition, chemical vapor deposition, rapid solidification process, and chemical processes, including the colloidal or sol–gel processes.

Ceramic nanocomposites are synthesized by various processes: sol–gel process, polymer precursor process, and powder process. Lastly, the production of polymer nanocomposites comprises in situ intercalation, melt intercalation, mixing in situ polymerization, template synthesis, and the sol–gel method. Nanoparticles manufactured from building blocks have attracted many researchers' attention because of their significant roles in biotechnology. Such nanostructures undergo self-assembly, which is a spontaneous process, and under which, nanostructures get mineralized into organized aggregates. Many biomolecules including proteins and peptides interact and spontaneously self-assemble, resulting in the fabrication of supramolecules with characteristic functionalities that make fibrous scaffolds, helical ribbons, and many more. Biomolecules like deoxyribose nucleic acid (DNA) can be combined or interacted with other substances or entities via different chemical self-assembly processes to form desired nanomaterials of significant importance. The distinctive and unique structure of DNA and its self-recognition property can be utilized to develop self-assembling DNA nanoparticles of various shapes using the "bottom-up" approach. The specific storage capacity of DNA is used in multiple applications, for example, the logic sensor for multiple sequence signals. Circular dichroism spectroscopy, fluorescence resonance energy transfer, laser scanning confocal microscopy, cryogenic transmission electron microscopy, small-angle X-ray scattering, and atomic force microscopy are used to study these nanostructures.

There are two main methods for the synthesis of nanoparticles that are bottom-up and the top-down approach. In a bottom-up approach, the materials are minimized up to an atomic level, which then undergoes a self-assembly process, leading to many significant nanoparticles with specific features and functionalities. Various physical forces act on the nanoscale during this process, thereby combining units into larger stable structures. In a top-down approach, larger initial systems are externally controlled, leading to nanostructure formation. Figure 14.1 shows the top-down and bottom-up processes of nanomaterial synthesis.

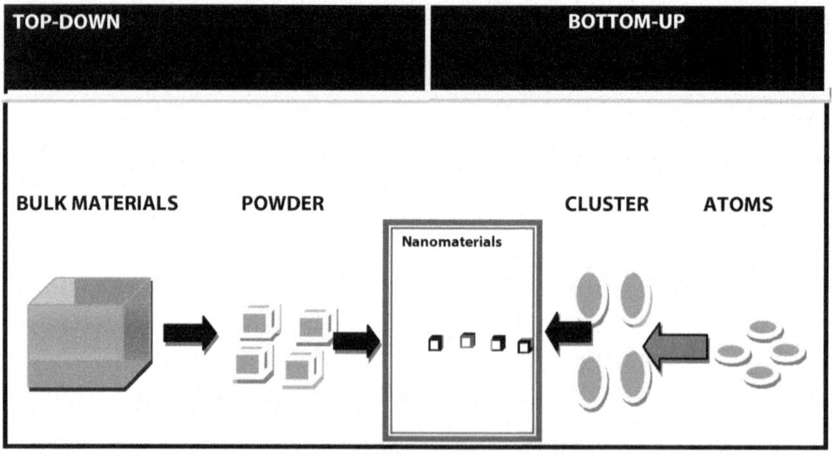

Figure 14.1 The top-down and bottom-up approaches of nanomaterials.

14.3.1 Chemical Methods

1. **Chemical reduction method:** The synthesis of nanoparticles was first reported by Michael Faraday in 1857 using the chemical reduction method. This method is the easiest and convenient for the synthesis of copper nanoparticles. Therefore, the synthesis of nano-sized copper metal particles with distinctive morphologies, shapes, and sizes can be achieved using this specific technique.
2. **Micro-emulsion/colloidal method:** The synthesis of nanoparticles using the micro-emulsion or colloidal form has already been reported. When appropriate amounts of surfactant, alcohol or amine-based co-surfactant, water, and oil are mixed, an emulsion of the clear and homogenous solution is formed, termed "a microemulsion." Thus, microemulsion uses a technique with which a nanoparticle can be synthesized by mixing two immiscible liquids as (water in oil) or (oil in water) or (water in water) or (water in supercritical carbon dioxide), forming thermodynamically stable dissemination with the help of a surfactant because the surfactant can reduce the tension between immiscible liquids. These microemulsions consist of surfactant aggregates, which are of 1–99.5 nm in dimension. Thus, the location of surfactant, oil, and water phases affects the geometry of aggregates. The system is said to be water in oil if the oil is in bulk quantity than water, and similarly, the microemulsion system is said to be oil in water, when the amount of water is in bulk than oil. Mixing of oil, water, and surfactant leads to the formation of a product called micelles, which reduced the free energy in the system. Hydrophobic head groups are pointed toward the water, the bulk matter, while the hydrophobic surfactants are pointed toward the center of aggregates.
3. **Sonochemical method:** In this method, powerful ultrasound radiations (20 kHz to MHz) are applied to the molecules to enhance the chemical reaction. A physical phenomenon called acoustic activation is responsible for the Sonochemical method. Initially, this method was used to synthesize iron oxide particles, but nowadays, this method is applied to produce different metals and metal oxides. This method is quite simple, convenient, and easy to control over the size of the nanoparticles that are being synthesized by using various precursors with a different concentration in the solution. When the powerful ultrasound radiations are projected into the solution, several chemical changes occur because of the acoustic cavitation phenomenon leading to the formation, growth, and collision of bubbles in the liquid solution. A technique called *sonolysis*, where sound waves of fixed frequency are allowed to pass through slurry or a solution of selected metal complex precursors, is included in this method. Meanwhile, in the solvent having a vapor pressure of a specific threshold, the alternating waves of expansion and compression cause cavities to grow and implode. The Sonochemical method has been found to have a general approach in the synthesis of nanoparticles by changing the reaction medium. The Sonochemical method employs both electrolytes as well as ultra-sonic pulses for the synthesis of nanoparticles.
4. **Electrochemical method:** Electricity is used as a driving or controlling force for the synthesis and production of nanoparticles. This is allowed by passing electricity between two electrodes separated by an electrolyte, which means the synthesis occurs at the

electrolyte–electrolyte interface. This technique is less expensive, most accessible, most straightforward, and most convenient. The most significant advantage of this method is the avoidance of vacuum systems and less contamination. In addition, it is an eco-friendly process and is being highly exploited by scientists and brought into industrial uses, but still, many aspects of the technique are unknown.

5. **Solvothermal decomposition:** In this method, the chemical reactions occur inside a sealed chamber as an autoclave, where the temperature is maintained above the boiling point of the solvents, respectively. Thus, when water is used, it is called the hydrothermal process. The main advantage of using this process is the presence of a single-phase where highly purified nanoparticles with the highest crystalline properties are synthesized. In addition, the technique is eco-friendly as well.

14.3.2 Physical Methods

1. **Pulse laser ablation:** In this method, a plasma is created by passing a high volt pulsed laser beam inside the chamber to strike the target in the material, which later gets converted into colloidal solutions of nanoparticles. Many factors affect the formation of the final product, such as the type of pulse, timing of pulse, number of pulses, and type of solvent.
2. **Mechanical/high ball milling method:** It is the solid-state technique used to characterize and synthesize nanoparticles. Benjamin first implemented the technique for the production of two alloys. In this process, a micron size raw material is brought down to undergo specific procedures and changes. Various mechanical mills are used for the synthesis of nanoparticles categorized sequentially according to their applications and capacities. The obstacle in the method is that ultra-thin particles cannot be synthesized because of mechanical limitations, or it takes a long time. However, low-cost production of nanoparticles can be formed by this method. The factors that affect the product are type of mills, milling speed, time, temperature, container, atmosphere, size and distribution of the grinding medium, and extent of filling the vial.
3. **Mechanical chemical synthesis:** This process involves the mechanical energy, which is induced by the chemical reaction. Chemical reactors are usually a mixture of oxides, chlorides, or metals that react during the heat treatment and form a composite powder into which ultra-fine nanoparticles are dispersed in a stable salt matrix. These nanoparticles are recovered by washing with suitable solvents following the selective removal of the matrix.
4. **Pulsed wired discharged method:** As one of the techniques to prepare or synthesize nanoparticles, it follows an entirely different mechanism than the above-mentioned chemical and physical methods. In this particular process, a metal wire is evaporated by a pulsed current to produce vapors, cooled down by ambient gas to form nanoparticles. Nitrite and metal oxides prepared by this method are found to be more productive and efficient. However, the technique is not commonly used in industrial processes because of its high cost and inability to use different metals. The method is observed to be best for high conductivity metals, which come in a thin wire.

14.3.3 Biological Methods

Biological or biogenic synthesis of nanoparticles is an eco-friendly method, which is cheaper than physical and chemical methods. The physical and chemical processes are time-consuming, need costly equipment, and release several toxic substances into the environment. Therefore, the biological approach is preferred over the physical and chemical methods. The natural techniques employ bacteria, fungi, and plant extracts for nonmaterial bio-synthesis [11]. Figure 14.2 shows the different methods of nanomaterial synthesis.

14.3.4 Role of Nanotechnology in Agriculture

In most developing countries, agriculture forms the backbone as it supports nearly 60% of the human population. Accordingly, many challenges, including the non-reasonable use of resources, climate change, and the more significant chemical fertilizers, have been faced by the agriculture sector [12]. In the current era, nanotechnology has developed as the most revolutionary technology and has also been considered an emerging scientific field. It deals with using the smallest possible particles, which are responsible for increasing agricultural

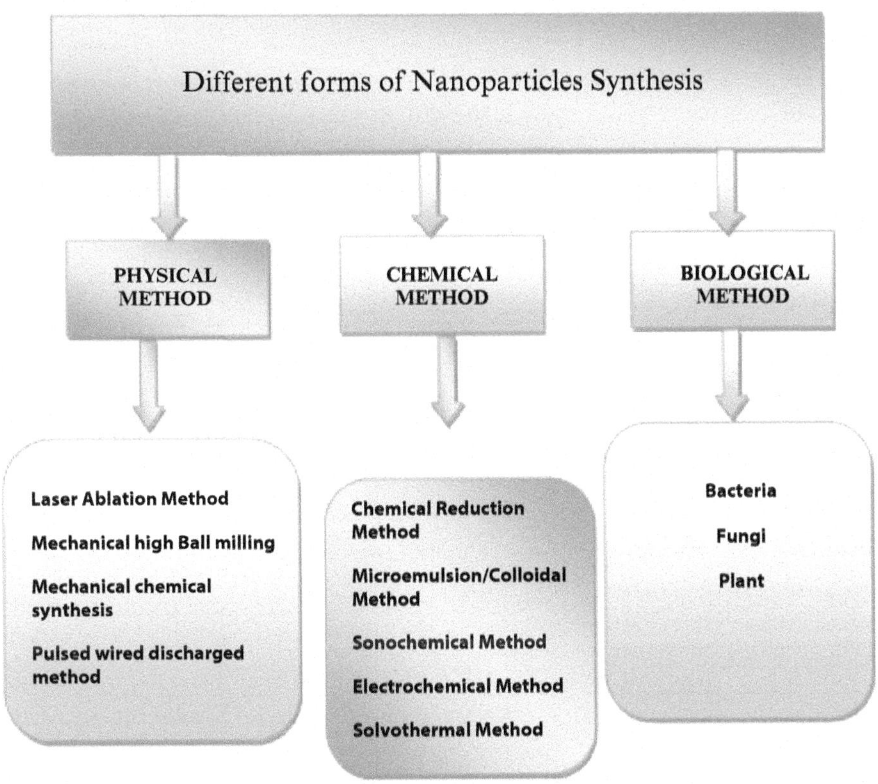

Figure 14.2 Different methods of nanoparticles synthesis.

productivity followed by better management and protecting the products exploited from plants and animals. To accomplish the growing population's upcoming needs, a great scope of novel applications is provided by nanotechnology because of the exclusive physicochemical characteristics of nanoparticles, such as higher reactive nature, more surface area, and appropriate pore size. Furthermore, applying nanotechnology directly in agricultural production generally embraces the improvement of genomes, the study of plant disease mechanisms, and the delivery of various agrochemicals [13].

The nano range includes many biological macromolecules (DNA, RNA, proteins) and other dynamic processes in nature. All around the globe, to defeat the food crisis, malnutrition, and to recover the scenario in agriculture and its secondary outfits for appropriate strategies, continuous efforts are being made by researchers [14, 15]. Nanotechnology forms connecting links between the agro-food industry and agricultural scientists, nano engineers, and molecular biology scientists as the chemical organizations of plants and animals are constructed at subatomic levels [16–18].

For agricultural applications, nanoparticles are supposed to possess the following properties:

1. In response to certain conditions, they provide definite concentration and meticulous release of pesticides or fertilizers [19].
2. Enhanced targeted activity [20].
3. Not spread to the ecosystem as compared to pesticides or fertilizers during usage and transportation.

Nanotechnology characterizes a field holding a significant promise [21–23]. Nanomaterials play a tremendous role in agriculture because of their dimensional size and properties, surface area, higher volume ratio, and classical optical properties. To prepare nanoparticles, a large variety of substances are used, such as ceramics, oxides, magnetic substances, emulsions, silicates, lipids, dendrimers, polymers, and semiconductor quantum dots (QDs) [24–26]. Size-dependent properties are displayed by metal nanoparticles, for instance, in agrochemical degradation, soil remediation and sensor development, semiconductor quantum dots, magnetic nanoparticles, and metal oxide nanoparticles possess great biotechnological applications.

In all the stages of agricultural fabrication, some of the main uses of nanomaterials are:

1. nano pesticides [27]
2. post-harvest technology [28]
3. weed management
4. nanosensors [29]
5. zinc nano-fertilizer used to enhance crop production of *Pennisetum americanum* [30]
6. plant growth regulators [31]
7. seed technology
8. food technology [32]
9. management of water [33]
10. management of soil [34]
11. aspects of agricultural engineering [35]
12. elements of plant protection (pesticides, weedicides, fungicides)

13. nano-fertilizers for balanced crop nutrition [13, 36]
14. nanobiotechnology
15. plant disease diagnostics.

The sensor systems, which are based on nanotechnology, hold the basic research level for agricultural applications. By early intervention, detection systems, such as diagnostic devices and sensors (for on-site monitoring of environmental conditions), can prove immensely helpful in the increase in productivity and decrease in the use of harmful agrochemicals, such as antibiotics, nutrients, and pesticides. Apart from detection, nanotechnology is very helpful in processing obstinate chemicals into innocuous ones, which often results in the formation or transformation of valuable components. Thus, nanotechnology can be brought into use to address the critical problems that the world has been facing in the given time regarding environmental pollution and its protection. Furthermore, it can streamline the technologies used in detecting, sensing, and remediating ecological problems.

14.4 Application of Nanotechnology in Plant Growth

Today, nanoparticles and nanotechnology have developed as the most potential and innovative branches of basic science and engineering technology. In day-to-day life, their incorporation has provided an essential diversity in every field of science and technology. To control the dimensions of the nanoparticles formed in plants during synthesis, naturally occurring chemicals with natural defense and biological activity called phytochemicals help themselves reduce agents during the synthesis period. Therefore, with no further use of reducing and capping materials, with a wide array of raw materials and large-scale production capabilities in a cost-effective environment, the biological route provides an eco-friendly process compared with the chemical and physical routes [37]. In various agricultural fields, nanotechnology has been widely employed, and one such application pertains to the usage of nanomaterials for protecting crops [38, 39]. Nanomaterials also act as magic bullets when it comes to the fabrication of dynamic chemicals and genes for upgradation of better yield and disease-free crops [40], for example, in the development of chemically formulated mesoporous silica nanoparticles, which function as gene shippers for the allocation of genes to plants similar to tobacco and corn. The target genes are then smoothly transferred to the host plant in an eco-friendly manner after the cell wall absorbs nanoparticles. Scientists are working closely to develop a powerful solution for protecting food and agricultural commodities from diverse pathogens throughout the globe. Nano-biotechnology can bring significant changes to agriculture; especially its contribution to plant health has been reported worldwide. For viable crop production, micronutrient and pesticide management are the main points of concern to scientists today. Plants are responsible for producing biologically efficient and harmless molecules, which are also non-persistent in soil and freshwater [41]. In several countries, the occurrence of a dreadful disease of tomato and lettuce is fusarium wilt. The disorders can be controlled to some extent by using specific resistant cultivars and chemicals. The conditions act by causing prolonged survival of fungus in the soil, which leads to a severe loss of soil production capacity. To control such plant pathogens, use of costly chemicals can be effectively replaced by nanomaterials. Nanomaterials arouse many essential facets of plant biology as the highly porous root and leaf surfaces of plants are the primary nutrient gateways [42, 43]. Subsequently, plant nutrient uptake through these gateways can be improved by applying nano-fertilizers either by enhancing the formation of new pores or by manipulating ion channels or endocytosis [44]. According to continuous research, it has been observed that the fertilizer nano-formulations safeguard balance in crop nutrition throughout the growth cycle, which ultimately helps develop the overall agricultural yield, thus encouraging farmers to make use of the products more lucratively. Besides the macronutrients, micronutrients, such as chlorine, iron, copper, zinc, boron, manganese, and molybdenum, are also responsible for an efficient increase in crop productivity. The deficiency of such nutrients decreases crop productivity and soil fertility and paves the way to adverse effects on human well-being through food consumption [45, 46]. In a nutshell, increasing the essential nutrients (element fertilization) is expected to improve crop productivity and soil fertility [47]. One of the significant nanotechnology applications includes nanoparticle-facilitated DNA gene transformation in plants to develop insect-resistant varieties, increased shelf life of products, and food dispensation and storage. The improvement in biomass-to-fuel production is also enhanced by nanotechnology. The applications of nanotechnologies intend to lessen the spraying of plant protection chemicals and improve plant yields in agriculture. For diagnosing and treating diseases in plants, nanocapsules, and nanoparticles are the best examples for use. In the science of plant breeding and plant genetics, nanotechnology-derived devices are also discovered. The introduction of new methods to a series of economic nanotech applications for boosted plant growth, seed propagation, and improvement besides acclimatization to the environment comes under a new podium of scientific revolution called nanoscience. Seed germination is one of the most sensitive phases in the life cycle of a plant, which is significantly affected by various parameters, such as environmental factors, moisture availability, genetic trait, and soil fertility [48]. Broader studies on this aspect of plant lifecycle have shown that nanomaterial application has a positive effect on seed germination and plant growth [49–52] and ultimately improve yields and product quality. Furthermore, various studies have shown that over the entire crop period of cultivation, the long-term accessibility of all the nutrients to the plant is crucial for promoting seed germination, flowering, fruiting, and plant growth and development [53]. Therefore, nanotechnology again plays an essential role in plant protection. Active ingredients in minimum effective concentration at the target sites are crucial for assurance of better plant protection from pest invasion and subsequent crop loss. In this regard, a new technology of plant protection called nano-formulation or encapsulation of pesticides involves a minimal number of particles that act as active ingredients of pesticides [54]. The nano-formulations of pesticides enable to strengthen the plant-based systemic acquired resistance (SAR) against pests. For example, the penetration of pesticides into plant and cell sap transportation is facilitated by silica nanosphere formulations, thereby exerting a systemic effect to control the harmony of chewing or sucking-type insects, such as aphids [55]. Presently, the use of numerous metallic and organic nanoparticles against a wide range of plant pathogens, insects, and pests as insecticides, pesticides, and insect repellants is in progress. To a greater extent, crop productivity is affected by various biotic factors, such as diseases and pests [56]. Farmers have been profoundly dependent on pesticides, which ultimately affect human health and environmental sustainability.

On the other hand, nanomaterials could successfully bring changes and decrease the risks of these pests and diseases, thereby minimizing yield losses and ecological hazards. This property of nanomaterials can be highlighted in the form of an example of the cotton plant (*Gossypium hirsutum*) in which biosynthesized AgNPs (obtained from the stem extract) have a vigorous antibacterial activity as evidenced by the zone of inhibition for the two major bacterial pathogens of Malvaceae and Brassicaceae family crops – *Xanthomonas axonopodis* and *Xanthomonas campestris*, respectively [57]. Hence, without disturbing environmental health, crop productivity can be improved or increased by using the calculated number of nanomaterials. Because of their eco-friendly nature and high capability, scientists have broadened the use of nanocomposites for plant protection [58]. Based upon various studies, it has been estimated that for efficient absorption, accumulation, translocation, and other effects on the growth and development of different plants, such as corn, wheat, spinach, cucumber, radish, tomato, soybean, alfalfa, onion, and pumpkin, metal- and carbon-based nanoparticles play a vital role [59, 60]. Therefore, for improving germination, photosynthetic activity, and mRNA expression, a remarkable role is played by the nanoparticles paving the way to the note-worthy positive changes in gene expression, which in turn can have a positive effect on the inclusive growth and yield of the crops [61, 43]. Nano-fertilizers deliver essential encapsulated substances/nutrients in a harmonized manner, that is, slowly and in a controlled way that it can assist in improving the efficacy of the nutrients for more than a month [62] and the proficient uptake of the encapsulated nutrients by the standing crops [63]. Such nanoparticles can be produced or synthesized by various nanotechnological methods, such as the bottom-up or the top-down approach, by exploring multiple organic and inorganic materials that are further loaded with various macro- and micronutrients that the plants can quickly take up. In the twenty-first century, nanotechnology has created a remarkable place, mainly in medical equipment, detergents, air and water purifiers, textiles, cosmetics, and medicines [64, 65]. The green synthesis route of nanoparticle synthesis became popular in the twenty-first century as inspired by the various texts available in Ayurveda. However, the production and synthesis of nanomaterials via multiple methods, which involves both physical and chemical processes and the formation of various toxic by-products, constitute some main drawbacks [66]. Figure 14.3 presents the role and use of nanomaterials for different agricultural purposes.

14.5 Role of Nanotechnology in Soil Microbiome Development

Nano agriculture has become a positive trend when it comes to soil development and its fertility. Maintaining soil microbiome and their habitat has become a debatable issue where scientists question and analyze problems and try to resolve them. Soil microbiome is significant as other nutrients are readily available in the soil and has tremendously improved in the past few decades. The considerable contribution of nanotechnology and its

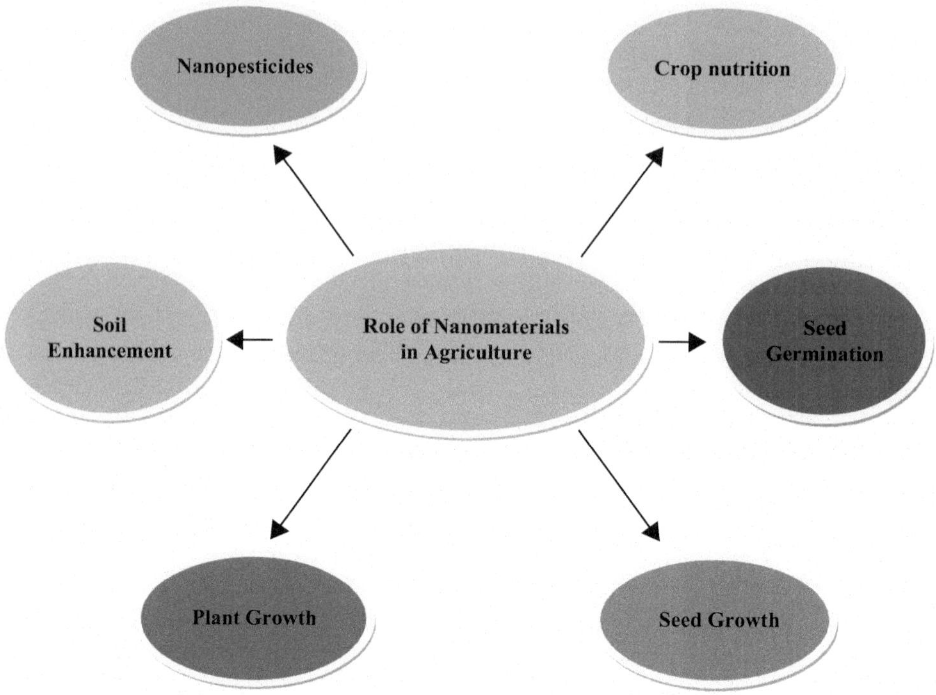

Figure 14.3 Role of nanomaterials in agriculture.

thriving aspect has brought laurels and pioneers to soil scientists. Scientists are learning to deal with the role and population of the microbiome more positively. Nanotechnology has helped maintain and develop the soil microbiome to a significant extent, which was undoubtedly not possible some 10 years ago. Millions of microbiomes have contributed to soil enrichment and fertility. Yet, specific microbes are selected by plants themselves for allowing the microbes to play a part in their growth and development. While phytopharmaceuticals tackle problems attributable to the soil microbiome, some other technologies as nano chemicals, nano agrochemicals, or nano pesticides have made a significant difference. These are active substances that are fabricated via nanotechnologies and their nanoformulations to form more efficient substances that increase soil characteristics and improve its quality. Nano-based agrochemicals enhance fertilization, pesticides, and finally, crop yield. Certain nanosensors have also been introduced lately, which are proving to be of great benefit for crops. Unlike nanomaterials of dimensions between 1 and 100 nm, the size of nano agrochemicals can exceed up to 100 nm threshold due to mixtures of chemicals.

Most nano pesticides, when being exposed to plants, become active inside the plant. However, as soon as they are told, they get scattered into the environment, disturbing the primary role of nano pesticides. Finally, the plant becomes a non-target organism. Thus, the impact of nano pesticides on these microbiomes as a non-target organism remains underestimated up to now.

14.6 Conclusions

Nanotechnology has tremendous uses in all aspects of life, whether science, physics, chemistry, or biology. Along with that, it has excellent roles in production, processing, storage, packaging, and all other agricultural aspects. The potential benefits of nanotechnology for agriculture, food, fisheries, and aquaculture have been felt and need to be analyzed and balanced against soil fertility, ecosystem, and workers' health. Nanotechnology is believed to have revolutionized agriculture and the food industry, for instance, by enhancing the ability of plants to absorb nutrients, while helping farmers detect diseases and control pests and practice other comprehensive farming techniques. Nanotechnology has always offered real solutions, such as assessing nutrient efficiency and improving crop varieties, and monitoring plant growth and development. In the development of the agricultural sector, nanotechnology acts as a lynchpin by increasing the volume of global crops production and yield and thereby can feed the whole world population in the following decades, and also supports agriculture through modern techniques and the increasing probability of products. It has already developed some promising results and applications in nano nutrients, plant protection (herbicides and pesticide), nano-packing, crop productivity implementation, and nanosensors. Nanomaterials used instead of pesticides and fertilizers are supposed to reduce the dosage and safeguard controlled slow delivery. In reducing environmental threats, the application of nanoparticles is the main contribution estimated for stabilizing biocontrol preparations. By using the unique properties of nanomaterials, such as the transformation of persistent chemicals into useful ones and sometimes functional components, nanosensors are developed, which have that excellent capability to detect pathogens even to the extent of parts per billion. In agriculture, the potential of nanotechnology is quite vast, but with regard to risk assessment, a few issues still ought to be considered. Agricultural techniques and practices of growing crops appreciate the structured, systematic, and organized application of a comprehensive assemblage of active compounds at variable dosages and frequencies. Recently, nanotechnology has gained attention from the world due to its many applications in different fields, such as agriculture, medicine, and the environment. The larger surface area possessed by the minute nanoparticles makes them appealing to deal with the challenges that are not yet met by different control methods provided by scientists earlier.

For the betterment of humankind, agricultural technologies should take massive advantage of these practical tools of nanotechnology, which can be used to deal with critical issues, such as pollution and other threats to agriculture. Nanotechnology applications are continuously being researched, giving rise to a new spectrum of discovery each day.

References

1 Feynman, R.P. (1960). There's plenty of room at the bottom. *Engineering and Science*. 23(5): 22–36.

2 Drexler, K.E. (1990). *Engines of Destruction, in Engines of Creation: The Coming Era of Nanotechnology*. 1986. New York, NY: Anchor Books. *Elimelech, M., and CR O'Melia, Effect of particle size on collision efficiency in the deposition of Brownian particles with electrostatic energy barriers. Langmuir, 6*: 1153-1163.

3 Faraday, M. (1857). X. The Bakerian Lecture—Experimental relations of gold (and other metals) to light. *Philosophical Transactions of the Royal Society of London.* 147: 145–181.

4 Rajput, N. (2015). Methods of preparation of nanoparticles-a review. *International Journal of Advances in Engineering & Technology* 7 (6): 1806.

5 Thodeti, S., Reddy, S.S., and Vemula, S. (2018). Synthesis and characterization of copper nanoparticles by chemical reduction method. *Research Journal of Science and Technology* 10 (1): 52–57.

6 Feng, P. and Cao, W. (2016). Properties, application and synthesis methods of nano-molybdenum powder. *Journal of Materials Science and Chemical Engineering* 4 (9): 36–44.

7 Bar-Yosef, O. (1998). On the nature of transitions: The middle to upper Palaeolithic and the Neolithic revolution. *Cam. Arch. Jnl* 8 (02 October): 141–146.

8 Watson, A. (1974). The Arab Agricultural Revolution and Its Diffusion, 700–1100. *The Journal of Economic History* 34 (1): 8–35.

9 Sapkota, T.B., Vetter, S.H., Jat, M.L., Sirohi, S., Shirsath, P.B., Singh, R., and Stirling, C.M. (2019). Cost-effective opportunities for climate change mitigation in Indian agriculture. *Science of the Total Environment* 655: 1342–1354.

10 Akoijam, R., Telem, R.S., and Marangmei, L. (2014a). Insect pest problems and their changing trends on crop losses. *Environment and Ecology* 32 (3): 837–841.

11 Sharmila, G., Thirumarimurugan, M., and Muthukumaran, C. (2019). Green synthesis of ZnO nanoparticles using *Tecoma castanifolia* leaf extract: Characterization and evaluation of its antioxidant, bactericidal and anticancer activities. *Microchemical Journal* Mar 1 145: 578–587.

12 Raliya, R., Saharan, V., Dimkpa, C., and Biswas, P. (2017). Nano fertilizer for precision and sustainable agriculture: Current state and future perspectives. *Journal of Agricultural and Food Chemistry* 66 (26): 6487–6503.

13 Abobatta, W.F. (2017a). Different Impacts of Nanotechnology in Agricultural sector development. Nano Technology Science and application-the Creative Researchers first scientific annual conference.

14 Bird, F.A., Pradhan, A., Bhavani, R.V., and Dangour, A.D. (2019). Interventions in agriculture for nutrition outcomes: A systematic review focused on South Asia. *Food Policy* 82: 39–49. doi: 10.1016/j.foodpol.2018.10.015.

15 Ngo, J. and Serra-Majem, L. (2019). Hunger and malnutrition. In: *Encyclopedia of Food Security and Sustainability*, (eds. P. Ferranti, E.M. Berry and J.R. Anderson), 315–335. Oxford: Elsevier. DOI: 10.1016/B978-0-08-100596-5.22064-7.

16 Chhipa, H. (2019). Applications of nanotechnology in agriculture (Chapter 6). In: *Methods in Microbiology*, (eds. V. Gurtler, A.S. Ball and S. Soni), 46, 115–142. Academic Press.

17 Duhan, J.S., Kumar, R., Kumar, N., Kaur, P., Nehra, K., and Duhan, S. (2017). Nanotechnology: The new perspective in precision agriculture. *Biotechnology Reports* 15: doi: 11–23. 10.1016/j.btre.2017.03.002.

18 Iavicoli, I., Leso, V., Beezhold, D.H., and Shvedova, A.A. (2017). Nanotechnology in agriculture: Opportunities, toxicological implications, and occupational risks. *Toxicology and Applied Pharmacology* 329: 96–111. doi: 10.1016/j.taap.2017.05.025.

19 Raliya, R., Biswas, P., and Tarafdar, J.C. (2015). TiO2 nanoparticle biosynthesis and its physiological effect on mung bean (*Vigna radiata* L.). *Biotechnology Reports* 5: 22–26.

20 Changmei, L., Chaoying, Z., Junqiang, W., Guorong, W., and Mingxuan, T. (2002). Research of the effect of nanometer materials on germination and growth enhancement of Glycine max and its mechanism. *Soybean Science* 21 (3): 168–171.

21 Baruah, S. and Dutta, J. (2009). Nanotechnology applications in pollution sensing and degradation in agriculture: A review. *Environmental Chemistry Letters* 7: 161–204.

22 Kuzma, J. (2007). Moving forward responsibly: Oversight for the nanotechnology-biology interface. *Journal of Nanoparticle Research: An Interdisciplinary Forum for Nanoscale Science and Technology* 9: 165–182.

23 Navrotsky, A. (2000). Technology and applications Nanomaterials in the environment, agriculture, and technology (NEAT). *Journal of Nanoparticle Research: An Interdisciplinary Forum for Nanoscale Science and Technology* 2: 321–323.

24 Niemeyer, C.M. and Doz, P. (2001). Nanoparticles, proteins, and nucleic acids: Biotechnology meets materials science. *Angewandte Chemie Int Ed* 40: 4128–4158.

25 Oskam, G. (2006). Metal oxide nanoparticles: Synthesis, characterization and application. *J Sol-Gel Sci Technol* 37: 161–164.

26 Puoci, F., Lemma, F., Spizzirri, U.G., Cirillo, G., Curcio, M., and Picci, N. (2008). Polymer in agriculture: A review. *American Journal of Agricultural and Biological Sciences* 3: 299–314.

27 Corradini, E., De Moura, M.R., and Mattoso, L.H.C. (2010). A preliminary study of the incorporation of NPK fertilizer into chitosan nanoparticles. *Express Polymer Letters* 4: 8.

28 Meetoo, D. (2011). Nanotechnology and the Food Sector: From the Farm to the Table. *Emirates Journal of Food and Agriculture* 23 (5): 387–403.

29 Das, M., Saxena, N., and Dwivedi, P.D. (2009). Emerging trends of nanoparticles application in food technology: Safety paradigms. *Nanotoxicology* 3 (1): 10–18.

30 Tarafdar, J.C., Raliya, R., Mahawar, H., and Rathore, I. (2014). Development of zinc nanofertilizer to enhance crop production in pearl millet (*Pennisetum americanum*). *Agricultural Research* 3 (3): 257–262.

31 Choy, J.H., Choi, S.J., Oh, J.M., and Park, T. (2007). Clay minerals and layered double hydroxides for novel biological applications. *Applied Clay Science* 36 (1-3): 122–132.

32 Yata, V.K., Tiwari, B.C., and Ahmad, I. (2018). Nanoscience in food and agriculture: Research, industries and patents. *Environmental Chemistry Letters* 16 (1): 79–84.

33 Bharathi, P., Balasubramanian, N., Anitha, S., and Vijayabharathi, V. (2016). Improvement of membrane system for water treatment by synthesized gold nanoparticles. *Journal of Environmental Biology* 37 (6): 1407.

34 Klingenfuss, F. (2014). Testing of Tio2 nanoparticles on wheat and microorganisms in a soil microcosm. Thesis for Master of Science in ecotoxicology, University of Gothenburg: 62.

35 González-Melendi, P., Fernández-Pacheco, R., Coronado, M.J., Corredor, E., Testillano, P.S., Risueño, M.C., and Pérez-de-luque, A. (2008). Nanoparticles as smart treatment-delivery systems in plants: Assessment of different techniques of microscopy for their visualization in plant tissues. *Annals of Botany* 101 (1): 187–195.

36 Janmohammadi, M., Amanzadeh, T., Sabaghnia, N., and Dashti, S. (2016). Impact of foliar application of nano micronutrient fertilizers and titanium dioxide nanoparticles on the growth and yield components of barley under supplemental irrigation. *Acta Agriculturae Slovenica* 107 (2): 265–276.

37 Nayak, D., Pradhan, S., Ashe, S., Rauta, P.R., and Nayak, B. (2015). Biologically synthesized silver nanoparticles from three diverse families of plant extracts and their anticancer activity against epidermoid A431 carcinoma. *Journal of Colloid and Interface Science* 457: 329–339. doi: 10.1016/j.jcis.2015.07.012.

38 Kah, M. (2015). Nanopesticides and nanofertilizers: Emerging contaminants or opportunities for risk mitigation? *Frontiers in Chemistry* 3.

39 Servin, A., Elmer, W., Mukherjee, A., De La Torre-roche, R., Hamdi, H., White, J.C., Bindraban, P., and Dimkpa, C. (2015). A review of the use of engineered nanomaterials to suppress plant disease and enhance crop yield. *Journal of Nanoparticle* 17: 1–21.

40 Pérez-de-luque, A. and Rubiales, D. (2009). Nanotechnology for parasitic plant control. *Pest Management Science* 65 (2): 540–545.

41 Isman, M.B. (2000). Plant essential oils for pest and disease management. *Crop Protect* 19: 603–608.

42 Eichert, T. and Goldbach, H.E. (2008). Equivalent pore radii of hydrophilic foliar uptake routes in stomatous and astomatous leaf surfaces-further evidence for a stomatal pathway. *Physiologia Plantarum* 132: 491–502.

43 Rico, C.M., Majumdar, S., Duarte-Gardea, M., Peralta-Videa, J.R., and Gardea-Torresdey, J.L. (2011a). Interaction of nanoparticles with edible plants and their possible implications in the food chain. *Journal of Agricultural and Food Chemistry* 59 (8): 3485–3498. doi: 10.1021/jf104517j.

44 Mastronardi, E., Tsae, P., Zhang, X., Monreal, C., and DeRosa, M.C. (2015). Strategic role of nanotechnology in fertilizers: Potential and limitations. In: *Nanotechnologies in Food and Agriculture*, (ed. M. Rai, C. Ribeiro, L. Mattoso and N. Duran), 25–67. Cham, Switzerland: Springer.

45 Monreal, C.M., DeRosa, M., Mallubhotla, S.C., Bindraban, P.S., and Dimkpa, C. (2016). Nanotechnologies for increasing the crop use efficiency of fertilizer-micronutrients. *Biology and Fertility of Soils* 52: 423–437.

46 Swaminathan, S., Edward, B.S., and Kurpad, A.V. (2013). Micronutrient deficiency and cognitive and physical performance in Indian children. *European Journal of Clinical Nutrition* 67: 467–474.

47 Li, C., Li, Y., Li, Y., and Fu, G. (2018). Cultivation techniques and nutrient management strategies to improve productivity of rain-fed maize in semi-arid regions. *Agricultural Water Management* 210: 149–157.

48 Manjaiah, K.M., Mukhopadhyay, R., Paul, R., Datta, S.C., Kumararaja, P., and Sarkar, B. (2019). Clay minerals and zeolites for environmentally sustainable agriculture. In: *Modified Clay and Zeolite Nanocomposite Materials* (ed. M. Mariano, S. Binoy and L. Alessio), 309–329. Amsterdam, The Netherlands: Elsevier.

49 Joshi, A., Kaur, S., Dharamvir, K., Nayyar, H., and Verma, G. (2018). Multi-walled carbon nanotubes applied through seed-priming influence early germination, root hair, growth and yield of bread wheat (*Triticum aestivum* L.). *Journal of the Science of Food and Agriculture* 98: 3148–3160.

50 Khodakovskaya, M.V. and Biris, A.S. (2010). Method of Using Carbon Nanotubes to affect Seed Germination and Plant Growth. *U.S. Patent Application* 13: 509–487.

51 Lahiani, M.H., Dervishi, E., Chen, J.H., Nima, Z., Gaume, A., Biris, A.S., and Khodakovskaya, M.V. (2013). Impact of carbon nanotube exposure to seeds of valuable crops. *ACS Applied Materials & Interfaces* 5: 7965–7973.

52 Srivastava, A. and Rao, D. (2014). Enhancement of seed germination and plant growth of wheat, maize, peanut and garlic using multiwalled carbon nanotubes. *European Chemical Bulletin* 3: 502–504.

53 Lateef, A., Nazir, R., Jamil, N., Alam, S., Shah, R., Khan, M.N., and Saleem, M. (2016). Synthesis and characterization of zeolite-based nanocomposite: An environment friendly slow-release fertilizer. *Microporous Mater* 232: 174–183.

54 Haq, I.U. and Ijaz, S. (2019). Use of Metallic Nanoparticles and Nanoformulations as Nanofungicides for Sustainable Disease Management in Plants. In: *Nanobiotechnology in Bioformulations*, (eds. R. Prasad, V. Kumar, M. Kumar and D. Choudhary), 289–316. Cham, Switzerland: Springer.

55 Li, Z.Z., Chen, J.F., Liu, F., Liu, A.Q., Wang, Q., Sun, H.Y., and Wen, L.X. (2007). Study of UV- shielding properties of novel porous hollow silica nanoparticle carriers for avermectin. *Pest Management Science* 63: 241–246.

56 Oerke, E.C. (2006). Crop losses to pests. *Journal of Agricultural Science* 144: 31–43.

57 Vanti, G.L., Nargund, V.B., Basavesha, K.N., Vanarchi, R., Kurjogi, M., Mulla, S.I., Tubaki, S., and Patil, R.R. (2019). Synthesis of *Gossypium hirsutum*-derived silver nanoparticles and their antibacterial efficacy against plant pathogens. *Applied Organometallic Chemistry* 33: e4630.

58 Gomathi, T., Rajeshwari, K., Kanchana, V., Sudha, P.N., and Parthasarathy, K. (2019). Impact of nanoparticle shape, size, and properties of the sustainable nanocomposites. In: *Sustainable Polymer Composites and Nanocomposites* (ed. Sabu Thomas Inamuddin, Raghvendra Kumar Mishra, Abdullah M. Asiri)), 313–336. Cham, Switzerland: Springer International Publishing.

59 Nair, R., Varghese, S.H., Nair, B.G., Maekawa, T., Yoshida, Y., and Kumar, D.S. (2010). Nanoparticulate material delivery to plants. *Plant Science* 179 (3): 154–163. doi: 10.1016/j.plantsci.2010.04.012.

60 Rai, P.K., Kumar, V., Lee, S., Raza, N., Kim, K.-H., Ok, Y.S., and Tsang, D.C.W. (2018). Nanoparticle-plant interaction: Implications in energy, environment, and agriculture. *Environment International* 119: 1–19. doi: 10.1016/j.envint.2018.06.012.

61 Hussain, I., Singh, A., Singh, N.B., Singh, A., and Singh, P. (2019). Plant-nanoceria interaction: Toxicity, accumulation, translocation and biotransformation. *South African Journal of Botany* 121: 239–247. doi: 10.1016/j.sajb.2018.11.013.

62 Juárez-Maldonado, A., González-Morales, S., Cabrera-de La Fuente, M., Medrano-Macías, J., and Benavides-Mendoza, A. (2018). Chapter 10 – Nanometals as promoters of nutraceutical quality in crop plants. In: *Handbook of Food Bioengineering* (ed. A. M. Grumezescu and A. M. B. T.-I. of N. in the F. I. Holban), 277–310. Academic Press. doi: 10.1016/B978-0-12-811441-4.00010-8.

63 Chen, J., Lü, S., Zhang, Z., Zhao, X., Li, X., Ning, P., and Liu, M. (2018). Environmentally friendly fertilizers: A review of materials used and their effects on the environment. *Science of the Total Environment* 613–614, 829–839. doi: 10.1016/j.scitotenv.2017.09.186.

64 Guo, Z. and Sadler, P.J. (1999). Metals in medicine. *Angewandte Chemie International Edition* 38 (11): 1512–1531. doi: 10.1002/(SICI)1521-3773(19990601)38:11<1512::AIDANIE1512>3.0.CO;2-Y.

65 Hofmann-Amtenbrink, M., Grainger, D.W., and Hofmann, H. (2015). Nanoparticles in medicine: Current challenges facing inorganic nanoparticle toxicity assessments and standardizations. *Nanomedicine: Nanotechnology, Biology, and Medicine* 11 (7): 1689–1694. doi: 10.1016/j.nano.2015.05.005.

66 Panigrahi, S., Kundu, S., Ghosh, S., Nath, S., and Pal, T. (2004). General method of Synthesis for metal nanoparticles. *Journal of Nanoparticle Research* 6 (4): 411–414.

15

Microbial Biofilms

Optimal Genetic Material Exchange in a Microbiome Environment

Niraj Singh[1] *and Pranjal Pratim Das*[2]

[1]*Department of Microbiology, The Assam Royal Global University, Guwahati, Assam, 781035*
[2]*Department of Biotechnology, Darrang College, Tezpur, Assam, India, 784028*

15.1 Introduction

A majority of the bacterial population can form "biofilms" when they are in contact with abiotic or biotic surfaces [1]. Bill Costerton, the "father of biofilms," has explained biofilm as "when microbial cells are confined in an own-generated, polymeric matrix" [2,3]. Microbes involved in biofilm formation also have been noted to have a potent ability to produce exopolymers [4]. Thus, biofilms are also defined as a well-organized group of microorganisms persisting within a self-secreted polymeric matrix. Generally, biofilms form a monolayer or multilayer structural setup through which bacterial cells can comfortably attach to the surface and the surrounding population of bacteria with the help of an extracellular matrix comprising polysaccharides, proteins, and DNA [5]. Bacterial communities present in the biofilm are distinct from planktonic ones in various ways, such as gene expression pattern and growth rate. In addition to prokaryotic microbes, bacteria, and archaea, some of the eukaryotes, for example, fungi *Aspergillus fumigates*, *Candida albicans*, *Fusarium oxysporum*, *Lomentosporaprolificans*, *Phytophthora parasitica*, *Scedosporiumapiospermum*, *Trichophyton mentagrophytes*, and *Trichoderma harzianum* [6,7] as well as microalgae such as *Binuclearia*, *Botryococcus*, *Chlorella*, *Cladophora*, *Fragilariopsiscylindrus*, *Microspora*, *Oscillatoria* spp., *Phormidium*, *Synedropsis* spp., and *Ulothrix* spp., are also known for biofilm formation and diverse potential applications [8–10].

In most biofilms, microorganisms contribute less than 10% of the total dry mass, whereas the rest of the 90% dry mass input is a matrix. The matrix is considered an extracellular secreted material manufactured by the organisms themselves in which the cells of biofilms are embedded. It comprises a cluster of different types of biopolymers, also known as extracellular polymeric substances (EPS) that establish the frame for the three-dimensional architecture of biofilms and act as a crucial player for adhesion and cohesion in biofilms [11]. Therefore, a proper understanding of biofilms development by microbes at the molecular level is fundamental. A few research studies have been carried out on a molecular level and some essential genes involved in diverse functions at different stages of biofilm formation have been reported (Table 15.1). Thus, the development of biofilms facilitates an entirely different lifestyle in the bacterial cell from the planktonic state.

Core Microbiome: Improving Crop Quality and Productivity, First Edition. Edited by
Javid A. Parray, Nowsheen Shameem, Elsayed Fathi Abd-Allah, and Mohammad Yaseen Mir.

Table 15.1 Genes involved in biofilm formation in agriculturally important bacteria.

Genes	Microbes	Functional attributes	References
calY	*Bacillus cereus*	Biofilm formation	[12]
espB	*Herbaspirillum seropedicae*	Production of EPS and biofilm formation	[13]
fadD	*Sinorhizobium meliloti*	Surface motility and biofilm formation	[14]
ftsE	*Bacillus velezensis*	Biofilm formation	[15]
IapD	*Pseudomonas fluorescens*	Biofilm formation	[16]
mtpABCDE	*Mesorhizobium tianshanense*	EPS and regulation	[17]
ndvB	*Rhizobium etli*	Formation of the cellular matrix in biofilm	[18]
nodD1ABC	*Sinorhizobium meliloti*	Biofilm formation	[19]
nod-gene	*Sinorhizobium fredii*	Transition of the biofilm structure	[20]
Pfl01_1508	*Pseudomonas fluorescens*	Motility and formation of biofilms	[21]
rhB	*Sinorhizobium meliloti*	Surface motility and biofilm formation	[22]
rirA	*Sinorhizobium meliloti*	Surface motility and biofilm formation	[22]
rpoA	*Rhizobium etli*	Biofilm formation	[18]
rpoE4	*Rhizobium etli*	Biofilm formation	[18]
sipW-tasA	*Bacillus cereus*	Biofilm formation	[12]
thuG	*Rhizobium etli*	Adhesion system in biofilm formation	[18]
yqxM	*Bacillus subtilis*	Formation of biofilms	[23]

EPS has also been known as "the dark matter of biofilms" due to the wide range of matrix biopolymers and difficult task analysis [24]. EPS varies between biofilms, and the variability of EPS depends on the type of microorganisms present, exerted shear forces, temperature, and nutrients availability. EPS was earlier signified as "extracellular polysaccharides," but in the recent past, it has been renamed. It became more accurately studied and confirmed that the matrix consists of nucleic acids, proteins, lipids, and other biopolymers, viz humic substances [11, 25, 26]. Biofilms facilitate an ideal habitat for developing syntrophic relationships between two metabolically active distinct bacterial cells to utilize certain substrates for energy production and life maintenance. In recent research and observations, eDNA has emerged as a crucial component of the biofilm matrix, which connects with other matrix components. Therefore, it behaves like a "glue" structure between cells that provide stability [27]. The biofilm matrix also serves as a recycling hub by containing allelements of different lysed cells. This also includes DNA, which may represent a pool of genes for horizontal gene transfer. Horizontal gene transfer is a well-known mechanism for developing new functions and adaptation in a bacterial system [28]. Horizontal gene transfer consists of three main primary mechanisms: transformation, conjugation, and transduction [28]. From the past several decades, horizontal gene transfer and biofilm have emerged as significant research areas in the field of environment, agriculture, and medical microbiology, leading to the acceptance of their crucial role in bacterial adaptation and evolution.

Interestingly, recent research and findings have indicated that biofilm and plasmid biology are intertwined through diverse, complicated interactions and vary from the genetic level up to the community level. For example, some recent scientific work has established that biofilm formation and horizontal gene transfer are interconnected [29]. Even though published research article investigating this interconnectedness is still scant, it has also been observed that flat gene transfer rates are typically higher in biofilm communities than planktonic states of the bacterial cell. In addition, biofilms also boost plasmid stability, and it promotes the host range of mobile genetic elements during horizontal gene transfer [30]. However, at present, detailed information regarding dynamic gene transfer through biofilms is limited. Still, very little is known and understood about when the gene transfer happens and how it propagates within a biofilm [31].

15.2 Role of Extracellular DNA (eDNA) in Bacterial Biofilms

In the bacterial biofilm matrix, exopolysaccharide is not alone as a significant component. In the recent past, the considerable role of nucleic acids, more specifically DNA, in the biofilm matrix has been noticed. However, the DNA presence was known in biofilms, while the active role of eDNA was not understood earlier in the context of biofilm formation. The crucial part of extracellular DNA (eDNA) in biofilm matrix was first time reported by Whitchurch and coworkers [32]. Since then, DNA has been found in many research as a significant player for biofilm development across different bacterial species (Table 15.2). In some bacteria, extracellular DNA (eDNA) quantified as the most abundant constituents of matrix polymer in biofilms [33, 34]. In addition, the DNA of some Gram-negative bacteria also were found to be associated with outer membrane vesicles [35, 36].

Interestingly, it was observed that outer membrane vesicles promote biofilm formation in *Pseudomonas putida* and *Helicobacter pylori* [37, 38]. Thus, nowadays, eDNA has emerged as a potent target for biofilm regulation and modification by targeting eDNA interactions pattern with other matrix components that could be amended to enhance bacterial susceptibility against antibacterial agents [39]. Apart from these, in many bacteria, eDNA is also utilized as a nutritional source [40].

Extracellular DNA also acts as a hub for horizontal gene transfer (HGT) that helps in natural competence and genetic recombination in bacterial cells to gain evolutionarily favorable properties. The active role of extracellular DNA, i.e. the facilitation of naturally occurring genetic transformation has been reported in Gram-positive and

Table 15.2 Role of eDNA in biofilm formation in Gram-positive and Gram-negative bacteria.

	Gram-positive bacteria	Reported observation	References
1	*Streptococcus mutans*	Cultures grown along with DNase have shown decreased biofilm phenotype	[41]
2	*Staphylococcus aureus*	The presence of DNase affects biofilm phenotype	[42]
3	*Staphylococcus haemolyticus*	DNase treatment culture has shown detachment from mature biofilm	[43]
4	*Streptococcus pyogenes*	DNase treatment depleted 24-h-old biofilms	[44]
5	*Listeria monocytogenes*	DNase treatment reduced initial attachment in biofilms	[45]
	Gram-negative bacteria	**Reported observation**	**References**
1	*Comamonas denitrificans*	Biofilm formation inhibited by DNase in the initial stage	[46]
2	*Neisseria meningitides*	Depletion in biofilm formation of 51 different meningococcal strains observed by DNase treatment	[47]
3	*Acinetobacter baumannii*	DNase treatment depleted biofilms performance up to 60%	[48]
4	*Shewanella oneidensis*	Disruption happens in all stages of biofilm formation under both static and flow conditions by DNase treatment	[49]
5	*Bordetella pertussis*	DNase treatment affects biofilms formed under both static and flow conditions	[50]

Gram-negative soil bacterium species [51]. However, genuine transformation is not only limited to soil bacterial populations. Most of the bacterial species on earth are accompanied by eDNA [52], and it is also reported that such eDNA are well suited for HGT [53]. It is well known that HGT occurs more efficiently at an enhanced rate [54, 55]. In addition to horizontal gene transformation, competence is also closely related to the release of eDNA and subsequent biofilm formation at the genetic level. Natural competence is a state of the bacterial cell in which they can uptake external DNA and recombine this outer DNA segment into the chromosome and thereby undergo natural transformation [56]. A wide range of bacterial species can develop natural competence and subsequently take up foreign DNA [57]. For example, in the recent past, a putative murein hydrolase-choline binding protein D (*CbpD*) was noticed on *Streptococcus pneumonia* as the leading player of competence-induced cell lysis, and along with DNA release, it strongly attenuated in a *cbpD* mutant [58]. Competence-induced release of DNA and natural genetic transformation are closely related phenomena that have evolved to enhance gene transfer efficiency. A study on bacterial species has shown that competence-related *comA*, *comB*, *comC*, *comD*, and *comE* genes are necessary for biofilm development [59]. In addition, the connecting link between competence, eDNA release, and biofilm formation of these genes have also been reported in many other studies related to streptococcal species [60–62].

15.3 Horizontal Gene Transfer and Bacterial Biofilm Interconnection

Horizontal gene transfer phenomenon in bacterial cells was first noticed in 1928 by Griffith in laboratory cultures of pneumococci cells [63]. Horizontal gene transfer is a crucial source of new genetic material for bacterial evolution. In the current research, it is well established that genetic material transfer occurs in bacterial biofilms with magnificent efficiency [55, 64]. Now it is also well identified that HGT and biofilm development are interconnected processes [29]. Although, most common HGT mechanisms are conjugation, transformation, and transduction, in addition to that recently, membrane vesicles (MVs) also have been observed as DNA reservoirs implicated in interspecies (Figure 15.1). Experimental findings have suggested that all three modes of genetic material transfer occur in planktonic cultures. Recent data have also revealed the occurrence of all three ways of HGT in bacterial biofilms. One of the essential agents of HGT is plasmids, a self-replicating mobile genetic element that can transfer into a closely or distantly related bacterial population. Nowadays, there is a consensus among scientists that biofilms used to stabilize cell-to-cell contact promote HGT through conjugation and are considered horizontal gene transfer "hot spots" [29, 65–68]. It is reported in the literature that some conjugative plasmids promote biofilm development by encoding

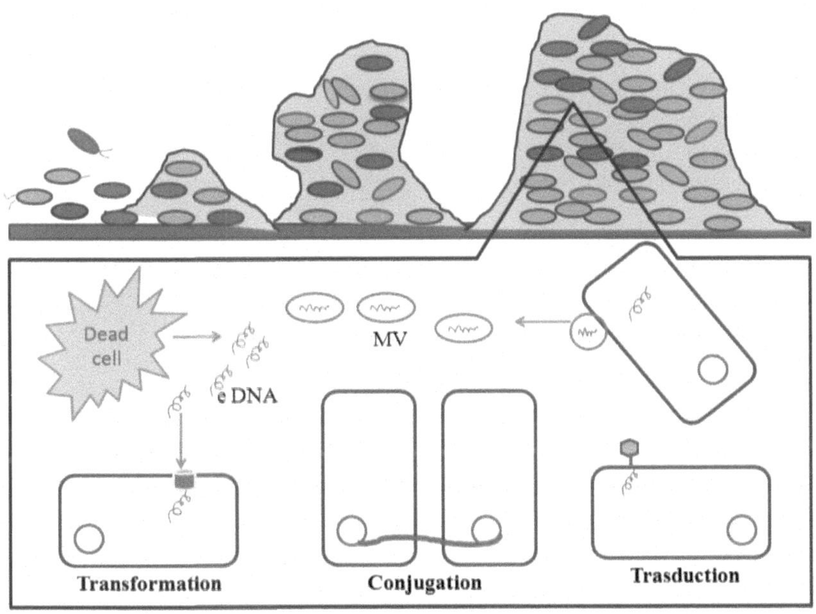

Figure 15.1 Horizontal gene transfer in biofilms.

biofilm-related proteins. For example, an *E. faecalis* conjugative plasmid known as pCF10 encodes three different cell-wall-anchoring proteins, viz*Prague, PrgB*, and *PC*, promoting in-between adhesion cells at an early stage of biofilm development [69]. In addition, *Klebsiellapneumonia* plasmid(pOLA52) possesses genes encoding type III fimbriae [70], and *Escherichia coli* is reported to have conjugative plasmids(F plasmid) promoting biofilm formation [71]. In addition to conjugation, natural transformation is closely connected to bacterial biofilm development in streptococci and *V. cholera* [72, 73], particularly eDNA being a principal constituent of the biofilm matrix released during the competence development. Apart from conjugation and transformation, recent past studies have demonstrated that phages can also play a crucial role in supporting the biofilm life cycle.

In many cases, phage mediates cell lysis and promotes eDNA production in the matrix, further strengthening biofilm structures [74]. In another case, *P. aeruginosa* phage Pf4 involves stabilizing and shaping mature biofilm structures [75]. More recently, MVs are reported for biofilms development in many bacteria such as *H. pylori* [38], *V. cholerae* [76], and *Pseudomonas putida* [37]. Membrane vesicles help theadhesion phenomena of bacterial cells to surfaces and cell-to-cell aggregation at the initial stages of biofilm development, probably by enhancing cell surface hydrophobicity. Transportation of various biomolecules such as DNA and RNA, are the most significant roles of MVs [77, 78]. With high capacity to carry DNA and their natural abundance, MVs, nowadays have been seeking close attention as potent agents for HGT [78, 79].

15.4 Different Modes of HGT in Biofilms

15.4.1 Conjugative Transfer of Genetic Material in Bacterial Biofilms

The conjugation phenomenon hardly occurs in motile planktonic cells as it demands direct physical contact between the donor and recipient cells. However, this drawback is easily overcome in biofilms where bacterial cells are attached through a matrix and located in very close contact. Indeed, much research also revealed that the conjugation frequency of biofilms enhanced significantly [29, 77]. Earlier, fewer experimental data also reportedly hinted that in bacterial biofilms, the DNA transfer rate was boosted through conjugation [30, 80]. Hausnerand Wuertzand Roberts and Mullany first reported the conjugative DNA transfer of plasmid in biofilms [80, 81]. Over and above, plasmids, while maintaining high transfer frequencies, were only able to persist in biofilms [82]. It is well explained that biofilms are very dense bacterial communities that accelerate the sharing of genetic elements [29]. Therefore, HGT through conjugation is a powerful technique to spread antibiotic resistance in biofilms beyond bacterial species boundary. Under experimental setup in laboratory conditions, in the case of *Staphylococcus aureus*'sbiofilms, it is observed that the transfer rate of a conjugative plasmid (pGO1) bearing gentamycin and trimethoprim resistance genes highly enhanced up to 16 000-fold as compared with the planktonic state of cells [83].

Nguyen and his coworkers studied the conjugative transfer of chromosomal DNA among *Mycobacterium smegmatis* strains [84]. Based on this investigation, they suggested the efficiency of DNA transfer in biofilm depends upon the expression of the *lsr2* gene. *Lsr2* protein plays a crucial role in biofilm formation in donor strain but not in the recipient bacterial strain. In addition, a study on *E. faecalis* strain's biofilms, invitrointraspecies conjugation assays have revealed that *E. faecium* were transferring tetracycline, doxycycline, streptomycin, gentamicin, and erythromycin resistance gene to human clinical strains with a very high rate. High transfer rates for both tetracycline and doxycycline resistance gene were quantified. As described above, higher HGT frequencies in biofilms are now well established.

Nevertheless, some reports on spatial constraints in biofilms may inhibit the spread of plasmids in a well-developed biofilm [85]. Króland his team have shown that the transfer of an *IncP-1* plasmid has spatial and nutritional constraints in *E. coli*'s biofilm aerobic zone [86]. Madsen and coworkers suggested the efficient transfer of some plasmids in a bacterial biofilm [29], and plasmid is likely to present during the initial phases of biofilm development. Biofilm is inherent to the typical bacterial lifestyle in most natural habitats. Based on research and findings, domain scientists concluded that the air–liquid interface could be a hot zone for HGT in biofilms [86]. However, apart from the air–liquid interphase, high efficiency was also noticed in the conjugative transfer under submerged biofilms.

Moreover, plasmid transfer frequencies in conjugation within bacterial bio films were also linked to substrate loading. Ma and Bryers estimated the transfer of *the TOL* plasmid of *Pseudomonas* in biofilms that limit nutrient concentration. They observed that plasmid transfer frequencies within bio films were linked to limiting substrate loading. Low concentrations of a limiting substrate, such as glucose, promotes thinner and porous biofilm cluster formation, allowing additional penetration of donor cells throughout the clusters with more exposure of recipient bacterial population to donor cells, ultimately boosting the plasmid transfer frequencies through conjugation [87].

15.4.2 Transformation in Biofilms

Aside from conjugation, HGT through transformation mode also occurs at higher rates in biofilms. Higher transformation efficiency is limited to small DNA fragments. Still, it also promotes a higher rate of transformation in significant DNA elements, such as plasmids, including those parts of the genome that do not have genes related to the mobilization process [88–90]. The development of natural competence is directly linked with DNA. That is based on research data, and it is believed that natural competence in biofilm is ignited by extracellular DNA [77]. The transformation efficiency of a bacterial cell depends on many factors, such as physicochemical components of the DNA molecules, for example, their size, conformation, and DNA concentration. Some other environmental factors, viz pH, temperature, UV light, salt, and extracellular nucleases present in the surroundings, also play a crucial role [91]. Furthermore, some genetic obstacles were also studied on successful transformation in cells, for example, the availability of restriction enzyme and the autonomous replication or integration ability of the incoming DNA into the recipient genome [91, 92].

The transformation process is not dependent on live donor cells, while in the case of transformation, DNA released during cell death acts as a significant source for gene transfer. Therefore, one of the crucial rate-limiting factors for the transformation process in biofilm is the lastingness of DNA in the biofilm and inside the cytoplasm of the transformed cell [91]. For the first time, Tribble and coworkers observed that in *P. gingivalis*'s biofilms, extracellular DNA acts as the significant DNA source for HGT and allelic exchange among different strains. They also established DNA exchange in biofilms by a transformation kind of process that plays a crucial role in the survival and persistence of *P. gingivalis* in a challenging environment [93]. In the recent past, research has demonstrated the strong linkage of the fratricide mechanism with intra-biofilm HGT; now, it is established that it plays an essential role in the efficient acquisition of donor DNA under the natural environment. Wei and Håvarstein have observed that competent biofilm cells of *S. pneumoniae* acquire a resistance marker gene more efficiently from neighboring cells than the growth medium [94].

Vibrio cholerae, a natural inhabitant of aquatic environments, coordinates many functions and behaviors with the help of a quorum-sensing system conserved among most members of the genus *Vibrio*. Mainly, the quorum-sensing system in *Vibrio* is mediated by two auto-inducers, i.e. CAI-1 and AI-2. The extracellular auto-inducer CAI-1 is secreted only by *Vibrio* cells, while the AI-2 extracellular auto-inducer is produced by many other bacteria [95]. During a marine biofilms study on chitinous surfaces, quorum-sensing-proficient *V. cholerae* cells develop a natural competency in chitinous surface biofilms to uptake DNA. Based on observed behavior, it was hypothesized that in a chitinous environmental biofilm, *V. cholerae* could switch to the competent state in the presence of auto-inducers derived from various multispecies bacterial consortium present in the surrounding environment [95]. Further, Antonova and Hammer also established that comEA transcription rate and horizontal uptake of DNA by *V. cholerae* are highly enhanced in response to purified extracellular auto-inducer CAI-1 and AI-2, and it is also observed in auto-inducers originating from other vibrios co-cultured with *V. cholerae* within a mixed-species biofilm [95]. Hence, there is an increasing amount of research supporting the evidence of transformation induced in biofilms.

15.4.3 Transduction in Biofilms

Transduction is a virus-mediated specific HGT process involving bacteriophages that have a narrow host range, but it acts as an actual gene-transfer vehicle. Bacteriophages have an incredible abundance and efficient insertion ability into chromosomes as a prophage without causing bacterial cell lysis. Therefore, it can easily create a modification in the genetic content or genome sequence of their hosts. Many bacteria contain a large number of prophages; some also encode virulence genes and play a significant contribution in the transduction mode of gene transfer, including the transfer of plasmid-borne genes.

Schmiegerand Schicklmaierhave summarized the current understanding of the role of temperate phages in the spreading of antibiotic resistance and gene transfer. In Salmonella, transformation and conjugation are the most common way of HGT. In this bacterium, phage-mediated transduction has been advised as the most crucial mode of HGT. In their study, the phages –ES18 and PDT17 transduce the gene related to antibiotic resistance in *S. Typhimurium*DT104, further reporting PDT17 phage integration into the genome of all known strains of DT104 [96]. Most of the other investigated bacterial strains of Salmonella have been found to carry complete prophase within their genomes. Many of them are also very capable of spontaneous transduction or transduction under induction conditions [97, 98].

In another study on transduction, the spreading of Shiga toxin (Stx)-encoding phages, which is the most common mechanism for the dissemination of Stx-encoding genes and the origination of a new strain of Stx producing *E. coli* (STEC), has been reported [99]. At two different temperatures, i.e. 20 C and 37 C, Solheim and coworkers studied the transfer of Stx-encoding phages to *E. coli* strain in biofilm formation. During the experiment, they observed that the infection rates were higher at 37 C in compression with 20 C. The study of Solheim and coworkers was the first reported on temperate phage-mediated HGT investigation in a laboratory-grown biofilm.

In the recent past, research on metagenomics analyses has revealed various antibiotic-resistant genes in a phage fraction that have been isolated from natural water samples, such as genes encoding resistance to aminoglycoside, β-lactam, macrolide, quinolone and sulphonamide, and tetracycline antibiotics from river water, sewages, and seawater [100–104]. Apart from these, phage-mediated gene transfer also has been reported in many other bacterial species under laboratory conditions [105]. For example, Solheim and coworkers have registered phage-mediated transfer of *cat* in case of *E. coli* biofilms in liquid media [99]. In another report, *Clostridium difficile* phage (phiC2) has been found to enhance dissemination of erythromycin resistance gene *erm*(B) (carried by a Tn*6215* transposon) between *C. difficile* bacterialcells [106]. Likewise, in *Acinetobacter baumannii*, a well-known multidrug-resistant pathogen, the transfer of *blaNDM-1* (carried within a Tn *125* transposon) through phage transduction mode is observed [107]. Asa commensal species present in waste bodies, *A. baumannii* is a crucial player in transporting the antibiotic resistance gene in-between the natural aqueous and artificial environments.

15.4.4 Membrane Vesicles-Mediated HGT in Biofilms

MVs are typical of small-sized (20–400 nmdiameter) lipid-bilayer-enclosed objects released from bacterial cells [78, 108]. Membrane vesicles were initially reported during the 1960s in gram-negative bacteria [109]; however, in a recently published work, it was also found in gram-positive bacteria capable of producing MVs [108–109]. Further, it has also been established that MVs are released from planktonic bacterial cells and work within biofilms. Membrane vesicles construction within biofilms has been reported in *Pseudomonas aeruginosa, Helicobacter pylori, E. coli, B. subtilis,* and *S. aureus* [38, 110, 112–114], while the first membrane vesicles-mediated gene transfer was reported for R-plasmid carrying *bla*in *Neisseria gonorrhoeae* [35]. From that first observation, many scientists have demonstrated the membrane vesicle-mediated spreading of antibiotic resistance gene in various bacterial species [79]. Apart from all these, the abundance of MVs in seawater samples, in coastal surface water (6×10^6 particles/ml), and Sargasso seawater samples (3×10^5 particles/ml), respectively, was also reported [115]. In further analysis of the seawater isolated membrane vesicle samples, a diverse pool of DNA has been identified. They have shown significant homology with the members of 33 different species, including *Bacteroidetes, Cyanobacteria, Firmicutes,* and *Proteobacteria* [115]. Apart from seawater sample, MVs are also found in river water sample [116]. Chiura and coworkers experimentally demonstrated the isolation of membrane vesicle from seawater and its ability to transfer auxotrophicmarker DNA to *E. coli* bacterial cellunder a laboratory environment [117].

This sketch represents the different HGT modes in a bacterial biofilm; the upper portion indicates the biofilm formation on the solid surface while the lower box portion describes the transformation, conjugation, transduction, and MVs process inside the structure of a bacterial biofilm.

15.5 Methods for Studying HGT in Biofilms

Several molecular-based techniques, such as quantitative PCR, have been used to track and quantify HGT. At present, high-throughput next-generation DNA sequencers (NGS) have been playing a crucial role by providing large amounts of whole-genome data of organisms, and this whole-genome sequence technique is beneficial for researchers to generate multi-species genomic data directly from natural environments (metagenome) [118]. The whole-genome sequence data are deposited on public databases, such as the comprehensive ARG database and CARD, which are readily available for further investigation and analysis [119]. Until today, many computational pipelines and software have been successfully developed to track HGT events [120]. Very recently, Song and his coworkers, including Li and Jiang, have developed MetaCHIP and LEMON, respectively. This recently developed program aims to predict HGT events in different bacterial communities based on metagenomic data generated by whole-genome sequencing [121, 122].

Moreover, combinations of experimental and bioinformatics methodologies may contribute significant help in discovering novel HGT pathways and mechanisms. For example, Jiang and Nonaka haverecruited integrative approaches to investigate the interspecies transfer of the antibiotic resistance gene through a combination of transposase/integrase-mediated transposition and homologous recombination [123, 124]. Furthermore, Li et al. and Qiu et al. established an innovative, integrative micro-fluidic system [31, 125] made up of flow cytometry, laser confocal microscopy, and micro-fluidics. In this approach, with the help of GFP-carrying conjugative plasmids as donor DNA, it can efficiently perform in situ tracking of HGT in the plasmid of viable biofilms through micro-fluidics devices. This approach provides temporal and spatial investigation and quantification of HGT in three-dimensional biofilms structures, while the rate can be measured with the help of flow cytometric data analysis. Based on this innovative technique, they have successfully calculated real-time HGT rates within activated sludge biofilms. Apart from it, they also claimed that their recently developed approach is capable of determining HGT rates within biofilms collected from diverse sources or environments.

15.6 Conclusion and Future Challenges

Conjugation, transformation, and transduction are the primary, well-known approaches for HGT. All three major gene transfer modes happen efficiently in planktonic cultures and microbial biofilms. While based on recent research and development, it has been well established that HGT occurs at higher frequencies within a microbial biofilm than in a planktonic state. This chapter has also highlighted the importance of MVs as a potential agent for HGT in microbial biofilms. However, the biological properties of MVsare still not adequately understood.

For the accurate assessment of antibiotic resistance emergence in different microbes in the environments, efficient investigation of the HGT pathway, rates, and mechanism is urgent. However, due to complex multiple HGT mechanisms, instrument efficiency, and experimental limitations, it remains a challenging job. At present, there is an intense need for some natural conditions mimicking a laboratory microcosm, biofilm model, or novel in situ tools, which must be effectively applied to measure nondestructive HGT events under varying real-life aquatic and other terrestrial habitats. Recent research and development indicate that the integrated (experimental and bioinformatics) approaches could be crucial tools to facilitate new evidence accumulation on HGT mechanisms, rate, and pathway. Over and above, the integrative approaches may be helpful to provide a unique insight and deeper understanding of the chronological distribution and the evolution of the resistance gene in the larger microbial community.

References

1 Donlan, R.M. and Costerton, J.W. (2002). Biofilms: Survival mechanisms of clinically relevant microorganisms. *Clinical Microbiology Reviews* 15 (2): 167–193.

2 Costerton, J.W., Stewart, P.S., and Greenberg, E.P. (1999). Bacterial biofilms: A common cause of persistent infections. *Science* 284 (5418): 1318–1322.

3 Al-Shabib, N.A., Husain, F.M., Ahmad, I., Khan, M.S., Khan, R.A., and Khan, J.M. (2017). Rutin inhibits mono and multispecies biofilm formation by foodborne drug-resistant Escherichia coli and Staphylococcus aureus. *Food Control* 79: 325–332.

4 Stanley, N.R. and Lazazzera, B.A. (2005). Defining the genetic differences between wild and domestic strains of Bacillus subtilis that affect poly-γ-DL-glutamic acid production and biofilm formation. *Molecular Microbiology* 57 (4): 1143–1158.

5 Karatan, E. and Watnick, P. (2009). Signals, regulatory networks, and materials that build and break bacterial biofilms. *Microbiology and Molecular Biology Reviews* 73 (2): 310–347.

6 Fanning, S., Xu, W., Solis, N., Woolford, C.A., Filler, S.G., and Mitchell, A.P. (2012). Divergent targets of Candida albicans biofilm regulator Bcr1 in vitro and in vivo. *Eukaryotic Cell* 11 (7): 896–904.

7 Villa, F., Cappitelli, F., Cortesi, P., and Kunova, A. (2017). Fungal biofilms: Targets for the development of novel strategies in plant disease management. *Frontiers in Microbiology* 8: 654.

8 Joint, I., Callow, M.E., Callow, J.A., and Clarke, K.R. (2000). The attachment of Enteromorpha zoospores to a bacterial biofilm assemblage.

9 Caramujo, M.-J., Mendes, C.R.B., Cartaxana, P., Brotas, V., and Boavida, M.-J. (2008). Influence of drought on algal biofilms and meiofaunal assemblages of temperate reservoirs and rivers. *Hydrobiologia* 598 (1): 77–94.

10 Kesaano, M. and Sims, R.C. (2014). Algal biofilm based technology for wastewater treatment. *Algal Research* 5: 231–240.

11 Flemming, H.-C. and Wingender, J. (2010). The biofilm matrix. *Nature Reviews Microbiology* 8 (9): 623–633.

12 Caro-Astorga, J., Pérez-García, A., de Vicente, A. and Romero, D. (2015). A genomic region involved in the formation of adhesin fibers in Bacillus cereus biofilms. *Frontiers in microbiology* 5: 745.

13 Balsanelli, E., Baura, V.A.D., Pedrosa, F.D.O., Souza, E.M.D. and Monteiro, R.A. (2014). Exopolysaccharide biosynthesis enables mature biofilm formation on abiotic surfaces by Herbaspirillum seropedicae. *PloS one* 9(10): e110392.

14 Amaya-Gómez, C.V., Hirsch, A.M. and Soto, M.J. (2015). Biofilm formation assessment in Sinorhizobium meliloti reveals interlinked control with surface motility. *BMC microbiology* 15(1): 1–14.

15 Li, Q., Li, Z., Li, X., Xia, L., Zhou, X., Xu, Z., Shao, J., Shen, Q. and Zhang, R. (2018). FtsEX-CwlO regulates biofilm formation by a plant-beneficial rhizobacterium Bacillus velezensis SQR9. *Research in microbiology* 169(3): 166–176.

16 Hinsa, S.M. and O'Toole, G.A. (2006). Biofilm formation by Pseudomonas fluorescens WCS365: a role for LapD. *Microbiology* 152(5): 1375–1383.

17 Wang, P., Zhong, Z., Zhou, J., Cai, T. and Zhu, J. (2008). Exopolysaccharide biosynthesis is important for Mesorhizobium tianshanense: plant host interaction. *Archives of microbiology* 189(5): 525–530.

18 Reyes-Pérez, A., del Carmen Vargas, M., Hernández, M., Aguirre-von-Wobeser, E., Pérez-Rueda, E. and Encarnacion, S. (2016). Transcriptomic analysis of the process of biofilm formation in Rhizobium etli CFN42. *Archives of microbiology* 198(9): 847–860.

19 Fujishige, N.A., Lum, M.R., De Hoff, P.L., Whitelegge, J.P., Faull, K.F. and Hirsch, A.M. (2008). Rhizobium common nod genes are required for biofilm formation. *Molecular microbiology* 67(3): 504–515.

20 Perez-Montan˜o, F., Jimenez-Guerrero, I., Del Cerro, P., Baena-Ropero, I., Lo´pez-Baena, F.J., Ollero, F.J., Bellogı´n, R., Lloret, J., Espuny, R. (2014). Thesymbiotic biofilm of Sinorhizobium fredii SMH12, necessary for successful colonization and symbiosis of Glycine max cv Osumi, is regulated by quorum sensing systems and inducing flavonoids via NodD1. *PLoS One* 9: e105901.

21 Mastropaolo, M.D., Silby, M.W., Nicoll, J.S. and Levy, S.B. (2012). Novel genes involved in Pseudomonas fluorescens Pf0-1 motility and biofilm formation. *Applied and environmental microbiology* 78(12): 4318–4329.

22 Amaya-Go´mez, C.V., Hirsch, A.M., Soto, M.J., (2015). Biofilm formation assessment in Sinorhizobium meliloti reveals interlinked control with surface motility. *BMC Microbiol* 15 (1): 58.

23 Rudrappa, T., Czymmek, K.J., Paré, P.W. and Bais, H.P. (2008). Root-secreted malic acid recruits beneficial soil bacteria. *Plant physiology* 148(3): 1547–1556.

24 Flemming, H.-C., Neu, T.R., and Wozniak, D.J. (2007). The EPS matrix: The "house of biofilm cells. *Journal of Bacteriology* 189 (22): 7945–7947.

25 Wingender, J., Neu, T.R., and Flemming, H.-C. (1999). What are bacterial extracellular polymeric substances? In: *Microbial Extracellular Polymeric Substances*, 1–19. Springer.

26 Neu, T.R. and Lawrence, J.R. (2010). Extracellular polymeric substances in microbial biofilms. In: *Microbial Glycobiology*, 733–758. Elsevier.

27 Peterson, B.W., Van Der Mei, H.C., Sjollema, J., Busscher, H.J., and Sharma, P.K. (2013). a different role of Edna in the viscoelastic relaxation of biofilms. *MBio* 4 (5): e00497–00413.

28 Lerminiaux, N.A. and Cameron, A.D. (2019). Horizontal transfer of antibiotic resistance genes in clinical environments. *Canadian Journal of Microbiology* 65 (1): 34–44.

29 Madsen, J.S., Burmølle, M., Hansen, L.H., and Sørensen, S.J. (2012). The interconnection between biofilm formation and horizontal gene transfer. *FEMS Immunology & Medical Microbiology* 65 (2): 183–195.

30 Sørensen, S.J., Bailey, M., Hansen, L.H., Kroer, N., and Wuertz, S. (2005). Studying plasmid horizontal transfer in situ: A critical review. *Nature Reviews Microbiology* 3 (9): 700–710.

31 Li, B., Qiu, Y., Zhang, J., Huang, X., Shi, H., and Yin, H. (2018). Real-time study of rapid spread of antibiotic resistance plasmid in biofilm using microfluidics. *Environmental Science & Technology* 52 (19): 11132–11141.

32 Whitchurch, C.B., Tolker-Nielsen, T., Ragas, P.C., and Mattick, J.S. (2002). Extracellular DNA required for bacterial biofilm formation. *Science* 295 (5559): 1487–1487.

33 Izano, E.A., Shah, S.M., and Kaplan, J.B. (2009). Intercellular adhesion and biocide resistance in nontypeable *Haemophilus influenzae* biofilms. *Microbial Pathogenesis* 46 (4): 207–213.

34 Matsukawa, M. and Greenberg, E. (2004). Putative exopolysaccharide synthesis genes influence *Pseudomonas aeruginosa* biofilm development. *Journal of Bacteriology* 186 (14): 4449–4456.

35 Dorward, D.W., Garon, C.F., and Judd, R.C. (1989). Export and intercellular transfer of DNA via membrane blebs of *Neisseria gonorrhoeae*. *Journal of Bacteriology* 171 (5): 2499–2505.

36 Garon, C., Dorward, D., and Corwin, M. (1989). Structural features of *Borrelia burgdorferi*—the Lyme disease spirochete: Silverstaining for nucleic acids. *Scanning Microscopy Supplement* 3: 109–115.

37 Baumgarten, T., Sperling, S., Seifert, J., Von Bergen, M., Steiniger, F., Wick, L.Y., and Heipieper, H.J. (2012). Membrane vesicle formation as a multiple-stress response mechanism enhances *Pseudomonas putida* DOT-T1E cell surface hydrophobicity and biofilm formation. *Applied and Environmental Microbiology* 78 (17): 6217–6224.

38 Yonezawa, H., Osaki, T., Kurata, S., Fukuda, M., Kawakami, H., Ochiai, K., Hanawa, T., and Kamiya, S. (2009). Outer membrane vesicles of *Helicobacter pylori* TK1402 are involved in biofilm formation. *BMC Microbiology* 9 (1): 197.

39 Okshevsky, M., Regina, V.R., and Meyer, R.L. (2015). Extracellular DNA as a target for biofilm control. *Current Opinion in Biotechnology* 33: 73–80.

40 Vorkapic, D., Pressler, K., and Schild, S. (2016). Multifaceted roles of extracellular DNA in bacterial physiology. *Current Genetics* 62 (1): 71–79.

41 Petersen FC, Tao L, Scheie AA. (2005). DNA binding-uptake system: a link between cell-to-cell communication and biofilm formation. *J Bacteriol* 187:4392–400.

42 Rice KC, Mann EE, Endres JL, et al. (2007). The cidA murein hydrolase regulator contributes to DNA release and biofilm development in Staphylococcus aureus. *PNAS* 104:8113–8.

43 Fredheim EGA, Klingenberg C, Rohde H, et al. (2009). Biofilm formation by Staphylococcus haemolyticus. *J Clin Microbiol* 47:1172–80.

44 Tetz GV, Artemenko NK, Tetz VV. (2009). Effect of DNase and antibiotics onbiofilm characteristics. *Antimicrob Agents Chemother* 53:1204–9.

45 Harmsen M, Lappann M, Knøchel S, Molin S. (2010). Role of extracellular DNA during biofilm formation by Listeria monocytogenes. *Appl Environ Microbiol* 76:2271–9.

46 Andersson S, Dalhammar G, Land C, Kuttuva Rajarao G. (2009). Characterization of extracellular polymeric substances from denitrifying organism Comamonas denitrificans. *Appl Microbiol Biotechnol* 82:535–43.

47 Lappann M, Claus H, Van Alen T, et al. (2010). A dual role of extracellular DNA during biofilm formation of Neisseria meningitidis. *Mol Microbiol* 75:1355–71.

48 Sahu PK, Iyer PS, Oak AM, et al. (2012). Characterization of eDNA from the clinical strain Acinetobacter baumannii AIIMS 7 and its role in biofilm formation. *Sci World J* 2012:10.

49 Godeke J, Paul K, Lassak J, Thormann KM. (2011). Phage-induced lysis enhances biofilm formation in Shewanella oneidensis MR-1. *ISME J* 5:613–26.

50 Conover MS, Mishra M, Deora R. (2011). Extracellular DNA is essential for maintaining Bordetella biofilm integrity on abiotic surfaces and in the upper respiratory tract of mice. *PLoS ONE* 6:e16861.

51 Lorenz, M.G., Gerjets, D., and Wackernagel, W. (1991). Release of transforming plasmid and chromosomal DNA from two cultured soil bacteria. *Archives of Microbiology* 156 (4): 319–326.

52 Lorenz, M.G. and Wackernagel, W. (1994). Bacterial gene transfer by natural genetic transformation in the environment. *Microbiology and Molecular Biology Reviews* 58 (3): 563–602.

53 Kaneko, S. and Itaya, M. (2010). Stable extracellular DNA: A novel substrate for genetic engineering that mimics horizontal gene transfer in nature. In: *Extracellular Nucleic Acids*, 39–53. Springer.

54 Björklöf, K., Nurmiaho-Lassila, E.L., Klinger, N., Haahtela, K., and Romantschuk, M. (2000). Colonization strategies and conjugal gene transfer of inoculated *Pseudomonas syringae* on the leaf surface. *Journal of Applied Microbiology* 89 (3): 423–432.

55 Molin, S. and Tolker-Nielsen, T. (2003). Gene transfer occurs with enhanced efficiency in biofilms and induces enhanced stabilization of the biofilm structure. *Current Opinion in Biotechnology* 14 (3): 255–261.

56 Seitz, P. and Blokesch, M. (2013). Cues and regulatory pathways involved in natural competence and transformation in pathogenic and environmental Gram-negative bacteria. *FEMS Microbiology Reviews* 37 (3): 336–363.

57 Lorenz, M.G. and Wackernagel, W. (1994). Bacterial gene transfer by natural genetic transformation in the environment. *Microbiological Reviews* 58 (3): 563–602.

58 Kausmally, L., Johnsborg, O., Lunde, M., Knutsen, E., and Håvarstein, L.S. (2005). Choline-binding protein D (CbpD) in Streptococcus pneumoniae is essential for competence-induced cell lysis. *Journal of Bacteriology* 187 (13): 4338–4345.

59 Li, Y.-H., Tang, N., Aspiras, M.B., Lau, P.C., Lee, J.H., Ellen, R.P., and Cvitkovitch, D.G. (2002). A quorum-sensing signaling system essential for genetic competence in *Streptococcus mutans* is involved in biofilm formation. *Journal of Bacteriology* 184 (10): 2699–2708.

60 Loo, C., Corliss, D., and Ganeshkumar, N. (2000). *Streptococcus gordonii* biofilm formation: Identification of genes that code for biofilm phenotypes. *Journal of Bacteriology* 182 (5): 1374–1382.

61 Petersen, F.C., Pecharki, D., and Scheie, A.A. (2004). Biofilm mode of growth of *Streptococcus intermedius* favored by a competence-stimulating signaling peptide. *Journal of Bacteriology* 186 (18): 6327–6331.

62 Steinmoen, H., Knutsen, E., and Håvarstein, L.S. (2002). Induction of natural competence in *Streptococcus pneumoniae* triggers lysis and DNA release from a subfraction of the cell population. *Proceedings of the National Academy of Sciences* 99 (11): 7681–7686.

63 Griffith, F. (1928). The significance of pneumococcal types. *Epidemiology & Infection* 27 (2): 113–159.

64 Björklöf, K. and Jørgensen, K.S. (2001). Applicability of non-antibiotic resistance marker genes in ecological studies of introduced bacteria in forest soil. *FEMS Microbiology Ecology* 38 (2–3): 179–188.

65 Aminov, R.I. (2011). Horizontal gene exchange in environmental microbiota. *Frontiers in Microbiology* 2: 158.

66 Balcázar, J.L., Subirats, J., and Borrego, C.M. (2015). The role of biofilms as environmental reservoirs of antibiotic resistance. *Frontiers in Microbiology* 6: 1216.

67 Fux, C.A., Costerton, J.W., Stewart, P.S., and Stoodley, P. (2005). Survival strategies of infectious biofilms. *Trends in Microbiology* 13 (1): 34–40.

68 Kolenbrander, P.E., Palmer, R.J., Periasamy, S., and Jakubovics, N.S. (2010). Oral multispecies biofilm development and the key role of cell-cell distance. *Nature Reviews Microbiology* 8 (7): 471–480.

69 Bhatty, M., Cruz, M.R., Frank, K.L., LaverdeGomez, J.A., Andrade, F., Garsin, D.A., Dunny, G.M., Kaplan, H.B., and Christie, P.J. (2015). Enterococcus faecalis PCF 10-encoded surface proteins PrgA, PrgB (aggregation substance) and PrgC contribute to plasmid transfer, biofilm formation and virulence. *Molecular Microbiology* 95 (4): 660–677.

70 Burmølle, M., Bahl, M.I., Jensen, L.B., Sørensen, S.J., and Hansen, L.H. (2008). Type 3 fimbriae, encoded by the conjugative plasmid pOLA52, enhance biofilm formation and transfer frequencies in Enterobacteriaceae strains. *Microbiology* 154 (1): 187–195.

71 Ghigo, J.-M. (2001). Natural conjugative plasmids induce bacterial biofilm development. *Nature* 412 (6845): 442–445.

72 Ibáñez De Aldecoa, A.L., Zafra, O., and González-Pastor, J.E. (2017). Mechanisms and regulation of extracellular DNA release and its biological roles in microbial communities. *Frontiers in Microbiology* 8: 1390.

73 Veening, J.-W. and Blokesch, M. (2017). Interbacterial predation as a strategy for DNA acquisition in naturally competent bacteria. *Nature Reviews Microbiology* 15 (10): 621.

74 Fernández, L., Rodríguez, A., and García, P. (2018). Phage or foe: An insight into the impact of viral predation on microbial communities. *The ISME Journal* 12 (5): 1171–1179.

75 Rice, S.A., Tan, C.H., Mikkelsen, P.J., Kung, V., Woo, J., Tay, M., Hauser, A., McDougald, D., Webb, J.S., and Kjelleberg, S. (2009). The biofilm life cycle and virulence of *Pseudomonas aeruginosa* are dependent on a filamentous prophage. *The ISME Journal* 3 (3): 271–282.

76 Altindis, E., Fu, Y., and Mekalanos, J.J. (2014). Proteomic analysis of Vibrio cholerae outer membrane vesicles. *Proceedings of the National Academy of Sciences* 111 (15): E1548–E1556.

77 Tashiro, Y., Uchiyama, H., and Nomura, N. (2012). Multifunctional membrane vesicles in *Pseudomonas aeruginosa*. *Environmental Microbiology* 14 (6): 1349–1362.

78 Toyofuku, M., Tashiro, Y., Hasegawa, Y., Kurosawa, M., and Nomura, N. (2015). Bacterial membrane vesicles, an overlooked environmental colloid: Biology, environmental perspectives and applications. *Advances in Colloid and Interface Science* 226: 65–77.

79 Domingues, S. and Nielsen, K.M. (2017). Membrane vesicles and horizontal gene transfer in prokaryotes. *Current Opinion in Microbiology* 38: 16–21.

80 Hausner, M. and Wuertz, S. (1999). High rates of conjugation in bacterial biofilms as determined by quantitative in situ analysis. *Applied and Environmental Microbiology* 65 (8): 3710–3713.

81 Roberts, A.P., Pratten, J., Wilson, M., and Mullany, P. (1999). Transfer of a conjugative transposon, Tn 5397 in a model oral biofilm. *FEMS Microbiology Letters* 177 (1): 63–66.

82 Lili, L.N., Britton, N.F., and Feil, E.J. (2007). The persistence of parasitic plasmids. *Genetics* 177 (1): 399–405.

83 Savage, V.J., Chopra, I., and O'Neill, A.J. (2013). *Staphylococcus aureus* biofilms promote horizontal transfer of antibiotic resistance. *Antimicrobial Agents and Chemotherapy* 57 (4): 1968–1970.

84 Nguyen, K.T., Piastro, K., Gray, T.A., and Derbyshire, K.M. (2010). Mycobacterial biofilms facilitate horizontal DNA transfer between strains of *Mycobacterium smegmatis*. *Journal of Bacteriology* 192 (19): 5134–5142.

85 Merkey, B.V., Lardon, L.A., Seoane, J.M., Kreft, J.U., and Smets, B.F. (2011). Growth dependence of conjugation explains limited plasmid invasion in biofilms: An individual-based modelling study. *Environmental Microbiology* 13 (9): 2435–2452.

86 Król, J.E., Nguyen, H.D., Rogers, L.M., Beyenal, H., Krone, S.M., and Top, E.M. (2011). Increased transfer of a multidrug resistance plasmid in *Escherichia coli* biofilms at the air-liquid interface. *Applied and Environmental Microbiology* 77 (15): 5079–5088.

87 Ma, H. and Bryers, J.D. (2013). Non-invasive determination of conjugative transfer of plasmids bearing antibiotic-resistance genes in biofilm-bound bacteria: Effects of substrate loading and antibiotic selection. *Applied Microbiology and Biotechnology* 97 (1): 317–328.

88 Hendrickx, L., Hausner, M., and Wuertz, S. (2003). Natural genetic transformation in monoculture Acinetobacter sp. strain BD413 biofilms. *Applied and Environmental Microbiology* 69 (3): 1721–1727.

89 Maeda, S., Ito, M., Ando, T., Ishimoto, Y., Fujisawa, Y., Takahashi, H., Matsuda, A., Sawamura, A., and Kato, S. (2006). Horizontal transfer of nonconjugative plasmids in a colony biofilm of *Escherichia coli*. *FEMS Microbiology Letters* 255 (1): 115–120.

90 Etchuuya, R., Ito, M., Kitano, S., Shigi, F., Sobue, R., and Maeda, S. (2011). Cell-to-cell transformation in *Escherichia coli*: A novel type of natural transformation involving cell-derived DNA and a putative promoting pheromone. *PLoS One* 6 (1).

91 Roberts, A.P. and Mullany, P. (2010). Oral biofilms: A reservoir of transferable. *Bacterial, Antimicrobial Resistance. Expert Review of Anti-infective Therapy* 8 (12): 1441–1450.

92 Ogunseitan, O. (1995). Bacterial genetic exchange in nature. *Science Progress* 78 (3): 183–204.

93 Tribble, G.D., Rigney, T.W., Dao, D.-H.V., Wong, C.T., Kerr, J.E., Taylor, B.E., Pacha, S., and Kaplan, H.B. (2012). Natural competence is a major mechanism for horizontal DNA transfer in the oral pathogen *Porphyromonasgingivalis*. *MBio* 3: 1.

94 Wei, H. and Håvarstein, L.S. (2012). Fratricide is essential for efficient gene transfer between pneumococci in biofilms. *Applied and Environmental Microbiology* 78 (16): 5897–5905.

95 Antonova, E.S. and Hammer, B.K. (2011). Quorum-sensing autoinducer molecules produced by members of a multispecies biofilm promote horizontal gene transfer to *Vibrio cholerae*. *FEMS Microbiology Letters* 322 (1): 68–76.

96 Schmieger, H. and Schicklmaier, P. (1999). Transduction of multiple drug resistance of *Salmonella enterica* serovar Typhimurium DT104. *FEMS Microbiology Letters* 170 (1): 251–256.

97 Mmolawa, P.T., Willmore, R., Thomas, C.J., and Heuzenroeder, M.W. (2002). Temperate phages in *Salmonella enterica* serovar Typhimurium: Implications for epidemiology. *International Journal of Medical Microbiology* 291 (8): 633–644.

98 Bossi, L., Fuentes, J.A., Mora, G., and Figueroa-Bossi, N. (2003). Prophage contribution to bacterial population dynamics. *Journal of Bacteriology* 185 (21): 6467–6471.

99 Solheim, H., Sekse, C., Urdahl, A.M., Wasteson, Y., and Nesse, L.L. (2013). Biofilm as an environment for dissemination of stx genes by transduction. *Applied and Environmental Microbiology* 79 (3): 896–900.

100 Colomer-Lluch, M., Calero-Cáceres, W., Jebri, S., Hmaied, F., Muniesa, M., and Jofre, J. (2014). Antibiotic resistance genes in bacterial and bacteriophage fractions of Tunisian and Spanish wastewaters as markers to compare the antibiotic resistance patterns in each population. *Environment International* 73: 167–175.

101 Lekunberri, I., Subirats, J., Borrego, C.M., and Balcázar, J.L. (2017). Exploring the contribution of bacteriophages to antibiotic resistance. *Environmental Pollution* 220: 981–984.

102 Lekunberri, I., Villagrasa, M., Balcázar, J.L., and Borrego, C.M. (2017). Contribution of bacteriophage and plasmid DNA to the mobilization of antibiotic resistance genes in a river receiving treated wastewater discharges. *Science of the Total Environment* 601: 206–209.

103 Wang, M., Xiong, W., Liu, P., Xie, X., Zeng, J., Sun, Y., and Zeng, Z. (2018). Metagenomic insights into the contribution of phages to antibiotic resistance in water samples related to swine feedlot wastewater treatment. *Frontiers in Microbiology* 9: 2474.

104 Yang, Y., Shi, W., Lu, S.-Y., Liu, J., Liang, H., Yang, Y., Duan, G., Li, Y., Wang, H., and Zhang, A. (2018). Prevalence of antibiotic resistance genes in bacteriophage DNA fraction from Funan River water in Sichuan, China. *Science of the Total Environment* 626: 835–841.

105 Von Wintersdorff, C.J., Penders, J., Van Niekerk, J.M., Mills, N.D., Majumder, S., Van Alphen, L.B., Savelkoul, P.H., and Wolffs, P.F. (2016). Dissemination of antimicrobial resistance in microbial ecosystems through horizontal gene transfer. *Frontiers in Microbiology* 7: 173.

106 Goh, S., Hussain, H., Chang, B.J., Emmett, W., Riley, T.V., and Mullany, P. (2013). Phage ϕC2 mediates transduction of Tn6215, encoding erythromycin resistance, between Clostridium difficile strains. *MBio* 4 (6).

107 Krahn, T., Wibberg, D., Maus, I., Winkler, A., Bontron, S., Sczyrba, A., Nordmann, P., Pühler, A., Poirel, L., and Schlüter, A.(2016). Intraspecies transfer of the chromosomal *Acinetobacter baumannii*blaNDM-1 carbapenemase gene. *Antimicrobial Agents and Chemotherapy* 60 (5): 3032–3040.

108 Toyofuku, M., Nomura, N., and Eberl, L. (2019). Types and origins of bacterial membrane vesicles. *Nature Reviews Microbiology* 17 (1): 13–24.

109 Brown, L., Wolf, J.M., Prados-Rosales, R., and Casadevall, A. (2015). Through the wall: Extracellular vesicles in Gram-positive bacteria, mycobacteria and fungi. *Nature Reviews Microbiology* 13 (10): 620–630.

110 Sugimoto, S., Okuda, K.-I., Miyakawa, R., Sato, M., Arita-Morioka, K.-I., Chiba, A., Yamanaka, K., Ogura, T., Mizunoe, Y., and Sato, C.(2016). Imaging of bacterial multicellular behaviour in biofilms in liquid by atmospheric scanning electron microscopy. *Scientific Reports* 6 (1): 1–13.

111 Toyofuku, M., Cárcamo-Oyarce, G., Yamamoto, T., Eisenstein, F., Hsiao, -C.-C., Kurosawa, M., Gademann, K., Pilhofer, M., Nomura, N., and Eberl, L. (2017). Prophage-triggered membrane vesicle formation through peptidoglycan damage in *Bacillus subtilis*. *Nature Communications* 8 (1): 1–10.

112 Murphy, K., Park, A.J., Hao, Y., Brewer, D., Lam, J.S., and Khursigara, C.M. (2014). Influence of O polysaccharides on biofilm development and outer membrane vesicle biogenesis in *Pseudomonas aeruginosa*PAO1. *Journal of Bacteriology* 196 (7): 1306–1317.

113 Nakao, R., Myint, S.L., Wai, S.N., and Uhlin, B.E. (2018). Enhanced biofilm formation and membrane vesicle release by *Escherichia coli* expressing a commonly occurring plasmid gene, kil. *Frontiers in Microbiology* 9: 2605.

114 Brown, L., Kessler, A., Cabezas-Sanchez, P., Luque-Garcia, J.L., and Casadevall, A. (2014). Extracellular vesicles produced by the Gram-positive bacterium *Bacillus subtilis* are disrupted by the lipopeptide surfactin. *Molecular Microbiology* 93 (1): 183–198.

115 Biller, S.J., Schubotz, F., Roggensack, S.E., Thompson, A.W., Summons, R.E., and Chisholm, S.W. (2014). Bacterial vesicles in marine ecosystems. *Science* 343 (6167): 183–186.

116 Roose-Amsaleg, C., Fedala, Y., Vénien-Bryan, C., Garnier, J., Boccara, A.-C., and Boccara, M. (2017). Utilization of interferometric light microscopy for the rapid analysis of virus abundance in a river. *Research in Microbiology* 168 (5): 413–418.

117 Chiura, H.X., Kogure, K., Hagemann, S., Ellinger, A., and Velimirov, B. (2011). Evidence for particle-induced horizontal gene transfer and serial transduction between bacteria. *FEMS Microbiology Ecology* 76 (3): 576–591.

118 Bragg, L. and Tyson, G.W. (2014). Metagenomics using next-generation sequencing. In: *Environmental Microbiology*, 183–201. Springer.

119 McArthur, A.G., Waglechner, N., Nizam, F., Yan, A., Azad, M.A., Baylay, A.J., Bhullar, K., Canova, M.J., De Pascale, G., and Ejim, L. (2013). The comprehensive antibiotic resistance database. *Antimicrobial Agents and Chemotherapy* 57 (7): 3348–3357.

120 Douglas, G.M. and Langille, M.G. (2019). Current and promising approaches to identify horizontal gene transfer events in metagenomes. *Genome Biology and Evolution* 11 (10): 2750–2766.

121 Song, W., Wemheuer, B., Zhang, S., Steensen, K., and Thomas, T. (2019). MetaCHIP: Community-level horizontal gene transfer identification through the combination of best-match and phylogenetic approaches. *Microbiome* 7 (1): 36.

122 Li, C., Jiang, Y., and Li, S. (2019). LEMON: A method to construct the local strains at horizontal gene transfer sites in gut metagenomics. *BMC Bioinformatics* 20 (23): 1–12.

123 Jiang, X., Ellabaan, M.M.H., Charusanti, P., Munck, C., Blin, K., Tong, Y., Weber, T., Sommer, M.O., and Lee, S.Y. (2017). Dissemination of antibiotic resistance genes from antibiotic producers to pathogens. *Nature Communications* 8 (1): 1–7.

124 Nonaka, L., Yamamoto, T., Maruyama, F., Hirose, Y., Onishi, Y., Kobayashi, T., Suzuki, S., Nomura, N., Masuda, M., and Yano, H. (2018). Interplay of a non-conjugative integrative element and a conjugative plasmid in the spread of antibiotic resistance via suicidal plasmid transfer from an aquaculture Vibrio isolate. *PloS One* 13 (6): e0198613.

125 Qiu, Y., Zhang, J., Li, B., Wen, X., Liang, P., and Huang, X. (2018). A novel microfluidic system enables visualization and analysis of antibiotic resistance gene transfer to activated sludge bacteria in biofilm. *Science of the Total Environment* 642: 582–590.

16

Rhizosphere Improvement

Role of Biotechnology and Microbioengineering

Afrozah Hassan and Irshad Ahmad Nawchoo

[1]*Department of Botany, University of Kashmir, Srinagar, 190006, J&K, India*

16.1 Introduction

The rhizosphere is a zone between growing roots and the minerals in the soil. It provides a particular ecosystem where chemical interactions, ecological feedback, and inter-organism communication take place. Based on root proximity, soil can be divided into three main zones: the ectorhizosphere, the portion of the soil under root influence, the rhizoplane or root surface, and soil that is devoid of plants [1]. The rhizoplane, the ectorhizosphere, and the root cortex are together called the rhizosphere. The German agronomist Hiltner first defined the rhizosphere, as the effect of the roots of legumes on the surrounding soil, in terms of higher microbial activity, due to the organic matter released by the roots [2]. It harbors overabundance of microbes and small soil-inhabiting animals [3]. Rhizosphere is a hot spot for numerous organisms and is considered as one of the most complex ecosystems on Earth [4–7].

The rhizosphere comprising soil pores that contain many bacteria and other microorganisms that feed on sloughed-off plant cells termed as "rhizodeposition" [8] and proteins and sugars released by roots are termed "root exudates" [9]. These exudates are passively "released" as overflow/waste products of plants [10–12]. This symbiosis influences plant growth and competition for resources and hence leads to more complex interactions. The rhizosphere also provides space to produce allelochemicals to control neighboring plants [13]. Much of the nutrient cycling and disease suppression by antibiotics required by plants occurs immediately adjacent to roots. This occurs due to root exudates and metabolic products of symbiotic and pathogenic communities of microorganisms [14].

Organisms found in the rhizosphere include bacteria, fungi, oomycetes, nematodes, protozoa, algae, viruses, archaea, and arthropods [15, 16,]. Plants may modulate the rhizosphere microbiome to their benefit by selectively stimulating microorganisms with traits that are beneficial to plant growth and health [17]. Hence rhizosphere is the most active portion of that frontier in which biogeochemical processes influence a host of landscape and global-scale processes. A better understanding of these processes is critical for maintaining the health of the planet and feeding the organisms that live on it [18]. Intense effort has been carried out to harness the root system of plants to increase yield potentials of staple food crops to meet the global food demand in the next 50 years [19].

Core Microbiome: Improving Crop Quality and Productivity, First Edition. Edited by
Javid A. Parray, Nowsheen Shameem, Elsayed Fathi Abd-Allah, and Mohammad Yaseen Mir.

16.2 Rhizosphere Improvement

The feeding of our ever-increasing population is one of the largest challenges for maintaining worldwide agricultural sustainability. Soil and soil microorganisms play an essential role in combating hunger. Soil being a critical component of Earth is not only sufficient for food production but also for maintaining worldwide environmental sustainability for the next generation [20–22]. Hence soil microorganisms play significant roles in agricultural productivity, mainly by improving plant nutrition, soil health, and soil quality [23, 24]. The function of the agricultural ecosystem depends largely on soil microbial dynamics [25]. These microorganisms maintain the organic matter and mineralization and thus regulate nutrient cycling. The beneficial saprophytic rhizosphere microbes improve plant performance by acting as decomposers and plant growth-promoting bacteria/rhizobacteria and assist in other plant growth-promoting activities. The efficient growth promoting bacteria must have the ability to survive and multiply in the rhizosphere with indigenous microbes [26].

The rhizosphere microorganisms improve nutrient-cycling processes that include nitrogen fixation, phosphorus mobilization, and the release of other nutrients into soil solution for plant nutrition [27, 28]. An increased knowledge of rhizospheric chemistry-based symbioses in plants could provide potential ways of developing sustainable agriculture with minimal risk to the environment. Therefore, rhizosphere microorganisms have a key role in improving nitrogen-use efficiency and crop sustainability simultaneously [29–34]. Rhizosphere-management strategies emphasize maximizing the efficiency of root and rhizosphere processes involved in nutrient mobilization, acquisition, and use by crops [35, 36]. The efficiency of rhizosphere processes is highly dependent on the combined influence of a soil's inherent fertility and the input of external nutrient resources. With increasing nutrient supply, the efficiency of rhizosphere processes can be increased. However, extreme application of fertilizers may lead to high concentrations of soluble nutrients in the root zone, which can also restrict root growth and rhizosphere efficiency [37]. In the Philippines, the use of landraces has been successful in targeting rhizosphere processes to improve phosphorus deficiency and tolerance of upland rice on volcanic soils [38]. Sustainable approaches aim to improve short-term crop yields and also to assure the maximum long-term performance, protecting the ecology of agricultural systems and the interests of the farmers. The so-called microbial technologies deal with restructuring the crop rhizosphere by inoculating crops with beneficial microorganisms and using cultural practices that enrich indigenous beneficial ones [39].

Treatments to achieve greater adventitious rooting, increased number of lateral roots, higher length, and density of root hairs are targets of many research projects in plant biology [40]. It is generally accepted that root activity modifies the habitat of microorganisms, and these, in turn, could trigger changes in the overall plant behavior. Some plant growth promoting bacteria stimulate plant growth directly by the production of substances that resemble plant hormones. However, to produce these stimulatory effects on plants, the first step is proper root colonization by bacteria that attach to the root surface, forming clumps or biofilms on it [41]. As a primary target, root is the organ that shows the first stimulating bacterial effects. This is particularly remarkable in plants inoculated with *Azospirillum* sp. [42].

16.3 Role of Microbiome Engineering in Rhizosphere Improvement

The composition, relative abundance, and spatiotemporal dynamics of the members of the rhizosphere microbiome will not only impact on plant growth but may also affect human health. Hence, there is a major interest in developing strategies that reshape the rhizosphere microbiome in favor of microorganisms that improve plant productivity and that prevent the proliferation of plant and human pathogens. Numerous studies conducted over the past three decades have clearly shown that the plant genotype and the soil type are two main drivers that shape the rhizosphere microbiome [43, 44]. This rhizosphere is mainly crammed with plant roots having a significant impact on microbial composition and diversity as well as their activities [45, 46].

For a span of decades, beneficial rhizospheric microbes have been obligatory in biogeochemical nutrient cycles that drive soil fertility and crop productivity [47]. Harnessing microbiome functions in the context of agricultural production hold great potential in providing solutions to key current challenges related to food security, land degradation, and crop yield [48, 49]. Indeed, microbes are critical drivers of soil functions and agricultural crop productivity [50, 51]. In managing the rhizosphere, colonized plant roots may be an effective vector for introduction of genetically engineered bacteria to detoxify organic pollutants [52]. The characteristics of ecosystems are determined by the core microbiome that they possess [53]. The unique characteristic of each ecosystem is a result of the complex network of interactions within microbial communities and between microorganisms and hosts, as well as responses to the metabolites that are produced by microbial communities [54]. Plant microbiome is represented as root-associated microbiome. It has been shown that root-associated microbial communities are diverse, comprising up to a few thousand bacterial and fungal taxa. These microbial community structures are affected by multiple factors, including abiotic environment, host genotypes, and microbe–microbe interactions. Functionally, the root-associated microbiome influences plant physiology and plays anti-pathogen roles [55].

Thus, engineering of root-associated microbiome can be used to alter microbiome composition and networking for improved phenotypes, such as plant growth, fitness, and health. To compete with conventional systems in terms of crop yield, organic farming uses several approaches, focusing on the improvement of soil physical structure and enhancement of enzymatic activity and beneficial microbes, such as PGPR and arbuscular mycorrhizal fungi [56]. This method aims to promote and enhance soil microbial diversity, biological cycles, and biological activity with minimal usage of off-farm inputs as in the case of chemical fertilizers and herbicides. For example, compared with soil fertilized chemically, long-term use of organic fertilizers increases richness, decreases evenness, reduces dispersion, and shifts the structure of the soil microbiome [57].

Plant-associated microbes and microbiomes provide important "ecosystem services" by promoting plant growth, controlling pathogens, and alleviating abiotic stresses [58–60]. It is now certain that the plant genotype plays a crucial role in the assembly and function of rhizospheric microbiomes and in benefiting from PGPR. For instance, *Arabidopsis* plants inoculated with bacterial biofertilizer, *Pseudomonas simiae* WCS417r produced up to fourfold more biomass in some accessions than did other genotypes where the effect was hardly noticeable [61]. The recent development in synthetic biological tools and gene-editing approaches offers a distinct path to engineer a microbiome with specific functions [62]. Therefore, engineering the rhizosphere of the plant is through manipulating plant traits and by crop breeding. An effective understanding of root architecture, host-specific root exudates, and other plant-related traits that select specific benign microbes can be used to help to restructure the plant for those traits and transfer into crops by using gene-editing tools as CRISPR [63]. The production of plant-growth regulators, mainly auxins, cytokinins, and gibberellins, is the most commonly invoked mechanism for PGP exerted by PGPR [64]. Auxin, cytokinin, gibberellin, abscisic acid, and jasmonate production have been reported in several associative and endophytic diazotrophic species of many genera, such as *Enterobacter, Bacillus, Achromobacter, Acetobacter, Burkholderia, Pseudomonas, Serratia, Xanthomonas, Azospirillum, Klebsiella, Gluconacetobacter, Azoarcus, Herbaspirillum*, and *Azotobacter* [65–67].

Auxins and cytokinins are important regulators of plant growth and development, regulating processes involved in the determination of the root architecture [68]. Experiments with Indole acetic acid (IAA)-attenuated mutant bacteria inoculated on wheat [69, 70] or those carried with dwarf rice or maize deficient in the production of physiologically active gibberellins [71] are strong evidence that the production of phytohormones by associated bacteria accounts for the phytostimulatory effects. Besides, the study of the expression profiles of inoculated plants would help to understand the complex metabolic changes produced upon inoculation. The transcript profile of in vitro grown sugarcane inoculated with *G. diazotrophicus* and *H. rubrisubalbicans* revealed differentially expressed genes related to auxins, gibberellins, and ethylene [72].

Some rhizosphere and endophytic bacteria produce the enzyme 1-aminocyclopropane-1-carboxylate deaminase (ACC deaminase). The activity of ACC deaminase can reduce ACC content from the ethylene biosynthesis pathway in plants [73]. ACC deaminase-containing rhizobacteria bound to a plant act as a sink for ACC, thereby lowering ethylene levels in plant tissues. The result of the functioning of this enzyme is an increase in the growth of plant roots and shoots and a reduction in the inhibitory effects of ethylene synthesis, especially during stressful conditions [74]. Sugarcane inoculation with bacteria *G. diazotrophicus* resulted in improved nitrogen uptake [75]. Besides, studies on rice inoculated with 10 different associative and endophytic diazotrophs, including *Paenibacillus* spp., *Bacillus* spp., *Burkholderia* spp., *Herbaspirillum* spp., and *Azorhizobium* spp. indicated that bacterial inoculation had a significant positive impact on nitrogen uptake and on shoot and root growth [76].

The ability of biocontrol agents to compete in the rhizosphere is crucial to produce their benefits. This competence comprises effective root colonization combined with the ability to survive and proliferate along growing plant roots over a considerable period, in the presence of the indigenous microflora [77, 78]. *Azospirillum*-inoculated in *T.* aestivum seedlings subjected to mild osmotic stress developed significant higher coleoptiles, with higher fresh weight and better water status than non-inoculated seedlings [79, 80]. A larger root system was evident in *Azospirillum*-inoculated wheat seedlings growing either under well-irrigated or water stress conditions [81]. It was also proved that part of the negative effects mild and severe salt stresses would cause on wheat seedlings were significantly reversed in *Azospirillum*-inoculated roots [82].

16.4 Role of Biotechnology in Rhizosphere Improvement

Bacteria inhabiting the rhizosphere beneficial to plants are called plant growth-promoting rhizobacteria [83]. In wild plant species, rhizosphere appears to be the best source for isolation of plant growth promoting rhizobacteria (Table 16.1) due to the co-evolution processes that have taken place over time [84–86]. The use of PGPR in agriculture is one of the most interesting alternatives for improving sustainable agricultural practices as well as for recovering degraded ecosystems [87]. A screening for plant growth promoting bacteria to improve the growth of *Cistus ladanifer* seedlings for reforestation of degraded Mediterranean ecosystems was carried out by Solano et al. [88], with the aim of identifying putative strains that are able to enhance the mycorrhization ability of *Cistus* with its spontaneous mycorrhizal fungus, *Amanita ponderosa*. The aim of identifying these strains was for two reasons: (i) the great economic interest in the edible fruiting body of *Amanita* and (ii) mycorrhization for soil recovery.

The influence of *rhizobacterium* and *B. pumilus* on *A. glutinosa* and on the growth of alder and its rhizosphere microbial community structure in two soils (soils A and B) was carried out by Ramos et al. [89]. The aim of this study was to show that alteration of native communities of soils is negatively related to changes in the microbial rhizosphere structure. The increases detected in shoot surface were significant only in soil A, while the root system was affected in both soils, revealing the ability to produce auxin-like compounds of this strain that elicited better growth of the root system.

A relative new perspective in the biotechnology of the rhizosphere is the metagenomic approach. Metagenomics allows the study of non-culturable, or difficult-to-culture, microorganisms. This approach gives a new estimation of microbial communities independent of their culture in the laboratory [90]. Strigolactones play an important role in harnessing plants subjected to drought and salt stress [91]. Under phosphorus- and nitrogen-limiting soil conditions, the biosynthesis and exudation of strigolactone is up-regulated [92,93], which is beneficial for the interaction of the plant with AMF (Table 16.2), but also for the symbiosis with *Rhizobium* spp., which induce root nodulation of legumes and are responsible for atmospheric nitrogen fixation [94, 95].

Table 16.1 Several rhizosphere microorganisms help in soil sustainability and nutrient acquisition. Data from Prasad et al. (2017) [131].

No.	Plant growth-promoting rhizobacteria	Reference
1	*Pseudomonas* and *Azotobacter*	[132]
2	*P. putida* strain R-168	[133]
3	*P. fluorescens* strain R-93	[134]
4	*P. fluorescens* DSM 50090	[134]
5	*P. fluorescens* strain	[135]
6	*A. lipoferum* DSM 169	[135]
7	*Rhizobium leguminosarum*	[136]
8	*P. putida* strain	[137]

Table 16.2 Endophytic bacteria in rhizosphere with biocontrol properties. Data from Mitter et al. [138].

No.	Species	Reference
1	*Chaetomium bastrychodes*	[138]
2	*C. cancriodeum*	[138]
3	*Aspergillus candidas*	[138]
4	*A. niger*	[138]
5	*A. oryzae*	[138]
6	*Fusarium oxysporium*	[138]
7	*Penicillium chrysogenum*	[138]
8	*Botrytis cinerea*	[138]
9	*Curvularia lunata*	[138]

The secretion of strigolactone into the rhizosphere is actively regulated. *A petunia* ABC transporter, PDR1, was identified to be a strigolactone cellular transporter, exporting orobanchol from petunia roots into the rhizosphere. Heterologous expression of PDR1 in Arabidopsis resulted in improved export of exogenously supplied synthetic strigolactone, GR24 [96].The main enzymes involved in the biosynthesis of flavonoids have been identified, such as chalcone isomerase (CHI), flavanone-3-hydroxylase (F3H), flavonol synthase (FLS), and anthocyanidin synthase (ANS) [97]. These genes form the key for manipulating flavonoid biosynthesis, which could have a great impact on the rhizosphere and hence, on agriculture. In *Medicago truncatula* and soybean, silencing of chalcone synthase (CHS) and isoflavone synthase (IFS) resulted in flavonoid deficiency, which prevented the formation of root nodules [98, 99]

In *M. truncatula*, silencing of the ABCG transporter MtABCG10 in hairy roots decreased the level of several isoflavonoids in the root exudate, which resulted in faster infection by the root pathogen *Fusarium oxysporum* [100].

The important recent advances in the elucidation of plant natural product pathways has resulted in the discovery of many of the enzymes and corresponding genes required for the biosynthesis and transport of a variety of rhizosphere signaling molecules, enabling the use of metabolic engineering to affect the rhizosphere [101]. Bacterial diversity in the rhizosphere is generally lower than in the bulk soil [102] and microbial community (Figure 16.1) composition is very different [103, 104], suggesting a strongly selective environment (Table 16.3).

There is sufficient evidence to show that the plant–microbe relationship is critical to health, productivity, and the overall condition of the plant [105]. These interactions are complex and often depend on plant species/cultivar, soil type, and environmental conditions, and anthropogenic effects[106].

Colonization and dominance of specific microbial species in the rhizosphere is critical for both pathogenic and beneficial soil microbes and will have an impact on disease incidence. Although a general increase in the abundance of microbes is always noted in the rhizosphere as compared with bulk soil [107], the main microbial strategy to enhance the

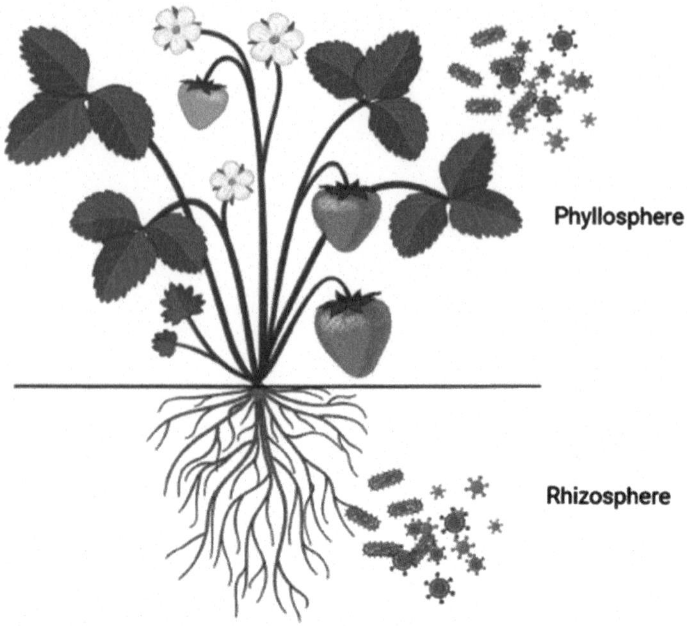

Figure 16.1 Plant–microbe interaction helps in growth promotion.

Table 16.3 Molecular studies on alleviating abiotic stresses in plants using PGPRs. Data from Ahkami et al. [139].

Bacterial species	Molecular mechanism identified	Reference
Azospirillum brasilense Sp245	Increase in nitrogen fixation and an increase in the expression of ethylene receptors	[140]
Dietzia natronolimnaea STR1	Confers salinity tolerance by modulating the transcriptional activity of salinity tolerance genes	[141]
Pseudomonas simiae strain AU	Facilitates induced systemic tolerance in soybean seedlings by producing a volatile organic blend	[142]
Enterobacter sp. UPMR18	Salt tolerance by increasing antioxidant activity and upregulation of ROS pathway genes	[143]
Burkholderia phytofirmans PsJN	Auxin and ethylene hormone homeostasis in plant growth promotion	[144]
Enterobacter sp. UPMR18	Salt tolerance by increasing antioxidant activity and upregulation of ROS pathway genes	[143]
Azospirillum brasilense Sp245	Increase in nitrogen fixation and an increase in the expression of ethylene receptors	[140]
Bacillus megaterium BP17	Identification of 150 genes differentially expressed in bacteria-treated plants	[145]
Bacillus licheniformis K11	Confers drought resistance by differential regulation of stress proteins and auxin and ACC deaminase production	[146]
Gluconacetobacter diazotrophicus PAL5	Activation the ABA-dependent signaling genes conferring drought tolerance	[147]

rhizosphere microbiome includes the direct inoculation of microorganisms, focusing on co-inoculation with several strains or mixed cultures of AMF. These ectomycorrhizal fungi PGPR and endophytes, enable plant growth and achieve niche saturation and competitive exclusion of pathogens [108].

The activation of specific functions are as equally important as the recruitment of the adequate microbiome for the plant. Quorum sensing is the mechanism used to regulate distinct microbial activities and is essential for within-species communication and for the crosstalk between species, which defines the relationship with the host plant [109, 110]. In the past decades, a number of genetic tools have been developed and employed to enhance productivity and reduce pest/pathogen damage [111, 112]. Several studies have investigated genotypic variation, which takes place, completely or partially, in the rhizosphere of cereals, particularly maize, wheat, and rice [113–115] have been targeted to nutrient efficiencies, most notably that of nitrogen, phosphorus, and potassium with most progress being made for potassium [116].

16.5 Molecular Approach

A study harnessing phytomicrobiome signaling for rhizosphere microbiome engineering carried out by Quiza et al. [117] reported that various approaches can be devised to engineer the plant microbiome, but one particularly promising approach is to take advantage of naturally evolving plant–microbe communication channels. Drought tolerance of wheat was found to have improved by rhizosphere bacteria from harsh environments studied by Timmusk et al. [118] who reported that gauging stress by elicited volatiles provides an effectual platform for rapid screening of potent bacterial strains. Besides, priming with isolates of rhizospheric bacteria from harsh environments is a promising novel way to improve plant water-use efficiency. These new advancements importantly contribute toward solving food-security issues in changing climates[118].

Hassan and Mathesius [119] studied the role of flavonoids in root–rhizosphere signaling: opportunities and challenges for improving plant–microbe interactions and reported that the manipulation of the flavonoid pathway to specifically synthesize certain products has been suggested as an avenue to improve root–rhizosphere interactions. Genetic engineering on targeted microbial species for agricultural use holds the potential of being effective, due to the direct introduction of individual, heterologous traits, into well-characterized microbes [120].

Mendes et al.'s [121] study on the influence of resistance breeding in common bean on rhizosphere microbiome composition and function network analyses showed a modular topology of the rhizosphere microbiome of the *Fox*-resistant cultivar, suggesting a more complex and highly connected bacterial community than in the rhizosphere of the *Fox*-susceptible cultivar. The breeding for *Fox* resistance in common bean may have co-selected for other unknown plant traits that support a higher abundance of specific beneficial bacterial families in the rhizosphere with functional traits that reinforce the first line of defense.

Rodriguez and Durán [122] studied the natural holobiome engineering by using native extreme microbiome to counteract the climate change effects and reported that the adaptation of plants to environmental changes not only suggests that the plants can adapt or migrate, but also can interact with their surrounding microbial communities to alleviate different stresses by natural microbiome selection of specialized strains, a phenomenon recently called "Cry for Help". From this way, plants have been co-evolved with their microbiota adapting to local environmental conditions to ensure the survival of the entire holobiome to improve plant fitness.

Igiehon and Babalola [123] studied the rhizosphere microbiome modulators: contributions of nitrogen-fixing bacteria toward sustainable agriculture reported that rhizosphere microbiomes that have been implicated to enhance plant growth and yield are modulated or influenced by a couple of environmental factors, such as soil type, plant cultivar, climate change, and anthropogenic activities. *Rhizobium, Pseudomonas, Azospirillum*, and *Bacillus* have been found to have a positive impact on crops by enhancing both above- and below-ground biomass and could therefore play a positive role in achieving sustainable agriculture.

Kwak et al. [124] study the rhizosphere microbiome structure alters to enable wilt resistance in tomato and reported that tomato variety (Hawaii 7996) is resistant to the soil-borne pathogen *Ralstonia solanacearum* besides Moneymaker, a tomato cultivar is susceptible to the pathogen. During the relative analyses of rhizosphere metagenomes from resistant and susceptible plants it was revealed that a flavobacterial genome was found to be more abundant in the rhizosphere microbiome of resistant plant than in that of the vulnerable plant. Therefore these findings divulge the role of native microbiota in protecting plants from microbial pathogens.

Dahlstrom et al. [125] studied keystone metabolites of crop rhizosphere microbiomes and reported redox-active metabolites, and in particular, phenazines, which are produced by many bacteria found in agricultural soils and have well-understood catalytic properties. These metabolites might affect microbial communities and ultimately agricultural productivity in two specific scenarios. Firstly, in the biocontrol of beneficial and pathogenic fungi in increasingly arid crop soils and, secondly, through promotion of phosphorus bioavailability and sustainable fertilizer use.

Lazcano et al. [126] studied the rhizosphere microbiome resistance in soil-borne pathogens and nutrient uptake of strawberry cultivars under field conditions and reported that microbial–root associations are significant to help plants cope with abiotic and biotic stressors. Management of these interactions offers an opportunity for improving the efficiency and sustainability of agricultural production.

De Vries et al. [127] studied harnessing rhizosphere microbiomes for drought-resilient crop production and reported that root-associated microbes can improve plant growth, and they offer the potential to increase crop resilience to future drought. This requires integrating ecology with plant, microbiome, and molecular approaches and is central to making crop production more resilient to future climate.

Hu et al. [128] studied *Streptomyces* spp. strain TOR3209, a rhizosphere bacterium promoting growth of tomato by affecting the rhizosphere microbial community, aiming at revealing the possible mechanism of its growth-promoting effect on tomato, and the correlations among *Streptomyces* spp.TOR3209. Analyses of illumina-sequencing and plate-coating revealed differences in rhizosphere microbial communities in different growth stages and distinct inoculation treatments.

A different strategy to mitigate climate change impacts on plant fitness is the utilization of beneficial mutualistic plant–microbe interactions in the rhizosphere. Such mutualism can offer enhancement of root nutrient uptake and biomass productivity, and potentially improved plant acclimation to abiotic stresses [129]. Identification of rhizospheric microbes capable of conferring stress tolerance to their plant hosts and employing symbiosis-based approaches to understanding and improving root biomass production, plant productivity, and soil carbon storage could significantly contribute to reducing the negative impact of abiotic stresses on plant ecosystem function [130, 131].

16.6 Conclusion

The role of rhizospheric microbes has been well recognized by many studies, although there is a dearth of data pertaining to the improvement of plant–microbe interactions. The improvement of soil is necessary for the growth of beneficial microbes and the management of rhizosphere is necessary for plant development, and hence, traditional rhizosphere-management practices should be incorporated to boost plant development. Moreover, rhizodeposition and plant growth regulators should be regulated. Through microbiome engineering, beneficial microorganisms should be inoculated in new crops for increasing crop productivity, efficiency, and resistance towards diseases. Besides, new research insights from molecular studies in identifying beneficial microbial species and rhizosphere engineering help in the enhancement of agricultural productivity for combating future climate challenges.

Acknowledgments

We are greatly thankful to Department of Botany, University of Kashmir, for providing the necessary facilities during this study.

References

1 Manthey, J.A., McCoy, D.L., and Crowley, D.E. (1994). Stimulation of rhizosphere iron reduction and uptake in response to iron deficiency in citrus rootstocks. *Plant Physiology and Biochemistry* 32: 211–215.
2 Lynch, J.M. (1990). In: *The Rhizosphere* (ed. J.M. Lynch), 458. Chichester: John Wiley & Sons, Ltd.
3 Dubey, A., Kumar, A., AbdAllah, E.F., Hashem, A., and Khan, M.L. (2019). Growing more with less: Breeding and developing drought resilient soybean to improve food security. *Ecological Indicators* 105: 425–437.
4 Hinsinger, P. and Marschner, P. (2006). Rhizosphere–perspectives and Challenges-a Tribute to Lorenz Hiltner 12–17 September 2004-Munich, Germany. *Plant and Soil* 283: 1–2.
5 Pierret, A., Doussan, C., Capowiez, Y., Bastardie, F., and Pagès, L. (2007). Root functional architecture: A framework for modeling the interplay between roots and soil. *Vadose Zone Journal* 6: 269–281.
6 Jones, D.L. and Hinsinger, P. (2008). The rhizosphere: complex by design.

7 Raaijmakers, J.M., Paulitz, T.C., Steinberg, C., Alabouvette, C., and Moënne-Loccoz, Y. (2009). The rhizosphere: A playground and battlefield for soil borne pathogens and beneficial microorganisms. *Plant and Soil* 321: 341–361.
8 Hütsch, B.W., Augustin, J., and Merbach, W. (2002). Plant rhizodeposition – An important source for carbon turnover in soils. *Journal of Plant Nutrition and Soil Science* 165 (4): 397–407.
9 Walker, T.S., Bais, H.P., Grotewold, E., and Vivanco, J.M. (2003). Root exudation and rhizosphere biology. *Plant Physiology* 132: 44–51.
10 Jones, D.L., Nguyen, C., and Finlay, R.D. (2009). Carbon flow in the rhizosphere: Carbon trading at the soil–root interface. *Plant Soil* 321: 5–33.
11 Hartmann, A., Schmid, M., Van Tuinen, D., and Berg, G. (2009). Plant-driven selection of microbes. *Plant Soil* 321: 235–257.
12 Dennis, P.G., Miller, A.J., and Hirsch, P.R. (2010). Are root exudates more important than other sources of rhizodeposits in structuring rhizosphere bacterial communities? *FEMS Microbiology Ecology* 72: 313–327.
13 Grayston, S.J., Wang, S., Campbell, C.D., and Edwards, A.C. (1998). Selective influence of plant species on microbial diversity in the rhizosphere. *Soil Biology and Biochemistry* 30: 369–378.
14 Ingham, E.R. (2011). Soil biology. *United States Department of Agriculture Natural Resources Conservation Service.Web, 19.*
15 Bonkowski, M., Villenave, C., and Griffiths, B. (2009). Rhizosphere fauna: The functional and structural diversity of intimate interactions of soil fauna with plant roots. *Plant Soil* 321: 213–233.
16 Buee, M., De Boer, W., Martin, F., van Overbeek, L., and Jurkevitch, E. (2009). The rhizosphere zoo: An overview of plant-associated communities of microorganisms, including phages, bacteria, archaea, and fungi, and of some of their structuring factors. *Plant Soil* 321: 189–212.
17 Cook, R.J., Thomashow, L.S., Weller, D.M., Fujimoto, D., Mazzola, M., Bangera, G., and Kim, D.S. (1995). Molecular mechanisms of defense by rhizobacteria against root disease. *Proceedings of National Academy Science USA* 92: 4197–4201.
18 Morrissey, J.P., Dow, J.M., Mark, G.L., and O'Gara, F. (2004). Are microbes at the root of a solution to world food production? Rational exploitation of interactions between microbes and plants can help to transform agriculture. *EMBO Reports* 5: 922–926.
19 Gewin, V. (2010). Food: An underground revolution. *Nature News* 466: 552–553.
20 Meena, O.P., Maurya, B.R., and Meena, V.S. (2013). Influence of K- solubilizing bacteria on release of potassium from waste mica. *Agriculture Sustainable Development* 1: 53–56.
21 Ahmad, M., Nadeem, S.M., Naveed, M., and Zahir, Z.A. (2016). Potassium-solubilizing bacteria and their application in agriculture. In: *Potassium Solubilizing Microorganisms for Sustainable Agriculture* (eds. V.S. Meena, B.R. Maurya, J.P. Verma and R.S. Meena), 293–313. New Delhi: Springer.
22 Zahedi, H. (2016). Growth-promoting effect of potassium-solubilizing microorganisms on some crop species. In: *Potassium Solubilizing Microorganisms for Sustainable Agriculture* (eds. V.S. Meena, B.R. Maurya, J.P. Verma and R.S. Meena), 31–42. New Delhi: Springer.
23 Barea, J.M., Pozo, M.J., Azcón, R., and Azcón-Aguilar, C. (2013). Microbial interactions in the rhizosphere. *Molecular Microbial Ecology of the Rhizosphere* 1: 29–44.

24 Lugtenberg, B. (2015). Life of microbes in the rhizosphere. In: *Principles of Plant-Microbe Interactions* (ed. B. Lugtenberg). Cham: Springer.

25 Barea, J.M., Azcón, R., and Azcón-Aguilar, C. (2004). Mycorrhizal fungi and plant growth promoting rhizobacteria. In: *Plant Surface Microbiology* (eds. A. Varma, L. Abbott, D. Werner and R. Hampp), 351–371. Heidelberg, Germany: Springer-Verlag.

26 Martínez-Viveros, O., Jorquera, M.A., Crowley, D.E., Gajardo, G.M.L.M., and Mora, M.L. (2010). Mechanisms and practical considerations involved in plant growth promotion by rhizobacteria. *Journal of Soil Science and Plant Nutrition* 10: 293–319.

27 Richardson, A.E., Barea, J.M., McNeill, A.M., and Prigent-Combaret, C. (2009). Acquisition of phosphorus and nitrogen in the rhizosphere and plant growth promotion by microorganisms. *Plant Soil* 321: 305–339.

28 Barea, J.M. and Richardson, A.E. (2015). Phosphate mobilisation by soil microorganisms. In: *Principles of Plant-Microbe Interactions* (ed. B. Lugtenberg), 225–234. Springer.

29 Bahadur, I., Maurya, B.R., Kumar, A., Meena, V.S., and Raghuwanshi, R., (2016). Towards the soil sustainability and potassium-solubilizing microorganisms. In: Meena, V.S., Maurya, B.R.,

30 Das, I. and Pradhan, M. (2016). Potassium-solubilizing microorganisms and their role in enhancing soil fertility and health. In: *Potassium Solubilizing Microorganisms for Sustainable Agriculture* (eds. V.S. Meena, B.R. Maurya, J.P. Verma and R.S. Meena), 281–291. India: Springer.

31 Dominguez-Nuñez, J.A., Benito, B., Berrocal-Lobo, M., and Albanesi, A. (2016). Mycorrhizal fungi: Role in the solubilization of potassium. In: *Potassium Solubilizing Microorganisms for Sustainable Agriculture* (ed. V. Meena, B. Maurya, J. Verma, R. Meena). New Delhi: Springer.

32 Verma, R., Maurya, B.R., and Meena, V.S. (2014). Integrated effect of bio-organics with chemical fertilizer on growth, yield and quality of cabbage (*Brassica oleracea* var capitata). *Indian Journal of Agricultural Science* 84: 914–919.

33 Yadav, B.K. and Sidhu, A.S. (2016). Dynamics of potassium and their bioavailability for plant nutrition. In: *Potassium Solubilizing Microorganisms for Sustainable Agriculture* (eds. V.S. Meena, B.R. Maurya, J.P. Verma and R.S. Meena), 187–201. India: Springer.

34 Yasin, M., Munir, I., and Faisal, M. (2016). Can Bacillus spp. enhance K+ uptake in crop species. In: *Potassium Solubilizing Microorganisms for Sustainable Agriculture* (ed. V. Meena, B. Maurya, J. Verma, R. Meena). New Delhi: Springer.

35 Shen, J., Li, C., Mi, G., Li, L., Yuan, L., Jiang, R., and Zhang, F. (2013). Maximizing root/rhizosphere efficiency to improve crop productivity and nutrient use efficiency in intensive agriculture of China. *Journal of Experimental Botany* 64: 1181–1192.

36 Jiao, X., Lyu, Y., Wu, X., Li, H., Cheng, L., Zhang, C., and Shen, J. (2016). Grain production versus resource and environmental costs: Towards increasing sustainability of nutrient use in China. *Journal of Experimental Botany* 67: 4935–4949.

37 Shen, J., Yuan, L., Zhang, J., Li, H., Bai, Z., Chen, X., and Zhang, F. (2011). Phosphorus dynamics: From soil to plant. *Plant Physiology* 156: 997–1005.

38 Wissuwa, M. and Ae, N. (2001). Genotypic variation for tolerance to phosphorus deficiency in rice and the potential for its exploitation in rice improvement. *Plant Breeding* 120: 43–48.

39 Roesch, L.F.W., de Quadros, P.D., Camargo, F.A., and Triplett, E.W. (2007). Screening of diazotrophic bacteria *Azopirillum* spp. for nitrogen fixation and auxin production in multiple field sites in southern Brazil. *World Journal of Microbiology and Biotechnology* 23: 1377–1383.

40 Molina-Favero, C., Creus, C.M., Lanteri, M.L., Correa-Aragunde, N., Lombardo, M.C., Barassi, C.A., and Lamattina, L. (2007). Nitric oxide and plant growth promoting rhizobacteria: Common features influencing root growth and development. *Advances in Botanical Research* 46: 1–33.

41 Salcedo, D.E., Lee, J.H., Ha, U.H., and Kim, S.P. (2015). The effects of antibiotics on the biofilm formation and antibiotic resistance gene transfer. *Desalination and Water Treatment* 54: 3582–3588.

42 Okon, Y. (1985). *Azospirillum* as a potential inoculant for agriculture. *Trends in Biotechnology* 3: 223–228.

43 Berg, G. and Smalla, K. (2009). Plant species and soil type cooperatively shape the structure and function of microbial communities in the rhizosphere. *FEMS Microbiology Ecology* 68: 1–13.

44 Bakker, M.G., Manter, D.K., Sheflin, A.M., Weir, T.L., and Vivanco, J.M. (2012). Harnessing the rhizosphere microbiome through plant breeding and agricultural management. *Plant and Soil* 360: 1–13.

45 Andersen, K.S. and Winding, A. (2004). Non-target effects of bacterial biological control agents on soil Protozoa. *Biology and Fertility of Soils* 40: 230–236.

46 de Vries, F.T. and Shade, A. (2013). Controls on soil microbial community stability under climate change. *Frontiers in Microbiology* 4: 265.

47 Sarkar, A., Saha, M., and Meena, V.S. (2017a). Plant beneficial rhizospheric microbes (PBRM): Prospects for increasing productivity and sustaining the resilience of soil fertility. In: *Agriculturally Important Microbes for Sustainable Agriculture* (eds. V.S. Meena et al), 1. Singapore: Springer.

48 Royal Agricultural Society of NSW. (2017). *Annual Report 2016/2017*. https://www.rasnsw.com.au/globalassets/document-library/annual-report/2016-2017_Annual_Report, Accessed date: 18 February 2019

49 National Academies of Sciences, Engineers and Medicine. (2018). *Science Breakthroughs to Advance Food and Agricultural Research by 2030*. National Academies Press. Accessed on 1 February, 2019

50 Nazaries, L., Pan, Y., Bodrossy, L., Baggs, E.M., Millard, P. et al. (2013). Microbial regulation of biogeochemcial cycles: Evidence from a study on methane flux and land-use change. *Applied and Environmental Microbiology* 79: 4031–4040.

51 Singh, B.K. and Trivedi, P. (2017). Microbiome and future for food and nutrient security. *Microbiome Biotechnology* 10: 50–53.

52 Brazil, G.M., Kenefick, L., Callanan, M., Haro, A., De Lorenzo, V., Dowling, D.N., and O'gara, F. (1995). Construction of a rhizosphere pseudomonad with potential to degrade polychlorinated biphenyls and detection of bph gene expression in the rhizosphere. *Applied and Environmental Microbiology* 61 (5): 1946.

53 Mueller, U.G. and Sachs, J.L. (2015). Engineering microbiomes to improve plant and animal health. *Trends in Microbiology* 23: 606–617.

54 Waldor, M.K., Tyson, G., Borenstein, E., Ochman, H. et al. (2015). Where next for microbiome research. *PLoS Biol* 13: e1002050.

55 Agler, M.T., Ruhe, J., Kroll, S., Morhenn, C., Kim, S.T., Weigel, D., and Kemen, E.M. (2016). Microbial hub taxa link host and abiotic factors to plant microbiome variation. *PLoS Biology* 14 (1): e1002352.

56 Mader, P., Fliessbach, A., Dubois, D., Gunst, L., Fried, P., and Niggli, U. (2002). Soil fertility and biodiversity in organic farming. *Science* 296(5573): 1694–1697.

57 Hartmann, M., Frey, B., Mayer, J., Mader, P., and Widmer, F. (2015). Distinct soil microbial diversity under long-term organic and conventional farming. *The ISME Journal* 9: 1177–1194.

58 AbRahman, S.F.S., Singh, E., Pieterse, C.M., and Schenk, P.M. (2018). Emerging microbial biocontrol strategies for plant pathogens. *Plant Science* 267: 102–111.

59 Sankar Ganesh, K., Sundaramoorthy, P., Nagarajan, M., and Xavier, R.L. (2017). Role of organic amendments in sustainable agriculture. In: *Sustainable Agriculture Towards Food Security* (ed. A. Dhanarajan). Singapore: Springer.

60 Saeid, A. and Chojnacka, K. (2019). Fertilizers: Need for new strategies. In: *Organic Farming* (ed. S. Chandran et al.), 91–116. Woodhead Publishing.

61 Wintermans, P.C., Bakker, P.A., and Pieterse, C.M. (2016). Natural genetic variation in Arabidopsis for responsiveness to plant growth-promoting rhizobacteria. *Plant Molecular Biology* 90: 623–634.

62 Sedeek, K.E.M., Mahas, A., and Mahfouz, M. (2019). Plant genome engineering for targeted improvement of crop traits. *Front Plant Science* 10: 114.

63 Wang, T., Zhang, H., and Zhu, H. (2019). CRISPR technology is revolutionizing the improvement of tomato and other fruit crops. *Horticulture Research* 6: 1–13.

64 Bashan, Y. and De-Bashan, L.E. (2010). How the plant growth-promoting bacterium Azospirillum promotes plant growth- a critical assessment. *Advances in Agronomy* 108: 77–136.

65 Ping, L. and Boland, W. (2004). Signals from the underground: Bacterial volatiles promote growth in *Arabidopsis*. *Trends in Plant Science* 9: 263–266.

66 Tsavkelova, E.A., Klimova, S.Y., Cherdyntseva, T.A., and Netrusov, A.I. (2006). Microbial producers of plant growth stimulators and their practical use: A review. *Applied Biochemistry and Microbiology* 42: 117–126.

67 Baca, B.E. and Elmerich, C. (2007). Microbial production of plant hormones. In: *Associative and Endophytic Nitrogen-fixing Bacteria and Cyanobacterial Associations, Nitrogen Fixation: Origins, Applications, and Research Progress*, 5 (ed. C. Elmerich, W.E. Newton). Dordrecht: Springer.

68 Overvoorde, P., Fukaki, H., and Beeckman, T. (2010). Auxin control of root development. *Cold Spring Harbor Perspectives in Biology* 2 (6): a001537.

69 Barbieri, P. and Galli, E. (1993). Effect on wheat root development of inoculation with an *Azospirillum brasilense* mutant with altered indole-3-acetic acid production. *Research in Microbiology* 144: 69–75.

70 Dobbelaere, S., Croonenborghs, A., Thys, A., Broek, A.V., and Vanderleyden, J. (1999). Phytostimulatory effect of *Azospirillum brasilense* wild type and mutant strains altered in IAA production on wheat. *Plant and Soil* 212: 153–162.

71 Castillo, P., Molina, R., Andrade, A., Vigliocco, A., Alemano, S., and Cassán, F.D. (2015). Phytohormones and other plant growth regulators produced by PGPR: The genus Azospirillum. In: *Handbook for Azospirillum* (ed. F. Cassán, Y. Okon, C. Creus), 115–138. Cham: Springer.

72 Nogueira, E.D.M., Vinagre, F., Masuda, H.P., Vargas, C., Pádua, V.L.M.D., Silva, F.R.D., and Hemerly, A.S. (2001). Expression of sugarcane genes induced by inoculation with *Gluconacetobacter diazotrophicus* and *Herbaspirillum rubrisubalbicans. Genetics and Molecular Biology* 24: 199–206.

73 Desbrosses, G., Contesto, C., Varoquaux, F., Galland, M., and Touraine, B. (2009). PGPR-Arabidopsis interactions is a useful system to study signaling pathways involved in plant developmental control. *Plant Signaling & Behavior* 4: 319–321.

74 Glick, B.R. (2004). Bacterial ACC deaminase and the alleviation of plant stress. *Advances in Applied Microbiology* 56: 291–312.

75 Suman, A., Gaur, A., Shrivastava, A.K., and Yadav, R.L. (2005). Improving sugarcane growth and nutrient uptake by inoculating *Gluconacetobacter diazotrophicus. Plant Growth Regulation* 47: 155–162.

76 Islam, M.R., Madhaiyan, M., Boruah, H.P.D., Yim, W.J., Lee, G.S., Saravanan, V.S., and Sa, T. (2009). Characterization of plant growth-promoting traits of free-living diazotrophic bacteria and their inoculation effects on growth and nitrogen uptake of crop plants. *Journal of Microbiology and Biotechnology* 19 (10): 1213–1222.

77 Parke, J.L. (1991). Root colonization by indigenous and introduced microorganisms. In: *The Rhizosphere and Plant Growth, Beltsville Symposia in Agricultural Research*, 14 (ed. D.L. Keister, P.B. Cregan). Dordrecht: Springer.

78 Lugtenberg, B.J., Dekkers, L., and Bloemberg, G.V. (2001). Molecular determinants of rhizosphere colonization by *Pseudomonas. Annual Review of Phytopathology* 39: 461–490.

79 Alvarez, M.I., Sueldo, R.J., and Barassi, C.A. (1996). Effect of *Azospirillum* on coleoptile growth in wheat seedlings under water stress. *Cereal Research Communications* 24: 101–107.

80 Creus, C.M., Sueldo, R.J., and Barassi, C.A. (1998). Water relations in *Azospirillum*-inoculated wheat seedlings under osmotic stress. *Canadian Journal of Botany* 76: 238–244.

81 Pereyra, M.A., Zalazar, C.A., and Barassi, C.A. (2006). Root phospholipids in *Azospirillum*-inoculated wheat seedlings exposed to water stress. *Plant Physiology and Biochemistry* 44: 873–879.

82 Creus, C.M., Sueldo, R.J., and Barassi, C.A. (1997). Shoot growth and water status in *Azospirillum*-inoculated wheat seedlings grown under osmotic and salt stresses. *Plant Physiology and Biochemistry* 35: 939–944.

83 Kloepper, J.W., Leong, J., Teintze, M., and Schroth, M.N. (1980). Enhanced plant growth by siderophores produced by plant growth-promoting rhizobacteria. *Nature* 286: 885–886.

84 Lucas García, J.A., Probanza, A., Ramos, B., and Gutierrez Mañero, F.J. (2001). Genetic variability of rhizobacteria from wild populations of four *Lupinus* species based on PCR-RAPDs. *Journal of Plant Nutrition and Soil Science* 164: 1–7.

85 Gutiérrez Mañero, F.J., Probanza, A., Ramos, B., Colón Flores, J.J., and Lucas García, J.A. (2003). Effects of culture filtrates of rhizobacteria isolated from wild lupine on germination, growth, and biological nitrogen fixation of lupine seedlings. *Journal of Plant Nutrition* 26: 1101–1115.

86 Barriuso Maicas, J., Pereyra de la Iglesia, M.T., Lucas García, J.A., Megías, M., Gutierrez Mañero, F.J., and Ramos Solano, B. (2005). Screening for PGPR to improve establishment of the symbiosis *Lactarius deliciosus–Pinus* sp. 1 50: 82–89

87 Vessey, J.K. (2003). Plant growth promoting rhizobacteria as biofertilizers. *Plant Soil* 255: 571–586.

88 Ramos Solano, B., de la Iglesia, M.P., Probanza, A., García, J.L., Megías, M., and Mañero, F.G. (2007). Screening for PGPR to improve growth of Cistus ladanifer seedlings for reforestation of degraded Mediterranean ecosystems. In: *First International Meeting on Microbial Phosphate Solubilization, Developments in Plant and Soil Sciences*, 102 (ed. E. Velázquez, C. Rodríguez-Barrueco). Dordrecht: Springer.

89 Ramos, B., García, J.A.L., Probanza, A., Domenech, J., and Mañero, F.J.G. (2003). Influence of an indigenous European alder (*Alnus glutinosa* (L.) Gaertn) rhizobacterium (*Bacillus pumilus*) on the growth of alder and its rhizosphere microbial community structure in two soils. *New Forests* 25: 149–159.

90 Rolf, D. (2005). The metagenomics of soil. *Nature Reviews Microbiology* 3: 470–478.

91 Van Ha, C., Leyva-González, M.A., Osakabe, Y., Tran, U.T., Nishiyama, R., Watanabe, Y., and Tran, L.S.P. (2014). Positive regulatory role of strigolactone in plant responses to drought and salt stress. *Proceedings of the National Academy of Sciences* 111 (2): 851–856.

92 Yoneyama, K., Xie, X., Kim, H.I., Kisugi, T., Nomura, T., Sekimoto, H., and Yoneyama, K. (2012). How do nitrogen and phosphorus deficiencies affect strigolactone production and exudation? *Planta* 235: 1197–1207.

93 López-Ráez, J.A. and Bouwmeester, H. (2008). Fine-tuning regulation of strigolactone biosynthesis under phosphate starvation. *Plant Signaling & Behavior* 3 (11): 963–965.

94 Akiyama, K., Matsuzaki, K.I., and Hayashi, H. (2005). Plant sesquiterpenes induce hyphal branching in arbuscular mycorrhizal fungi. *Nature* 435: 824–827.

95 Foo, E., Yoneyama, K., Hugill, C.J., Quittenden, L.J., and Reid, J.B. (2013). Strigolactones and the regulation of pea symbioses in response to nitrate and phosphate deficiency. *Molecular Plant* 6: 76–87.

96 Kretzschmar, T., Kohlen, W., Sasse, J., Borghi, L., Schlegel, M., Bachelier, J.B., and Martinoia, E. (2012). A petunia ABC protein controls strigolactone-dependent symbiotic signalling and branching. *Nature* 483: 341–344.

97 Saito, K., Yonekura-Sakakibara, K., Nakabayashi, R., Higashi, Y., Yamazaki, M., Tohge, T., and Fernie, A.R. (2013). The flavonoid biosynthetic pathway in Arabidopsis: Structural and genetic diversity. *Plant Physiology and Biochemistry* 72: 21–34.

98 Wasson, A.P., Pellerone, F.I., and Mathesius, U. (2006). Silencing the flavonoid pathway in *Medicago truncatula* inhibits root nodule formation and prevents auxin transport regulation by rhizobia. *The Plant Cell* 18: 1617–1629.

99 Subramanian, S., Stacey, G., and Yu, O. (2006). Endogenous isoflavones are essential for the establishment of symbiosis between soybean and *Bradyrhizobium japonicum*. *The Plant Journal* 48: 261–273.

100 Banasiak, J., Biała, W., Staszków, A., Swarcewicz, B., Kępczyńska, E., Figlerowicz, M., and Jasiński, M. (2013). A *Medicago truncatula* ABC transporter belonging to subfamily G modulates the level of isoflavonoids. *Journal of Experimental Botany* 64: 1005–1015.

101 Naqvi, S., Farré, G., Sanahuja, G., Capell, T., Zhu, C., and Christou, P. (2010). When more is better: Multigene engineering in plants. *Trends in Plant Science* 15: 48–56.

102 Marilley, L. and Aragno, M. (1999). Phylogenetic diversity of bacterial communities differing in degree of proximity of *Loliumperenne* and *Trifolium repens* roots. *Applied Soil Ecology* 13: 127–136.

103 Smalla, K., Wieland, G., Buchner, A., Zock, A., Parzy, J., Kaiser, S., and Berg, G. (2001). Bulk and rhizosphere soil bacterial communities studied by denaturing gradient gel electrophoresis: Plant-dependent enrichment and seasonal shifts revealed. *Applied and Environmental Microbiology* 67: 4742.

104 Peiffer, J.A., Spor, A., Koren, O., Jin, Z., Tringe, S.G., Dangl, J.L., and Ley, R.E. (2013). Diversity and heritability of the maize rhizosphere microbiome under field conditions. *Proceedings of the National Academy of Sciences* 110: 6548–6553.

105 Baudoin, E., Benizri, E., and Guckert, A. (2003). Impact of artificial root exudates on the bacterial community structure in bulk soil and maize rhizosphere. *Soil Biology and Biochemistry* 35: 1183–1192.

106 Oburger, E., Dell'mour, M., Hann, S., Wieshammer, G., Puschenreiter, M., and Wenzel, W.W. (2013). Evaluation of a novel tool for sampling root exudates from soil-grown plants compared to conventional techniques. *Environmental and Experimental Botany* 87: 235–247.

107 Bakker, M.G., Manter, D.K., Sheflin, A.M., Weir, T.L., and Vivanco, J.M. (2012). Harnessing the rhizosphere microbiome through plant breeding and agricultural management. *Plant and Soil* 360: 1–13.

108 Bünemann, E.K., Schwenke, G.D., and Van Zwieten, L. (2006). Impact of agricultural inputs on soil organisms- a review. *Soil Research* 44: 379–406.

109 Soto, M.J., Sanjuán, J., and Olivares, J. (2006). Rhizobia and plant-pathogenic bacteria: Common infection weapons. *Microbiology* 152: 3167–3174.

110 Straight, P.D. and Kolter, R. (2009). Interspecies chemical communication in bacterial development. *Annual Review of Microbiology* 63: 99–118.

111 Qaim, M. and Zilberman, D. (2003). Yield effects of genetically modified crops in developing countries. *Science* 299: 900–902.

112 Godfray, H.C.J., Beddington, J.R., Crute, I.R., Haddad, L., Lawrence, D., Muir, J.F., and Toulmin, C. (2010). Food security: The challenge of feeding 9 billion people. *Science* 327: 812–818.

113 Fageria, N.K., Wright, R.J., and Baligar, V.C. (1988). Rice cultivar evaluation for phosphorus use efficiency. *Plant and Soil* 111: 105–109.

114 Wissuwa, M. and Ae, N. (2001). Further characterization of two QTLs that increase phosphorus uptake of rice (*Oryza sativa* L.) under phosphorus deficiency. *Plant and Soil* 237: 275–286.

115 Osborne, L.D. and Rengel, Z. (2002). Screening cereals for genotypic variation in efficiency of phosphorus uptake and utilization. *Australian Journal of Agricultural Research* 53 (3): 295–303.

116 Rengel, Z. and Marschner, P. (2005). Nutrient availability and management in the rhizosphere: Exploiting genotypic differences. *New Phytologist* 168: 305–312.

117 Quiza, L., St-Arnaud, M., and Yergeau, E. (2015). Harnessing phytomicrobiome signaling for rhizosphere microbiome engineering. *Frontiers in Plant Science* 6: 507.

118 Timmusk, S., Abd El-Daim, I.A., Copolovici, L., Tanilas, T., Kännaste, A., Behers, L., and Niinemets, Ü. (2014). Drought-tolerance of wheat improved by rhizosphere bacteria from harsh environments: Enhanced biomass production and reduced emissions of stress volatiles. *PloS One* 9 (5): e96086.

119 Hassan, S. and Mathesius, U. (2012). The role of flavonoids in root–rhizosphere signalling: Opportunities and challenges for improving plant–microbe interactions. *Journal of Experimental Botany* 63: 3429–3444.

120 Ali, Z., Abulfaraj, A., Idris, A., Ali, S., Tashkandi, M., and Mahfouz, M.M. (2015). CRISPR/Cas9-ediated viral interference in plants. *Genome Biol* 16: 238.

121 Mendes, L.W., Raaijmakers, J.M., de Hollander, M., Mendes, R., and Tsai, S.M. (2018). Influence of resistance breeding in common bean on rhizosphere microbiome composition and function. *The ISME Journal* 12: 212–224.

122 Rodriguez, R. and Durán, P. (2020). Natural holobiome engineering by using native extreme microbiome to counteract the climate change effects. *Frontiers in Bioengineering and Biotechnology* 8.

123 Igiehon, N.O. and Babalola, O.O. (2018). Rhizosphere microbiome modulators: Contributions of nitrogen fixing bacteria towards sustainable agriculture. *International Journal of Environmental Research and Public Health* 15 (4): 574.

124 Kwak, M.J., Kong, H.G., Choi, K., Kwon, S.K., Song, J.Y., Lee, J., and Kim, J.F. (2018). Rhizosphere microbiome structure alters to enable wilt resistance in tomato. *Nature Biotechnology* 36: 1100–1109.

125 Dahlstrom, K.M., McRose, D.L., and Newman, D.K. (2020). Keystone metabolites of crop rhizosphere microbiomes. *Current Biology* 30: R1131–R1137.

126 Lazcano, C., Boyd, E., Holmes, G., Hewavitharana, S., Pasulka, A., and Ivors, K. (2021). The rhizosphere microbiome plays a role in the resistance to soil-borne pathogens and nutrient uptake of strawberry cultivars under field conditions. *Scientific Reports* 11: 1–17.

127 de Vries, F.T., Griffiths, R.I., Knight, C.G., Nicolitch, O., and Williams, A. (2020). Harnessing rhizosphere microbiomes for drought-resilient crop production. *Science* 368 (6488): 270–274.

128 Hu, D., Li, S., Li, Y., Peng, J., Wei, X., Ma, J., and Wang, Z. (2020). Streptomyces sp. strain TOR3209: A rhizosphere bacterium promoting growth of tomato by affecting the rhizosphere microbial community. *Scientific Reports* 10 (1): 1–15.

129 Vurukonda, S.S.,.K.P., Vardharajula, S., Shrivastava, M., and SkZ, A. (2016). Multifunctional *Pseudomonas putida* strain FBKV2 from arid rhizosphere soil and its growth promotional effects on maize under drought stress. *Rhizosphere* 1: 4–13.

130 Coleman-Derr, D. and Tringe, S.G. (2014). Building the crops of tomorrow: Advantages of 29 symbiont-based approaches to improving abiotic stress tolerance. *Front Microbiology* 5: 283.

131 Prasad, M., Chaudhary, M., Choudhary, M., Kumar, T.K., and Jat, L.K. (2017). Rhizosphere microorganisms towards soil sustainability and nutrient acquisition. In: *Agriculturally Important Microbes for Sustainable Agriculture* (ed. V. Meena, P. Mishra, J. Bisht, A. Pattanayak). Singapore: Springer.

132 Rokhzadi, A., Ahmad, A., Farrokh, D., Ghorban, N.M., and Eslam, M. (2008). Influence of plant growth-promoting rhizobacteria on dry matter accumulation and yield of chickpea (*Cicerarietinum* L.) under field conditions. *American-Eurasian Journal of Agricultural & Environmental Sciences* 3: 253–257.

133 Gholami, A., Shahsavani, S., and Nezarat, S. (2009). The effect of plant growth promoting rhizobacteria (PGPR) on germination, seedling growth and yield of maize. *World Academy of Science Engineering and Technology* 49: 19–24.

134 Kavino, M., Harish, S., Kumar, N., Saravanakumar, D., and Samiyappan, R. (2008). Induction of systemic resistance in banana (*Musa spp.*) against banana bunchy top virus (BBTV) by combining chitin with root-colonizing *Pseudomonas fluorescens* strain CHA0. *European Journal of Plant Pathology* 120: 353–362.

135 Niranjan, S.R., Deepak, S.A., Basavaraju, P., Shetty, H.S., Reddy, M.S., and Kloepper, J.W. (2003). Comparative performance of formulations of plant promoting rhizobacteria in growth promotion and suppression of downy mildew in pearl millet. *Crop Protection* 22: 579–588.

136 Afzal, A. and Asghari, B. (2008). Rhizobium and phosphate solubilizing bacteria improve the yield and phosphorus uptake in wheat. *International Journal of Agriculture Biology* 10: 85–88.

137 Gravel, V., Antoun, H., and Tweddell, R.J. (2007). Growth stimulation and fruit yield improvement of greenhouse tomato plants by inoculation with *Pseudomonas putida* or *Trichoderma atroviride*: Possible roe of Indole Acetic Acid (IAA). *Soil Biology and Biochemistry* 39: 1968–1977.

138 Mitter, B., Petric, A., SG Chain, P., Trognitz, F., Nowak, J., Compant, S., and Sessitsch, A. (2013). Genome analysis, ecology, and plant growth promotion of the endophyte *Burkholderia phytofirmans* strain PsJN. *Molecular Microbial Ecology of the Rhizosphere* 1: 865–874.

139 Ahkami, A.H., White III, R.A., Handakumbura, P.P., and Jansson, C. (2017). Rhizosphere engineering: Enhancing sustainable plant ecosystem productivity. *Rhizosphere* 3: 233–243.

140 Vargas, L., de Carvalho, T.L.G., Ferreira, P.C.G., Baldani, V.L.D., Baldani, J.I., and Hemerly, A.S. (2012). Early responses of rice (*Oryza sativa* L.) seedlings to inoculation with beneficial diazotrophic bacteria are dependent on plant and bacterial genotypes. *Plant and Soil* 356: 127–137.

141 Bharti, N., Pandey, S.S., Barnawal, D., Patel, V.K., and Kalra, A. (2016). Plant growth promoting rhizobacteria *Dietzia natronolimnaea* modulates the expression of stress responsive genes providing protection of wheat from salinity stress. *Scientific Reports* 6: 34768.

142 Vaishnav, A., Kumari, S., Jain, S., Varma, A., and Choudhary, D.K. (2015). Putative bacterial volatile-142.mediated growth in soybean (*Glycine max* L. Merrill) and expression of induced proteins under salt stress. *Journal of Applied Microbiology* 119: 539–551.

143 Habib, S.H., Kausar, H., and Saud, H.M. (2016). Plant growth-promoting rhizobacteria enhance salinity stress tolerance in Okra through ROS-scavenging enzymes. Biomed Research International.

144 Poupin, M.J., Timmermann, T., Vega, A., Zuniga, A., and Gonzalez, B. (2013). Effects of the plant growth-promoting bacterium *Burkholderia phytofirmans* PsJN throughout the life cycle of *Arabidopsis thaliana*. *Plos One* 8.

145 Vibhuti, M., Kumar, A., Sheoran, N., and Al, E. (2017). Molecular basis of endophytic *Bacillus megaterium*-induced growth promotion in *Aabidopsis thaliana*: Revelation by microarray-based gene expression analysis. *Journal of Plant Growth Regulation* 36: 118.

146 Lim, J.H. and Kim, S.D. (2013). Induction of drought stress resistance by multi-functional PGPR *Bacillus licheniformis* K11 in pepper. *The Plant Pathology Journal* 29: 201–208.

147 Vargas, L., Santa Brigida, A.B., Mota Filho, J.P., De Carvalho, T.G., Rojas, C.A., Vaneechoutte, D., van Bel, M., Farrinelli, L., Ferreira, P.C., Vandepoele, K., and Hemerly, A.S. (2014). Drought tolerance conferred to sugarcane by association with *Gluconacetobacter diazotrophicus*: A transcriptomic view of hormone pathways. *PLoS One* 9: e114744.

17

Exploring Biological Agents and Core Microbiomes as a Tool for Reclamation of Abandoned Mines

Seema B. Sharma and Rupak Dey

Department of Earth and Environmental Science, KSKV Kachchh University, Mundra Road, Bhuj 370 001, Gujarat, India

17.1 Introduction

Bioremediation is a waste-management technique that involves using organisms to remove or neutralize pollutants from a contaminated site, such as abandoned mines and industrial areas. It is also regarded as a treatment (Figure 17.1) that uses naturally occurring organisms to break down hazardous substances into less toxic or non-toxic substances.

Abandoned mines are sites where mining activities have occurred in the past, but though they were closer to active mine sites, reclamation did not occur or was incomplete. These abandoned mines result in an environmental imbalance. One of the significant outstanding ecological problems related to mining is the consequent abandonment of mines that are used as massive waste disposal sites. A technique called bioremediation ought to be induced and practiced to mitigate environmental problems [1].

17.1.1 Bioremediation Types

Bioremediation technologies can be generally carried out in two ways: (i) in situ and (ii) ex situ. In situ bioremediation involves treating the contaminated material at the site, e.g., abandoned mines. Ex situ treatment involves the removal of contaminated material to be treated elsewhere. Some examples of bioremediation related to technologies are phytoremediation, bioventing, bioleaching, and land farming.

17.2 Abandoned Mines and Bioremediation: Problems and Solutions

The most critical environmental impacts of mining projects are discussed, along with remedial measures.

17.2.1 Impacts on Water Resources

The most significant impact of a mining project is that it affects the water quality and water resources within the project area. Mining sites also imbalance groundwater

Core Microbiome: Improving Crop Quality and Productivity, First Edition. Edited by Javid A. Parray, Nowsheen Shameem, Elsayed Fathi Abd-Allah, and Mohammad Yaseen Mir.

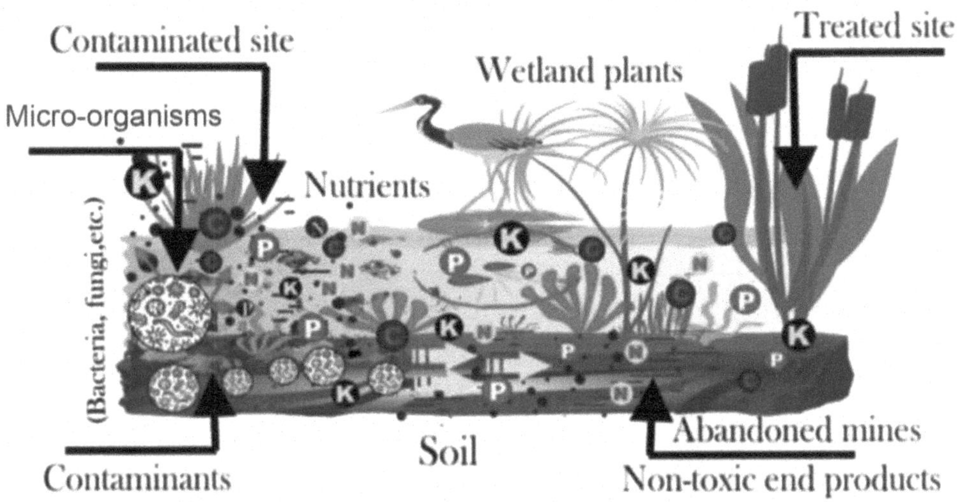

Figure 17.1 Bioremediation process.

supplies and groundwater recharge systems beneath the earth's surface, making the groundwater unfit for human consumption and thus affecting aquatic life and terrestrial wildlife [2]. Owing to water contamination, the close bonding between water, marine life, and terrestrial wildlife are prone to dislink, without creating a strong bonding between them; wetlands have to be built artificially as wetlands serve as transition zones between aquatic and terrestrial environments and provide a dynamic link between two Ecosystems that are an effective nutrient sink and absorber of organic and inorganic pollutants.

Aquatic plants can be used for treating contaminated water [3]. They are divided into (a) floating plants and (b) submerged plants.

Floating plants meet their need for CO_2 and O_2 directly from the atmosphere. Such plants derive their mineral needs from water.

Submerged plants can absorb O_2, CO_2, and minerals directly from water columns [4].

Planting native plants, such as sunflower, can prevent water contamination to some extent in an area [5]. Moreover, water lilies and irises are two great pond plants that also filtrate water [6].

Another technique called phytoremediation can also be used to protect water resources [8]. Phytoremediation involves the direct use of green plants and their associated microorganisms to stabilize or reduce contamination in soils (Table 17.1), sediments, surface water, or groundwater [7].

17.2.2 Erosion of Soil and Mine Wastes into Surface Water

Soil erosion is removing the upper layer of the soil, usually carried out by five to six erosive agents, such as water, snow, air, plants, and animals. In mining areas, soil erosion is the major threatening factor because of the removal of forest vegetation, thus

Table 17.1 Types of phytoremediation that can be used to treat water resources affected by various elements. Modified from Pivetz [8].

Type	Process involved	Pollutants treated
Phyto-volatilization	Plants take up heavy metals and organic compounds, bind or modify them and release the by-products into the atmosphere via transpiration	Mercury, selenium, and chlorinated solvents (tetrachloromethane and trichloromethane)
Rhizo-filtration	Plant roots absorb, precipitate, and concentrate heavy metals from contaminated liquid effluents and degrade organic compounds	Cadmium, cobalt, nickel, mercury, chromium, zinc, radioactive isotopes, and phenolic compounds
Phyto-stabilization	Plants with tolerance for metals are used to reduce the mobility of metals and prevent their passage to groundwater or the air	Lagoons are rid of mineral deposits proposed for phenolic and chlorinated compounds
Phyto-extraction	Plants are used to concentrate metals in parts harvestable (leaves and roots) in this process	Cadmium, cobalt, chromium, nickel, mercury, lead, selenium, and zinc
Phyto-stimulation	Root exudates promote the development of microorganisms (bacteria and fungi) capable of biodegrading compounds	Petroleum and polyaromatic hydrocarbons, benzene, toluene, atrazine, etc.
Phyto-decomposition	Both aquatic and terrestrial plants capture organic compounds and store them or decompose them to less toxic or non-toxic by-products	Munition products (TNT, DNT, RDX, nitrobenzene, nitrotoluene), atrazine, chlorinated solvents, DDT, phosphate pesticides, phenols, and nitriles, etc.

increasing the rate of erosion and transferring mine wastes through erosion into the surface water, polluting it for the successful operation of mining activities. This affects the quality of water.

Many large areas of land are disturbed by mining operations. Erosion can be a significant factor at complex rock mining sites, such as marble and granite, and to overcome such a high rate of erosion, a process called reclamation has to be followed from the beginning of the mining operation so that air and water quality mostly can be protected. Reclamation has to be done to prevent future environmental damage and to provide human safety. Reclamation or soil/land quality can be improved by creating wetlands and flood plains [9], and by doing this, the pH of streams and lands can be improved so that the whole area can be covered with vegetation and the beauty of nature can be maximized by planting crops and seeding. Artificial fishery ponds can also be created to improve the land area and control the rate of erosion to some extent.

For mines associated and found in connection with the seas or any water bodies (Figure 17.2), profound cement mixing reclamation has to be carried out to protect the marine ecology from being polluted and the organisms that sustain their life on and inside the water bodies. This process is also called deep cement mixing (DCM). The main advantage of DCM is that it does not require removing the contaminated mud, squeeze polluted water but treating the mud in place (Figure 17.3) with cement without any impact on the environment [10]. This process is an in situ bio-remediation process, i.e. treating the contaminated material at the site.

17.2.3 Mine Dewatering and Its Impact

Mine dewatering means the removal of groundwater from a mine. Dewatering an entirely or partially flooded mine can create several environmental issues, such as dewatering, that may lead to acid mine drainage, contaminating the surface water source. It also leads to the development of a cone of depression, which is a significant factor in lowering groundwater in the surrounding areas affecting the lives of humans who live nearby mine areas. A cone of depression is caused due to dewatering and extensive pumping from a

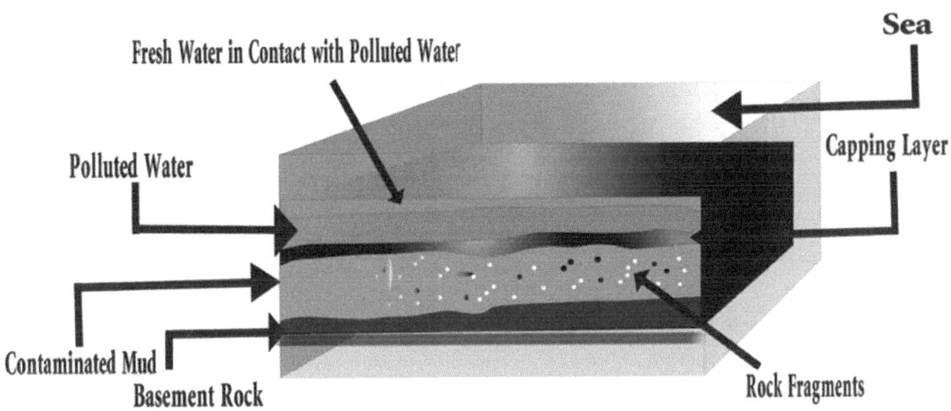

Figure 17.2 Mining area in connection with sea.

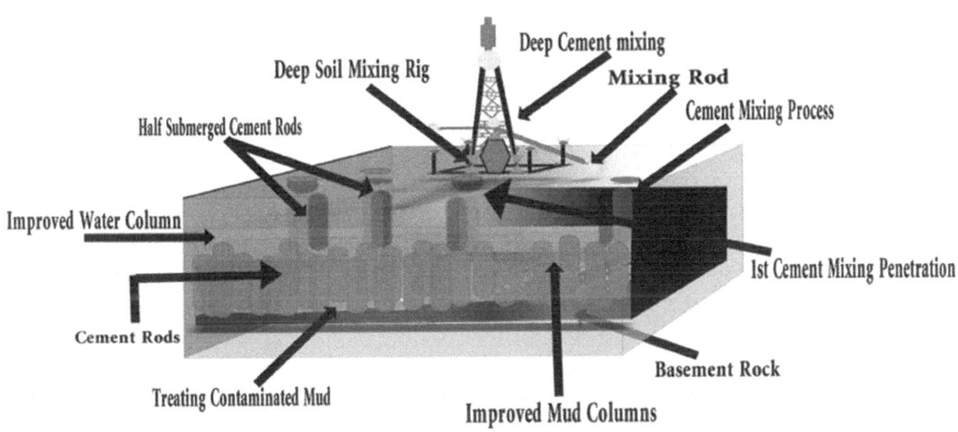

Figure 17.3 Deep-sea cement mixing.

well in a water table [11], and it is a serious problem with regard to the environment because dewatering lowers the water table near the well, due to which the natural direction of groundwater flow is influenced. This might even reduce the water availability of the specified area.

The only remedial measure that can be taken here is that dewatering should not influence or cause to build a cone of depression. It should be adequately maintained so that the natural groundwater direction does not get distributed. A mitigation measure that can be taken is to construct a wetland as long as dewatering occurs so that groundwater recharge occurs to some extent [12].

17.2.4 Mine Projects and Air Quality

Air, by volume, comprises 78% nitrogen, 21% of oxygen, 0.93% of argon, 0.04% carbon dioxide, and small amounts of other gases. It is also composed of a variable quantity of water vapor. Mining projects often imbalance the above composition and affect air quality. Air quality is the composition and condition of the air in our surroundings. Good air quality means that the air is clean, clear, and free from pollutants, such as smoke, dust, and other gaseous materials in the air. Poor air quality can harm human health and the environment. During each mining cycle, airborne emissions occur during exploration, development, construction activities, and operational activities. Their waste pipes and waste products are easily dispersed and carried away by the wind and severely affect air quality. These pollutants enter the atmosphere that causes specific physical and chemical changes and generates hazardous air pollutants, such as carbon monoxide, sulfur dioxide, and nitrogen oxides.

Trees and forests are to be constructed in the mining and abandoned mine sites for constantly cleaning, clearing, and removing pollutants from the air as proper oxygen levels are necessary for all human beings and animals.

Planting trees can break down most pollutants, such as sulfur dioxide and nitrogen oxides [13]. Moreover, planting large trees, such as pine, bamboo, and palm, can absorb tiny particles of pollution that have been generated from mining sites and other activities. These small, polluted particles are converted to non-harmful compounds by the tremendous power of such types of tress.

Special efforts are to be taken in coal-mining zones in generating substantial vegetation cover as coal mine areas generate much higher dust emissions followed by erosion and subsidence than any other mine site.

17.2.5 Public Health

Abandoned mines are a dangerous threat to public health as they may contain deadly gases in standing, contaminated, and polluted water, resulting in a lack of oxygen. The impacts of abandoned mine sites on human beings include altered landscape for habitation, subsidence, degradation of groundwater; lands are no longer usable due to their low pH value and air pollution from dust or toxic gases. Due to the adverse effects of mining activities, deficiency-related diseases, such as hypertension, anemia, diabetes, pneumonia, typhoid, asthma, and malaria, may occur in the human population [14]. The areas in touch with coal mines are more severe at risk of diseases, such as asthma.

Abandoned mine sites can also cause respiratory diseases because of the dust intake by the public in the residing area. In addition, the clouds of dust may contain silica. Once silica goes inside the body, it does not tend to come out of the body.

17.2.6 Climate Change

Climate change is a global problem, and mines accelerate climate change within the area of concern. Mining often involves the use of large trucks, different types of machinery used in blasting and exploration, and loaders running around the area, emitting CO_2, which is the leading cause for such places that are particularly vulnerable being emitters of greenhouse gases.

17.3 Organisms that Can Help Transform Abandoned Mine Sites

Human beings pollute and contaminate soil and water on the planet. Luckily, some natural organisms can help us clean those contaminants or any other toxic wastes. Some organisms, which are discussed next, owing to their functional characteristics and physiology, can help clean up contaminated sites.

17.3.1 Fungi

Fungi can be used as bio-remediators, such as oysters and mushrooms. They increase and act as great decomposers in burrowing organic matter, such as trees, and inorganic materials, such as rocks and mine wastes [15]. In the 1990s, a mycologist named Paul Edward Stamets developed a strain of oyster mushrooms with enzymes called peroxidases [16], which can break down hydrocarbon molecules in oil-contaminated soil or oil dirt areas of a mine. This is one of the best decomposers for oil because oil-contaminated sites can have toxic effects on the environment.

17.3.2 Sunflower

These plants can suck up most radioactive contaminants, such as strontium-90 and caesium-137, from contaminated sites [17]. Sunflower proliferates in almost all types of climates; growing sunflower on abandoned mine sites can help absorb those radiometric isotopes in their stems and leaves. The main reason for planting such flowers is that when a nuclear disaster or blasting site is operated in mine sites, they tend to emit harmful ionizing radiation for decades. These species mainly help accumulate radioactive isotopes into their stems, less effectively though, in clayey soil, because it is harder to extract than in other types of soils. But we can induce bacteria because clay mineral serves as a store house of food for them in the ground for them to continue their activity that is beneficial for us. They can also be used to clean and clear contaminants of groundwater near mineral processing plants.

17.3.3 Bivalves

Bivalves are aquatic mollusk with a tight body enclosed within a hinged shell, e.g., mussels and oysters. They can be used to filter the water around them because they are planktonic.

Planktonic organisms generally float, or they are tiny free-floating organisms. They also serve in sucking chemicals, such as weed killers, flame retardants, and pharmaceuticals, in places abounding in chemical contaminants, such as triclocarban. Bivalves have the ability to reduce 80% of one contaminant within 72 hours [18], and they can also be used to treat infections caused by *E. coli* bacteria.

17.3.4 Bonfire Moss

Bonfire moss or *Funaria hygrometrica* is a type of water moss that usually grows on moist, shady, and damp soil. It can also be found in rock crevices and in places where recent fire activity has taken place. This water moss can treat areas where the water is contaminated with lead [19]. They can also be used to treat heavy metals. Polygalacturonic acid binds the lead atoms, and because of this ability, they can be used as a part of water treatment systems.

17.3.5 Alpine Pennycress

They are an ornamental plant species with white flowers; when such plants grow in soil contaminated with heavy metals, they can concentrate a group of such metals into their leaves, which is why such plants are called hyperaccumulators. There are more than 100 hyperaccumulator species globally, but alpine pennycress is one of the special plants among them, because it can help treat both cadmium and zinc together [20].

17.3.6 Tobacco

Tobacco plants engineered with pentaerythritol tetranitrate (PETN) reductase gene and bacterial gene can be used and grown in soil to absorb and break down explosive nitrogen compounds [21]. Nitrogen-based volatile compounds are associated and found in industrial and mining areas. By growing such types of plants, explosive nitrogen compounds can be mitigated with the help of microbes.

17.3.7 Wave Moth Caterpillars

Wave moth caterpillars can be used to treat areas where soil is polluted with plastics [22], such as abandoned mines. Nearly a double century of wax moth caterpillars can eat a tenth of a gram of polyethene within 12 hours. Thus, they can be used as good decomposers for eliminating plastics.

17.3.8 Earthworms

Earthworms are suitable for the soil. They fertilize the ground by taking waste into their body and excrete rich nutrients into the soil. They can also remove toxic waste from contaminated soil. Therefore, earthworms can be used in highly polluted land to enrich soil fertility and nutrients for crop production.

17.3.9 Microorganisms

Microorganisms are very useful to human beings, animals, and plants. For example, some organisms are scavengers. Their work is to decompose dead bodies of plants and animals into essential nutrients, which are very useful for the growth and development of plants and animals. Some beneficial soil microorganisms along with their distribution are mentioned below.

- *Actinomycetes*: These are widely distributed in soil and found in abundance, next to bacteria. Their population increases with soil depth. These organisms are also responsible for peat (first stage of the coal-forming process) formation by converting dead plant matter deep under roots. The primary function of Actinomycetes is to degrade and decompose [23]. Organic residues added to the soil are first attacked by bacteria and fungi and later by Actinomycetes because they are slow in activity and growth than bacteria and fungi.
- *Bacteria*: Bacteria are the most abundant and predominant organisms inhabiting the soil. They are divided into three groups: spherical, rod-shaped, and spirilla (cells with long wavy chains). Most probably, bacteria are found on the top surface of the soil. Some bacteria are helpful in nitrogen-fixing, such as *Azotobacter*, *Azospirillum*, and *Gluconobacter* [24]. In addition, bacteria, such as *Thiobacillus* and *Bacillus megaterium* are beneficial in solubilizing phosphate and are sulfur oxidizers [25].

17.4 Core Microbiomes as Tools for Abandoned Mine Reclamation

It would not be an exaggeration to say that microbiomes stand out as the most potential biological agents in mine reclamation methods through biological agents. Although the concept of core microbiomes has emerged in the past few decades, the way it is steadily growing in research, we soon see a future of these microbiomes as the promising agents of change. After the first comprehensive definition given by Whipps in 1988, the term "microbiome" has now evolved to include a theatre of activity along with a suite of microorganisms that share similar habitats. This inclusion of the "activity sphere" is essential for understanding stability, plasticity, and functioning across complex microbial assemblages [26]. Unfortunately, microbiomes have been underexplored for their possibility as agents of mine site remediation.

Engineered microbial species such as *Thiobacillus ferrooxidans* can remediate soil contaminated with toxic metals like mercury and cadmium. After the advent of high-throughput techniques, such as 16S rRNA gene amplicon sequencing, where the microbiomes of complex environments have been simplified, microbial community profiling of mined sites has currently become possible. The genes that provide resistance to various toxic elements, heavy metals, and plant growth promotion have now been elucidated using comparative genomic analysis [27]. The remediation mechanisms at the molecular level are better understood for rhizobacteria than for fungi [28], owing to their increased popularity.

17.5 The Future of Bioremediation

This chapter attempted to summarize the different methodologies that can be put into action to nullify harmful post-mining effects through bioremediation techniques. Several attempts have been carried out in other parts of the globe toward the same end. Bioremediation is the safest and most promising method to curb the environmental implications caused by abandoned mines, alleviate the hazardous effects, and achieve the higher goal of ecological sustainability.

References

1 Lee, J.H. (2013). An overview of phytoremediation as a potentially promising technology for environmental pollution control. *Biotechnology and Bioprocess Engineering* 18 (3): 431–439.

2 Egboka, B.C., Nwankwor, G.I., Orajaka, I.P., and Ejiofor, A.O. (1989). Principles and problems of environmental pollution of groundwater resources with case examples from developing countries. *Environmental Health Perspectives* 83: 39–68.

3 Singh, D., Tiwari, A., and Gupta, R. (2012). Phytoremediation of lead from wastewater using aquatic plants. *Journal of Agricultural Technology* 8 (1): 1–11.

4 Srivastava, J., Gupta, A., and Chandra, H. (2008). Managing water quality with aquatic macrophytes. *Reviews in Environmental Science and Bio/Technology* 7 (3): 255–266.

5 Jadia, C.D. and Fulekar, M.H. (2008). Phytoremediation: The application of vermicompost to remove zinc, cadmium, copper, nickel and lead by sunflower plant. *Environmental Engineering & Management Journal (EEMJ)* 7 (5).

6 Rahmati, M., Weihermüller, L., Vanderborght, J., Pachepsky, Y.A., Mao, L., Sadeghi, S.H., ... Toth, B. (2018). Development and analysis of the Soil Water Infiltration Global database.

7 Garbisu, C., Allica, J.H., Barrutia, O., Alkorta, I., and Becerril, J.M. (2002). Phytoremediation: A technology using green plants to remove contaminants from polluted areas. *Reviews on Environmental Health* 17 (3): 173–188.

8 Pivetz, B.E. (2001). *Phytoremediation of Contaminated Soil and Ground Water at Hazardous Waste Sites. US Environmental Protection Agency*. Office of Research and Development, Office of Solid Waste and Emergency Response.

9 Moreno, D., Pedrocchi, C., Comín, F.A., García, M., and Cabezas, A. (2007). Creating wetlands for the improvement of water quality and landscape restoration in semi-arid zones degraded by intensive agricultural use. *Ecological Engineering* 30 (2): 103–111.

10 Chu, J. and Guo, W. (2016). Land reclamation using clay slurry or in deep water: Challenges and solutions. *Japanese Geotechnical Society Special Publication* 2 (51): 1790–1793.

11 Hoque, M.A., Hoque, M.M., and Ahmed, K.M. (2007). Declining groundwater level and aquifer dewatering in Dhaka metropolitan area, Bangladesh: Causes and quantification. *Hydrogeology Journal* 15 (8): 1523–1534.

12 Gregoire, C., Elsaesser, D., Huguenot, D., Lange, J., Lebeau, T., Merli, A., ... Wanko, A. (2009). Mitigation of agricultural nonpoint-source pesticide pollution in artificial wetland ecosystems – a review. In: *Climate Change, Intercropping, Pest Control and Beneficial Microorganisms*, 293–338. Dordrecht: Springer.

13 Fenger, J. (2009). Air pollution in the last 50 years – From local to global. *Atmospheric Environment* 43 (1): 13–22.

14 Ross, M.H. and Murray, J. (2004). Occupational respiratory disease in mining. *Occupational Medicine* 54 (5): 304–310.

15 Suwarniati, S. and Hanum, U. (2018). The effect of arbuscular michorriza fungi and organic fertilizer on the growth of sunflower (*Helianthus annuus* L.) on critical land of former iron ore mine in Lhoong–Aceh Besar. *Journal Natural* 18 (2): 77–84.

16 Kumar, R. and Kaur, A. (2018). Oil Spill Removal by Mycoremediation. In: *Microbial Action on Hydrocarbons* (ed. V. Kumar, M. Kumar, R. Prasad). Singapore: Springer.

17 Sinha, R.K., Valani, D., Sinha, S., Singh, S., and Herat, S. (2009). Bioremediation of contaminated sites: A low-cost nature's biotechnology for environmental clean up by versatile microbes, plants & earthworms. *Solid waste management and environmental remediation*, 978–1.

18 Cossu, C., Doyotte, A., Jacquin, M.C., Babut, M., Exinger, A., and Vasseur, P. (1997). Glutathione reductase, selenium-dependent glutathione peroxidase, glutathione levels, and lipid peroxidation in freshwater bivalves, *Unio tumidus*, as biomarkers of aquatic contamination in field studies. *Ecotoxicology and Environmental Safety* 38 (2): 122–131.

19 Pradhan, A., Kumari, S., Dash, S., Biswal, D.P., Dash, A.K., and Panigrahi, K.C. (2017, August). Heavy metal absorption efficiency of two species of mosses (*Physcomitrella patens* and *Funaria hygrometrica*) studied in mercury treated culture under laboratory condition. In *IOP Conference Series: Materials Science and Engineering* 225 (1): 012225. IOP Publishing.

20 Comis, D. (1996). Green remediation: Using plants to clean the soil. *Journal of Soil and Water Conservation* 51 (3): 184–187.

21 Cluis, C. (2004). Junk-greedy greens: Phytoremediation as a new option for soil decontamination. *BioTeach Journal* 2 (6): 1–67.

22 Worm, B., Lotze, H.K., Jubinville, I., Wilcox, C., and Jambeck, J. (2017). Plastic as a persistent marine pollutant. *Annual Review of Environment and Resources* 42: 1–26.

23 Crawford, D.L. and Sutherland, J.B. (1979). Role of actinomycetes in the decomposition of lignocellulose. *Dev. Ind. Microbiol.;(United States), 20.*

24 Ahmad, I., Ahmad, F., and Pichtel, J. (Eds.) (2011). *Microbes and Microbial Technology: Agricultural and Environmental Applications*. Springer Science & Business Media.

25 Dixit, S., & Kuttan, K. K. A. A review: Role of phosphorus solubilising and sulfur oxidizing bacteria in mine reclamation. *International Journal of Agricultural Science and Research (IJASR)* 5(2): 33–40.

26 Berg, G., Rybakova, D., Fischer, D., Cernava, T., Vergès, M.C., Charles, T., Chen, X., Cocolin, L., Eversole, K., Corral, G.H., Kazou, M., Kinkel, L., Lange, L., Lima, N., Loy, A., Macklin, J.A., Maguin, E., Mauchline, T., McClure, R., Mitter, B., and Schloter, M. (2020). Microbiome definition re-visited: Old concepts and new challenges. *Microbiome* 8 (1). 103. https://doi.org/10.1186/s40168-020-00875-0.

27 Xie, P., Hao, X., Herzberg, M., Luo, Y., Nies, D.H., and Wei, G. (2015). Genomic analyses of metal resistance genes in three plant growth promoting bacteria of legume plants in Northwest mine tailings, China. *Journal of Environmental Sciences* 27: 179–187. ISSN 1001-0742. https://doi.org/10.1016/j.jes.2014.07.017.

28 Dixit, R., Wasiullah, M.D., Pandiyan, K., Singh, U.B., Sahu, A., Shukla, R., Singh, B.P., Rai, J.P., Sharma, P.K., Lade, H., and Paul, D. (2015). Bioremediation of heavy metals from soil and aquatic environment: An overview of principles and criteria of fundamental processes. *Sustainability* 7 (2): 2189–2212. https://doi.org/10.3390/su7022189.

18

Mycorrhizal Strategy for the Management of Hazardous Chromium Contaminants

Abeer Hashem[1,2], Nowsheen Shameem[3], Javid A. Parray[3], and Elsayed Fathi Abd-Allah[4]*

[1]*Botany and Microbiology Department, College of Science, King Saud University, P.O. Box. 2460, Riyadh 11451, Saudi Arabia*
[2]*Mycology and Plant Disease Survey Department, Plant Pathology Research Institute, ARC, Giza, Egypt.*
[3]*Centre of Research for Development (CORD) University of Kashmir 190006 Jammu & Kashmir India*
[4]*Plant Production Department, College of Food and Agricultural Sciences, King Saud University, P.O. Box. 2460, Riyadh 11451, Saudi Arabia*
[*]*Botany and Microbiology Department, College of Science, King Saud University, P.O. Box. 2460, Riyadh 11451, Saudi Arabia. habeer@ksu.edu.sa*

18.1 Introduction

Heavy metals (HMs) are particularly dense elements with metallic characteristics that are damaging even in small quantities [1]. Different metals, such as mercury (Hg), cadmium (Cd), arsenic (As), chromium (Cr), and lead (Pb), are highly mobile and are absorbed by plants, including non-essential HMs that are toxic to plants and soil. Other HMs, i.e., copper (Cu), Cr, zinc (Zn), selenium (Se), molybdenum (Mo), tin (Sn), and nickel (Ni), which are responsible for causing pollution and damaging human health have also been investigated [2,3]. The quantity of metals in the root zone and in the atmosphere is generally evaluated by obtaining metal concentration values. Small quantities of toxic metals are present in the subsoil and atmosphere, but their concentrations can increase based on industrial and natural anthropogenic sources, weathering of minerals, sludge sources, water pollution, and excessive use of fertilizers [4]. Plant metabolism, soil fertility, terrestrial ecosystems, and human health are all harmed by these heavy metals [5]. Through soil–plant–food interactions, HMs are transmitted to the human body and cause severe abnormalities [6]. Heavy metals affect plant growth and productivity through a variety of mechanisms by (i) intruding into the function of essential nutrients for plant development; (ii) changing phenotyping and physiological behavior; (iii) affecting microbial activity and growth, which is likely to cause disturbances in the soil, and in turn, reduce the interaction of microbes with rhizobia, and (iv) disturbing soil physical and chemical characteristics [7]. The majority of heavy metals found on the surface of the soil are naturally occurring; however, the situation deteriorates the environment and plant growth habitats when concentrations of such metals increase [8]. High concentrations of HMs cause injury in plant tissues, by changing the permeability of the cell membrane, reduced nutrients uptake, decreased photosynthetic activity, impaired plant growth, altering biological pathways, and ultimately, complete failure of the crop [1]. Furthermore, it also affects free radicals in the root zone, i.e., superoxide radical (O^{2-}), hydrogen peroxide (H_2O_2), and hydroxyl radical (−OH), affects cell organelles, and, as a result, cell death occurs and affects the DNA [9]. The

Core Microbiome: Improving Crop Quality and Productivity, First Edition. Edited by Javid A. Parray, Nowsheen Shameem, Elsayed Fathi Abd-Allah, and Mohammad Yaseen Mir.

capability of HMs to cause injury depends on the stress of specific metals, the time, and bioavailability of the stress exposure in plant organisms [1,10]. Plants have indigenous mechanisms of defense, i.e., their own antioxidant enzymes that tolerate HM stress. Heavy metals result in reactive oxygen species (ROS), which means that these enzymes do not work at their highest concentration. Many physical and chemical strategies have been used to reduce HM ion levels in soil [11]. Meanwhile, biological approaches (plants and microorganisms) demonstrated an innovative option as the natural qualities of the environment are cost-effective and can be easily implemented and maintained [7]. In the environment, plants cooperate continuously with a vast group of rhizosphere-containing microbes and build mutually advantageous alliances. As beneficial fungi, arbuscular mycorrhizal fungi (AMF or AM fungi) develop a mutual association with plant roots that receive the products of photosynthesis made by plants and then assist them by improving absorption of nutrients, protecting them against HM toxicity, improving the accumulation of biomass, and improving photosynthesis [12,13]. Approximately 80% of terrestrial plants have symbiotic relationships with fungi, where shrubs and hyphae are formed and fungal hyphae percolate into the cortical cells of plant roots [14]. Furthermore, AM fungus improved medicinal plant development and nutrient absorption, speeding up the buildup of medicinal compounds and increasing resilience to abiotic conditions [15].

They also make HMs unmovable by attaching to the cortical region of the roots, obstructing transfer into aerial portions of the plant, and preventing injury to leaf tissues. Belonging to phylum Glomeromycota, AMF are a form of endomycorrhizal fungi that provide plant growth nutrients by enhancing water intake and modulating stomatal conductivity [16]. Furthermore, fallow practices have an indirect impact on AM diversity because they alter soil chemical characteristics by producing large amounts of C and N, which influences the growth of different AM species in the soils [17]. Multifunctional role was reported by AMF released glomalin. Exogenous glomalin-related soil protein (GRSP) induced varied impacts on growth response, and the effect was linked with the element, as well as changes in auxins and root shape was occurs [18]. As one of the greatest biological tools, AMF can detoxify the stress caused by HMs to boost plant development and shoot biomass by decreasing HM stress activation in the antioxidant–enzyme system following immobilization of the fungal structure, precipitation and rhizosphere chelation, and vacuole repossession (Mishra et al. 2021) as well as enhanced phosphates, antioxidant enzyme activity and photosynthetic activity and resistance against *Fusarium oxysporum* L [19].

Similarly, endophytes can boost plant nutrition while also improving agronomic performance, with arbuscular mycorrhizal fungus providing the highest benefits to their hosts. A fungus that looks like an arbuscular mycorrhizal fungus (AMF) *Piriformospora* indica colonized the roots of the trifoliate orange (*Poncirus trifoliata*) and showed beneficial impacts on nutrient uptake. The results suggested that P. *indica* showed a greater sensitivity to sugars and fatty acids than *F. mosseae* [20]. The application of AMF provides an environmentally benign strategy and is a useful element in achieving sustained production by increasing soil health and protecting plants from biotic and abiotic conditions [13]. This study thus mainly aims at examining the impact of AMF in preserving an environment under a Cr-contaminated ecosystem. Furthermore, it also discusses the hazardous effects of Cr on plants and how AMF help plants adopt certain mechanisms to withstand Cr stress. Meanwhile, it emphasizes several mechanisms by which AMF minimize the effect of Cr on plants.

18.2 Impact of Heavy Metals

Heavy metals, including As, Ni, Hg, Pb, Cd, Cu, and Cr, are designated as toxic compounds by the Agency for Toxic Substances and Disease Registry (ATSDR) and has hazardous effects on human health as the 1st, 2nd, 3rd, 7th, 17th, 58th, and 120th most dangerous elements, respectively [21]. Several cases of nasal and skin irritation were reported by employees at a chromate production plant [22]. Furthermore, studies have also reported that Cr has a hazardous effect on the gastrointestinal cavity on consumption of water from sources contaminated with Cr [23]. Chromium is the biggest contaminant hazardous to metabolic activity and plant production. The antioxidant defense system restricts the biological pathway and decreases plant growth. Increased Cr pollution of agricultural soil has harmed crop development and productivity, and it has become a global issue [24]. Leather and paint industries are the primary sources of harmful Cr (VI) contamination in the atmosphere, accounting for approximately 40% of total Cr lines in the biosphere [25]. Indian tanning industries solely discharge between 2 000 and 32 000 tons of Cr each year into the atmosphere [26]. There are two types of Cr that are stable: trivalent (III) and hexavalent (VI) forms [27]. The trivalent type is less toxic due to its high bioavailability characteristics, while the hexavalent type is more stable, less mobile, and forms insoluble hydroxide/oxide above pH 5.5 [28]. It is mostly found in soil and aquatic habitats linked to organic materials [29]. The hexavalent (VI) type is usually present as oxyanions of chromates (CrO42) or dichromats (Cr_2O72) in combination with oxygen. In a dynamic equilibrium governed by redox/oxide, dissolution/precipitate, and adsorption/desorption, the two major forms of Cr can interchange and coexist. In the rhizosphere, where many bacterial reductase enzymes exist and are secreted from roots, the redox reaction is highly active [30]. Photosynthesis, nutrient uptake, root damage, and, eventually, plant mortality are all symptoms of Cr toxicity [31]. Production of ROS, enzyme cofactors replacement, and transcription factors; antioxidative enzymes inhibition; redox cellular discrepancy; ionic transport restriction; DNA injury, and oxidation of protein are all well-known phytotoxic symptoms [32]. As a result of ROS precipitation in the form of free radical of lipid membranes and proteins, oxidative degradation occurs [33]. Plants have several antioxidant defense systems in place to protect their cells from ROS, one of which is the accumulation of a variety of tiny organic molecules known as suitable solutes [34]. Similar substances and compatible solutes in the form of amino acids, glycine betaine (GB), and polyols that help to prevent metal stress are also produced [35]. It has been shown that when plants are exposed to abiotic stress, their levels of GB increase [36]. As a quaternary NH_4 molecule, GB is found in plants, animals, and other organisms [37] and acts as an osmoprotectant, inhibiting the generation of ROS. It protects plants from oxidative stress by increasing proline levels and antioxidant enzymes, such as catalase (CAT), peroxidase (POX), and superoxide dismutase (SOD). It is a highly efficient osmo-regulating agent [30]. Its concentration fluctuates greatly among plant species.

18.3 Mechanism to Ameliorate Metal Toxicity

Many plant species, in both normal and severe situations, do not accumulate GB. The natural deposition of GB in some plants is insufficient to prevent from stress. Exogenous use of GB may assist in decreasing negative effects in such situations [38]. In plants, HMs pose long-term adverse consequences after they penetrate the soil. In this context, AMF is an important bioagent because it can improve ecosystem efficiency by building fungal structures, i.e., arbuscules, and assisting in the interchange of inorganic and mineral ingredients and providing the strength needed for plants to grow. They are also a biofilter for HMs, thereby assisting in their mitigation. They support vigorous plant growth under adverse conditions by enabling increasing water intake, photosynthetic rate, and nutrition [39]. As HMs are generally immobilized in fungal hyphae that live in mutual relationships with plants, their accessibility to plants is reduced by chelation and blocking them in the cell wall, cytoplasm, and vacuole, thereby reducing metal toxicity in plants [40,41]. Previous research results have shown that AMF have a significant effect on plant growth and productivity [42]. Inoculating potted *Camellia* with AMF has been proven to stimulate plant growth, improve nutrient absorption, nutritional quality, and increase resilience to drought, salinity, and heavy metal pollution [18]. By adopting a variety of techniques, AMF have been shown to alter the absorption, transport, and accumulation of HMs as well as improve plant resistance to HM stress. Strategies involve conservation of HM in mycorrhizal and external hyphae, stimulation of nutrient absorption, vacuolin sequestration of HM, binding of HMs to the fungal cell wall, protection and remediation of the reaction center, increased response to the antioxidant plant, and chelation of HMs on land. Higher-plant roots colonized by AMF allow plants to retain HMs in roots and external mycelium. A system for reducing HM toxicity in plants has been demonstrated by AMF by keeping HMs in mycorrhizal structures, such as the fungal mycelium and vesicles, where large concentrations of HMs are present, preventing their translocation toward the upper part of plant tissue [43]. Fungal vesicles of mycorrhizal plants are like plant vacuoles, which have been linked to hazardous chemical accumulation. Because of their larger surface area, external mycelial roots have a greater impact on the translocation of metals in soil under Cr stress. This enables metals to be more efficient than plant roots. The AMF retention mechanism reduces HMs translocation and protects leaf tissue from damage by keeping HMs in root zones and helping with Phyto stabilization. Because metals are accommodated in the fungal mycelium, HM tolerance of mycorrhizal plants is associated with mycelial biomass yield and fungal growth. In addition, with the growth of HMs, the vesicles in fungi also rise [44].

18.4 Amelioration of Cr Toxicity by AMF

The AMF have been identified as biological mediators and are capable of enhancing plant tolerance against HMs toxicity [45]. Under variable water levels, AMF inoculation has also been shown to boost antioxidant activity in wetland plants under Cr stress in prior investigations [46]. In semi-aquatic systems, AMF favorably influenced wetland plant growth under HM stress, and set the basis for using AMF to purify wastewater in real wetland systems. The reaction of wetland plants to HM stress at different amounts of water is still unknown, as does the response of wetland plants to varying HM concentrations [47,48]. In roots, AMF can encourage HM accumulation (Zn, Cr, Fe, Cu, and Mn) and impede HM transmission from shoots to roots [49]. According to [50], AMF improved plant tolerance to HM, owing to their ability to chelate, sequestrate, deactivate, and remove sensitive sections of hazardous HMs, as well as the control of the heavy metal acquisition and discharge in metal transport systems [50]. Previous research has shown that fungal structures (arbuscules, intraradical mycelium) in AMF-colonized roots can cumulate Cr and limit Cr translocation into the cytoplasm, hence reducing Cr toxicity in plants [51,27]. It has been observed that Cr accumulates more in the fungal structure (intraradical) than in root cells [51]. The causes for the heavy accumulation of Cr in intraradical fungal structures are as follows: (i) Cr can be absorbed and translocated by the intraradical mycelium of the fungal structure [52] and (2) Cr can be absorbed directly from the rhizosphere by root cells and transmitted to intraradical fungal structures [53]. Similarly, another study was used biochar, made from royal poinciana pods were used to remove Cr (VI) from aqueous solution. The FTIR and XRD analysis results suggested that biochar have (OH) and (O) containing groups including various other functional groups and minerals were present. According to the findings, biochar made from royal poinciana pods could be a promising solution for eliminating harmful Cr (VI) in wastewater treatment [54]. The use of calcium in conjunction with the AMF resulted in a considerable decrease in lipid peroxidation and the formation of hydrogen peroxide, hence mediating the mitigation of cadmium-induced oxidative stress in *Bassia indica* (Hashem et al. 2016).

AMF inoculation in *Melastoma malabathricum* resulted in enhanced soil Cr assimilation by the roots and enhanced plant growth [55]. Similar results were achieved on AMF application on the distribution of Cr and bioavailability under various water depths in semi-aquatic regions. Results demonstrated that Cr accumulation by *Iris pseudacorus* roots in the AMF symbiosis increased by 10–20%. Furthermore, because of Cr contamination, the Cr content in the shoots of AMF-inoculated *I. pseudacorus* were lower by about 21.7–68.4% than in the non-inoculated controls. Furthermore, AMF inoculation reduced Cr concentrations from water, and it was up to 35–23%, respectively, under low Cr stress (5 mg kg^{-1}) compared with non-inoculated treatments. In addition, Cr (VI) concentrations in substrates with Cr contents of 5 mg kg^{-1} were lower by about 27.0–44.0% than in non-inoculated controls. Hence, AMF lowered Cr bioavailability from water soil solution under variable water depth, resulting in a 4.1–5.6% reduction in Cr acid-soluble state [56].

18.5 Chromium-Tolerance Mechanism

Through direct and indirect mechanisms, AMF symbiosis can boost plant performance in the presence of Cr pollution: Directly, AMF symbiosis controls Cr actions in soil or plant roots, influencing its uptake and toxicity; indirectly, AMF symbiosis increases plant tolerance to Cr stress through improved plant nutrition or modulation of the host plant under Cr stress. Exudation of low molecular weight acids or glomalin-related soil protein (GRSP) called glycoprotein with repeated monomeric structures by AM fungi influences the rhizosphere micro-environment, changing the bioavailability of metals. Exudates or GRSP may also form complexes with Cr chelates, affecting metal speciation extractable in acid [57]. Similarly, AMF symbiosis increased organically combined Zn concentrations and decreased crystalline and residual Zn concentrations under Pb- and Zn-polluted soil [58] and increased interchangeable Pb but decreased Pb and Fe/Mn oxidized Pb and carbonated Pb in the rhizosphere [58,59]. Glycoprotein has a lot of potential when it comes to complexing HMs, such as Cr, Pb, Cu, and Zn [60], as concentrations of glycoprotein absorb Cr from metal-contaminated soil. [61], also suggested that trivalent (III) Cr complexes with T-GRSP represent the possibilities of the role of GRSP in soil Cr stability (T-GRSP). Rhizosphere GRSP exhibited a favorable connection with soil concentration [62]. Instead of electrostatic adsorption, the strong chemical bonding can be ascribed to glycoprotein complexation by metals. González-Chávez et al. [60] discovered GRSP to be strongly binding with Cu, which even 50 ml of citric acid, acid chloride, and boric acid could not break down. However, the classic extraction approach has its disadvantages since extraction may also alter metals. The use of sophisticated spectroscopy-based X-ray, i.e., synchrotron X-ray fluorescence spectroscopy [XRF), and X-ray absorption spectroscopy, in situ, can overcome the limitations posed by conventional approaches.

18.6 Cr Stress Amelioration with AMF Symbioses

[57], suggested that 20 mg kg^{-1} Cr (VI] amendment reduced colonization intensity t in the roots of dandelion from 75% to 50% and, similarly, in *Bermuda grass* from 30% to 16%. However, some research findings have also revealed that Cr contamination had no effect on AMF colonization [63]. Metal stress has significant impact on AMF colonization depending on the extent of pollution and specific metal types in the soil. Low levels of heavy metals may promote AMF symbiosis, whereas excessive levels of metal concentration inhibit growth [64,27]. Cr additions below 10 mg kg^{-1} Cr (VI) had no effect on mycorrhizal colonization (and even boosted M percent value) in our dandelion study; however, Cr levels above 20 mg kg^{-1} Cr(VI) dramatically decreased M percent value [57]. [65], discovered that mild Cr-polluted soil boosted *Plantago lanceolata* mycorrhizal association and execution [65]. Furthermore, the trivalent and hexavalent Cr types have had a different impact on AMF symbiosis development [66,67]. In a soil trial, [68], discovered that Cr (VI) had a greater impact on mycorrhizal colonization and arbuscule abundance than Cr (III), with 10 mg kg^{-1} Cr(III) only decreasing arbuscular abundance percentage from 97% to 81%, but 1 mg kg^{-1} Cr(VI) decreasing its concentration from 92% to 12%. There is no question that Cr(VI) is substantially more harmful than Cr(III) in the formation of AMF symbiosis because Cr(VI) generates more oxidative stress significantly [69].

In line with Wu et al.'s [57] suggestion, oxidation level of changing metal types produces variances in movement and bioavailability, impacting AMF colonization differentially. For example, [70], discovered that EDTA-NH_4OAc and $Ca[NO_3)_2$ extractible HMs had no effect on mycorrhizal association, whereas [70], discovered that ammonium nitrate extraction linked to Cd and Zn had a detrimental effect on AMF colonization. Cr might also have similar trends. Another study suggested that Fulvic acid (FA] increased plant growth, biomass, and photosynthetic pigments while decreasing oxidative stress, lipid peroxidation, and Cr buildup in stressed plants. It is also determined that the use of FA in crops irrigated with tannery effluent is one of the most effective ways for reducing Cr toxicity in plants [71].

18.7 Growth, Cr Absorption, and Photosynthate Impact on AMF: Cr Toxicity and Its Effect on the Physiology of Plants with AMF

Recent studies suggested that AMF enhance plant development under Cr-toxic soils. For example, in soil polluted with hexavalent levels of Cr, AMF colonization significantly improved plant development in *dandelion* (Cr-sensitive) and *Bermuda grass* (Cr-tolerant) and the mycorrhizal dependency increased with rising Cr levels [57]. The following study recorded best results in Cr-toxic soil when *Helianthus annuus* L. was in association with *Rhizophagus irregularis* (previously known as *Glomus intraradices*) [67], *Cannabis sativa* with *Glomus mosseae* AMF [72], mesquite plants, and *Glomus deserticola* [66]. *P. lanceolata* linked to *Rhizophagus irregularis* recorded best results and improved yield of crops [65] and fruits quality [73]. Furthermore, when both plants were exposed to Cr stress, AMF-colonized plants, and as a result, the physiological characteristics were improved under Cr-toxic condition compared with control. Under hexavalent type Cr toxicity, mycorrhizal dandelion recorded higher chlorophyll content in leaves than in the control condition [57], although amylase activity was decreased in leaves under Cr stress after AMF inoculation [66]. Mycorrhizal plants had higher chlorophyll content, lower MDA content, and lower guaiacol peroxidase (GPX) and ascorbate peroxidase (APX) activity than nonmycorrhizal plants with Cr contamination, according to certain research [67,74]. Cr speciation in polluted soil influences the perception of Cr effects on plant physiological characteristics. In Cr (III)-treated plants, AMF symbiosis improved *A. prolifera* growth and gas exchange, but not under hexavalent Cr stress [75].

18.8 Plant Cr Absorption and Partitioning as a Result of AMF Symbiosis

Symbiotic AMF affect plant Cr absorption, deposition, and translocation of photosynthetes, in addition to improving growth, development, and physiology characteristics under Cr-toxic condition. Previous research has shown that AMF symbiosis reduces Cr levels in cell organs [66,67,57]. The result of AMF symbiosis boosting plant biomass might be attributed to "growth dilution," which is analogous to the effects of AMF symbiosis on plant concentrations of other heavy metals as Cu, As, Zn, and Cd [76]. However, because AMF-inoculated plants exhibited substantially more biomass than control plants, AMF boosted total plant Cr intake in most circumstances. Plant Cr partitioning is also influenced by AMF symbiosis, which boosts Cr content in roots while reduced it in shoots in *Medicago truncatula* (Wu et al. 2018). Furthermore, Cr movement from the lower part to higher part was reduced, and toxicity was also found to be reduced in plant shoots [57]. Similarly, from investigations on other HMs, as Cu, Zn, Pb, and As, it has also been found that AMF inoculation enhanced metal solidity in roots while limited metal movement in shoots [77]. Heavy metal immobilization by fungi in roots reduces Cr accumulation in shoots by depositing and making complexes in the root zone and reducing Cr toxicity in leaves. However, in some cases, AMF symbiosis was found to improve metal, including Cr, translocation from roots to shoots [72], implying that AMF tasks for Cr partitioning may differ between plants and fungi species.

18.9 Factors Affecting AMF Functioning in Relation to Plant Cr Uptake and Tolerance

The form of phosphorus (P) available for plant uptake is an important factor that limits crop growth and development and it is affected by Cr toxicity at different levels. Furthermore, plant and AMF symbiosis improves the physicochemical qualities of soil that determine all its elements and the role of AMF in the amelioration of the Cr content in plants. One of the most important tasks of AMF symbiosis is to promote plant P uptake, while soil P availability has a major impact on mycorrhizal formation and development [78,79]. Phosphorus addition inhibited AMF function on Cr tolerance in *Medicago truncatula* [80]. Similar results were recorded by [81], who found that AMF have significant impact on plant growth as tolerance increased with decreasing P levels. Enhanced P in soil reduces the development of symbiotic structures (i.e., arbuscules) and associated mycorrhizal activities, according to these findings. Furthermore, enhanced P feeding may boost resistance against Cr by direct contact of P with compounds (Arshad et al. 2016), reducing the reliance of plants on AMF inoculation for survival when subject to metal toxicity. In addition to Cr not being required for plant growth, even at relatively low concentrations in soil, AMF reduces whatever Cr absorption occurs in plants. Plant uptake comprises significant mineral nutrients as Cu and Zn; however, AMF prevent metal uptake. Meta-analysis results predicted that AMF symbiosis normally boosts plant uptake of Cu and Zn at low concentrations but lowers its elevated levels in Cr-toxic soil [82]. The role of AMF in metal absorption in plants differs according to the plant family and their tolerance to heavy metals. Although AMF symbiosis dramatically reduced Cr quantity in dandelion shoots and roots, it had no effect on *Bermuda grass* roots or shoots [57]. Similarly, AMF symbiosis boosted Pb translocation factor in *Kummerowia striata*, *Ecrusgallivar mitis*, and *Ixeris denticulate* L., but had no effect on *Trifolium repens* and *Lolium perenne* L. [64]. Types of AMF and plant symbiosis do not have a strong effect on correlation, despite its dependence on the type of fungus and plant relationship to show species specificity, i.e., one species may have more symbiosis compared with others and a few AMF species [42]. However, it cannot be stated that such AMF have similar association functions with plant organs. In plant metal uptake mechanism, different partitioning has occurred with different AMF strains and may also work inversely. For example, *Glomus mosseae* and *Glomus unicatum* reduced metal concentration in a metal toxic environment; however, the *Glomus constrictum* growth and yield function have not been affected by these metal ions [83]. Moreover, the use and tolerance of various isolates of the same AMF strains may differ. *Tripholium* was extracted from Cd toxic soils, while *G. mosseae* was extracted from non-polluted soils and recorded no Cd absorbance in plants [84]. On the basis of the research on As and Cd, different AMF species are expected to have a different impact on uptake and tolerance of plant Cr. Physical–chemical features of the soil may also affect AMF functioning under different soil pH, which influences AMF functions by regulating the establishment of AMF symbiosis, with AM fungi and their relationship with plants under varying pH conditions [85]. *Glomus* spp. have indicated a preference to grow under pH 5–9, whereas *Gigaspora* spp. and *Acaulospora* have shown better growth and performance under lower pH (Yu and Zhao 2008).

18.10 Effects of AMF on Cr Translocation by Roots

It has been observed that AM fungi affect metal translocation and distribution, hence reducing the phytotoxicity of metals. [66], discovered that mycorrhizal inoculation improved the vascular system by reducing the deposition of Cr and inhibited the translocation of Cr into springs. [51], suggested that AMF modified the translocation pathway in roots, which resulted in CR distribution mainly in mycorrhizal root epidermis, compared with nonmycorrhizal roots in which Cr had been distributed mainly in a vassal bundle with the use of a synchrotron-based XRF microscope. It has been demonstrated that mycorrhizal symbiosis can limit transportation of Cr from lateral roots to major roots. Chromium has been reported to be translocated via vascular bundles from a lower to a higher zone (shoot] [86]. This could explain why the Cr translocation factor is so important to limit toxicity with AMF inoculation [51]. Moreover, similar results were obtained when the Cr portion was substituted by AMF association structures within roots by utilizing a soft-ray X-ray microscope-based synchrotron scanning transmission [51]. The Cr concentration moved from the fungal cells to the cytoplasm of the plant cells and reduced Cr phytotoxicity in plant tissue while increased and retained deposition in mycelium roots. Further research has shown that as shown in synchrotron-based XAS, Cr may be complexed with phosphate and histidine inside fungal structures [51]. With the support of P plant acquisition, the symbioses of AMF are widely known to help P intake in plants due to mycelium and help absorb P and transport it to the upper part [87]. Hence, the preservation of Cr in fungal structures by P can be indicated by precipitated Cr. In addition to Cr, several cation metals [Cu, Cd, Pb, and Zn) have also been discovered to be precipitated in the fungal root structure. The concentration of the metals, translocation, and their movement can be accessed by different techniques. [88], and [89], as disclosed by the Synchrotron Micro-XRF findings, found that Cd accumulated primarily in fungal structures within mycorrhizal roots. Some investigations even have discovered that U was detected inside mycorrhizal roots in intraradical mycelia or in vesicles [90]. [91], suggested that Ni has been collected inside AMF roots and Ni concentration and movement was confirmed with the particulate-induced X-ray emissions hyperaccumulator technique. In mycorrhizal maize, roots produced under Zn/Pb toxic soil suggested that the primary immobilization of Zn has been noticed [92]. Collectively, these investigations have shown that mycorrhizal heavy metal stabilization may be a frequent plant response to withstand HM toxicity in soil. Cr processing in plants can be affected by redox modifications by AMF, which decreased Cr from the hexavalent to trivalent (VI–III) form, when the roots transaction analysis was done. Cr(III) was identified in a majority of the investigations as a significant HM or plant Cr in plant roots as trivalents exist in the form of carboxyl groups [93]. The Cr(III) spectrum inside the roots may also be increased by the decrease of the Cr(III)-systeine and Cr(III)-acetate proportions, while raising the Cr(III) and Cr(III)-histidine ratios [80]. The variants in Cr root speciation by AMF can be caused by the direct participation of fungal roots or by promotion metabolism mechanism. In general, Cr tended to be more stable in mycorrhizal roots, thereby improving root stability and helping to improve plant Cr tolerance. In other words, it was more stable than in nonmycorrhizal roots. Similar AMF symbiosis roles were also identified for As in the metal processing of plants. For example, AMF symbiosis can help plant shoots generate dimethyl arsenic acid (DMA) [94]. [83], also discovered that AMF symbiosis can prevent As(VI) and reduction to As(III) and lower concentrations, possibly due to decreased enzyme activity.

References

1 Morkunas, I., Woźniak, A., Mai, V.C., Rucińska-Sobkowiak, R., and Jeandet, P. (2018). The role of heavy metals in plant response to biotic stress. *Molecules* 23 (9): 2320.

2 Clemens, S. (2006). Toxic metal accumulation, responses to exposure and mechanisms of tolerance in plants. *Biochimie* 88 (11): 1707–1719.

3 Dago, À., González, I., Ariño, C., Díaz-Cruz, J.M., and Esteban, M. (2014). Chemometrics applied to the analysis of induced phytochelatins in Hordeum vulgare plants stressed with various toxic non-essential metals and metalloids. *Talanta* 118: 201–209.

4 Gupta, N., Khan, D., and Santra, S. (2010). Determination of public health hazard potential of wastewater reuse in crop production. *World Review of Science, Technology and Sustainable Development* 7 (4): 328–340.

5 Grimm, N.B., Foster, D., Groffman, P., Grove, J.M., Hopkinson, C.S., Nadelhoffer, K.J., ... Peters, D.P. (2008). The changing landscape: Ecosystem responses to urbanization and pollution across climatic and societal gradients. *Frontiers in Ecology and the Environment* 6 (5): 264–272.

6 Liu, X., Song, Q., Tang, Y., Li, W., Xu, J., Wu, J., ... Brookes, P.C. (2013). Human health risk assessment of heavy metals in soil–vegetable system: A multi-medium analysis. *Science of the Total Environment* 463: 530–540.

7 Miransari, M. (2017). Arbuscular mycorrhizal fungi and heavy metal tolerance in plants. In: *Arbuscular Mycorrhizas and Stress Tolerance of Plants*, 147–161. Springer.

8 Singh, P.C., Srivastava, S., Shukla, D., Bist, V., Tripathi, P., Anand, V., ... Srivastava, S. (2018). Mycoremediation mechanisms for heavy metal resistance/tolerance in plants. In: *Mycoremediation and Environmental Sustainability*, 351–381. Springer.

9 Igiri, B.E., Okoduwa, S.I., Idoko, G.O., Akabuogu, E.P., Adeyi, A.O., and Ejiogu, I.K. (2018). Toxicity and bioremediation of heavy metals contaminated ecosystem from tannery wastewater: A review. *Journal of Toxicology* 2018.

10 Zhang, J., Yang, R., Chen, R., Peng, Y., Wen, X., and Gao, L. (2018). Accumulation of heavy metals in tea leaves and potential health risk assessment: A case study from Puan County, Guizhou Province, China. *International Journal of Environmental Research and Public Health* 15 (1): 133.

11 Singh, J.S. (2019). *New and Future Developments in Microbial Biotechnology and Bioengineering: Microbes in Soil, Crop and Environmental Sustainability*. Elsevier.

12 Chen, S., Zhao, H., Zou, C., Li, Y., Chen, Y., Wang, Z., Jiang, Y., Liu, A., Zhao, P., Wang, M. et al. (2017). Combined inoculation with multiple arbuscular mycorrhizal fungi improves growth, nutrient uptake and photosynthesis in cucumber seedlings. *Front. Microbiol.* 8: 2516.

13 Mitra, D., Uniyal, N., Panneerselvam, P., Senapati, A., and Ganeshamurthy, A. (2019). Role of mycorrhiza and its associated bacteria on plant growth promotion and nutrient management in sustainable agriculture. *International Journal of Life Sciences and Applied Sciences* 1 (1): 1–10.

14 Verma, R., Tapwal, A., Kumar, D., Parkash, V., and Puri, S. (2019). Vesicular arbuscular mycorrhizal diversity in some important ethnomedicinal plants of Western Himalaya. *Med. Plants* 11: 279–285.

15 Rui-Ting, S., Zhang, Z.-Z., Nong, Z., Srivastava, A., Kuča, K., Abd-Allah, E. F.,. Qiang-Sheng, W (2021). A review of the interaction of medicinal plants and A review of the interaction of medicinal plants and arbuscular arbuscular mycorrhizal fungi in the rhizosphere. *Notulae Botanicae Horti Agrobotanici Cluj-Napoca* 49 (3).

16 Schüßler, A. (2002). Molecular phylogeny, taxonomy, and evolution of Geosiphon pyriformis and arbuscular mycorrhizal fungi. In: *Diversity and Integration in Mycorrhizas*, 75–83. Springer.

17 Jemo, M., Dhiba, D., Hashem, A., Abd_Allah, E.F., Alqarawi, A.A., and Tran, L.-S.P. (2018). Mycorrhizal fungal community structure in tropical humid soils under fallow and cropping conditions. *Scientific Reports* 8 (1): 1–17.

18 Liu, R.-C., Gao, W.-Q., Srivastava, A.K., Zou, Y.-N., Kuča, K., Hashem, A., ... Wu, Q.-S. (2021a). Differential effects of exogenous glomalin-related soil proteins on plant growth of trifoliate orange through regulating auxin changes. *Frontiers in Plant Science*, 12.

19 Hashem, A., Akhter, A., Alqarawi, A.A., Singh, G., Almutairi, K.F., and Abd_Allah, E.F. (2021). Mycorrhizal fungi induced activation of tomato defense system mitigates Fusarium wilt stress. *Saudi Journal of Biological Sciences* 28 (10): 5442–5450.

20 Meng, -L.-L., Liu, R.-C., Yang, L., Zou, Y.-N., Srivastava, A.K., Kuča, K., ... Wu, Q.-S. (2021). The Change in Fatty Acids and Sugars Reveals the Association between Trifoliate Orange and Endophytic Fungi. *Journal of Fungi* 7 (9): 716.

21 Agency for Toxic Substances and Disease Registry (ATSDR). CERCLA Priority List. Available online: https://www.atsdr.cdc.gov/spl/resources/ATSDR_2017_SPL_Support_Document.pdf (accessed on 2 March 2019).

22 Gibb, H.J., Lees, P.S., Pinsky, P.F., and Rooney, B.C. (2000). Clinical findings of irritation among chromium chemical production workers. *American Journal of Industrial Medicine* 38 (2): 127–131.

23 Pizarro, F., Olivares, M., Uauy, R., Contreras, P., Rebelo, A., and Gidi, V. (1999). Acute gastrointestinal effects of graded levels of copper in drinking water. *Environmental Health Perspectives* 107 (2): 117–121.

24 Adrees, M., Ali, S., Rizwan, M., Ibrahim, M., Abbas, F., Farid, M., ... Bharwana, S.A. (2015). The effect of excess copper on growth and physiology of important food crops: A review. *Environmental Science and Pollution Research* 22 (11): 8148–8162.

25 Barnhart, J. (1997). Occurrences, uses, and properties of chromium. *Regulatory Toxicology and Pharmacology* 26 (1): S3–S7.

26 Nigam, H., Das, M., Chauhan, S., Pandey, P., Swati, P., Yadav, M., and Tiwari, A. (2015). Effect of chromium generated by solid waste of tannery and microbial degradation of chromium to reduce its toxicity: A review. *Adv Appl Sci Res* 6 (3): 129–136.

27 Zhang, T., Liang, J., Wang, M., Li, D., Liu, Y., Chen, T.H., and Yang, X. (2019a). Genetic engineering of the biosynthesis of glycinebetaine enhances the fruit development and size of tomato. *Plant Science* 280: 355–366.

28 He, Z., Gao, F., Sha, T., Hu, Y., and He, C. (2009). Isolation and characterization of a Cr (VI)-reduction Ochrobactrum sp. strain CSCr-3 from chromium landfill. *Journal of Hazardous Materials* 163 (2-3): 869–873.

29 Ertani, A., Mietto, A., Borin, M., and Nardi, S. (2017). Chromium in agricultural soils and crops: A review. *Water, Air, & Soil Pollution* 228 (5): 190.

30 Kumar, P. (2020). Exogenous AMF with Glycine Betaine Reduces Cr (VI) Uptake and Alleviates Chromium Toxicity by Supressing the Oxidative Stress in Three Genetically Different Sorghum (Sorghum bicolor L.) Cultivars.

31 Hussain, A., Rizwan, M., Ali, Q., and Ali, S. (2019). Seed priming with silicon nanoparticles improved the biomass and yield while reduced the oxidative stress and cadmium concentration in wheat grains. *Environmental Science and Pollution Research* 26 (8): 7579–7588.

32 Huang, W., Yang, X., Yao, S., LwinOo, T., He, H., Wang, A., ... He, L. (2014). Reactive oxygen species burst induced by aluminum stress triggers mitochondria-dependent programmed cell death in peanut root tip cells. *Plant Physiology and Biochemistry* 82: 76–84.

33 Cuypers, A., Karen, S., Jos, R., Kelly, O., Els, K., Tony, R., ... Jan, C. (2011). The cellular redox state as a modulator in cadmium and copper responses in Arabidopsis thaliana seedlings. *Journal of Plant Physiology* 168 (4): 309–316.

34 Ashraf, M. and Foolad, M.R. (2007). Roles of glycine betaine and proline in improving plant abiotic stress resistance. *Environmental and Experimental Botany* 59 (2): 206–216.

35 Hanson, A. (1992). Compatible solute synthesis and compartmentation in higher plants. In: *Water and Life*, 52–60. Springer.

36 Dhir, B., Nasim, S., Samantary, S., and Srivastava, S. (2012). Assessment of osmolyte accumulation in heavy metal exposed Salvinia natans. *International Journal of Botany*.

37 Chen, T.H. and Murata, N. (2008). Glycinebetaine: An effective protectant against abiotic stress in plants. *Trends in Plant Science* 13 (9): 499–505.

38 Hossain, M.A. and Fujita, M. (2010). Evidence for a role of exogenous glycinebetaine and proline in antioxidant defense and methylglyoxal detoxification systems in mung bean seedlings under salt stress. *Physiology and Molecular Biology of Plants* 16 (1): 19–29.

39 Birhane, E., Sterck, F.J., Fetene, M., Bongers, F., and Kuyper, T.W. (2012). Arbuscular mycorrhizal fungi enhance photosynthesis, water use efficiency, and growth of frankincense seedlings under pulsed water availability conditions. *Oecologia* 169 (4): 895–904.

40 Ouziad, F., Hildebrandt, U., Schmelzer, E., and Bothe, H. (2005). Differential gene expressions in arbuscular mycorrhizal-colonized tomato grown under heavy metal stress. *Journal of Plant Physiology* 162 (6): 634–649.

41 Punamiya, P., Datta, R., Sarkar, D., Barber, S., Patel, M., and Das, P. (2010). Symbiotic role of Glomus mosseae in phytoextraction of lead in vetiver grass [Chrysopogon zizanioides (L.)]. *Journal of Hazardous Materials* 177 (1-3): 465–474.

42 Smith, S.E. and Read, D.J. (2008). The symbionts forming arbuscular mycorrhizas. *Mycorrhizal Symbiosis* 2: 13–41.

43 Vogel-Mikuš, K., Pongrac, P., Kump, P., Nečemer, M., and Regvar, M. (2006). Colonisation of a Zn, Cd and Pb hyperaccumulator Thlaspi praecox Wulfen with indigenous arbuscular mycorrhizal fungal mixture induces changes in heavy metal and nutrient uptake. *Environmental Pollution* 139 (2): 362–371.

44 Yang, Y., Han, X., Liang, Y., Ghosh, A., Chen, J., and Tang, M. (2015). The combined effects of arbuscular mycorrhizal fungi (AMF) and lead (Pb) stress on Pb accumulation, plant growth parameters, photosynthesis, and antioxidant enzymes in Robinia pseudoacacia L. *PloS One* 10 (12): e0145726.

45 Vivas, A., Barea, J., and Azcón, R. (2005). Interactive effect of Brevibacillus brevis and Glomus mosseae, both isolated from Cd contaminated soil, on plant growth, physiological mycorrhizal fungal characteristics and soil enzymatic activities in Cd polluted soil. *Environmental Pollution* 134 (2): 257–266.

46 Hu, S., Hu, B., Chen, Z., Vosátka, M., and Vymazal, J. (2020). Antioxidant response in arbuscular mycorrhizal fungi inoculated wetland plant under Cr stress. *Environmental Research* 191: 110203.

47 Li, H., Luo, N., Zhang, L.J., Zhao, H.M., Li, Y.W., Cai, Q.Y., and Mo, C.H. (2016). Do arbuscular mycorrhizal fungi affect cadmium uptake kinetics, subcellular distribution and chemical forms in rice? *Science of the Total Environment* 571: 1183–1190.

48 Xu, Z., Wu, C., Lv, Y., Meng, F., and Ban, Y. (2021). Effects of Aeration on the Formation of Arbuscular Mycorrhiza under a Flooded State and Copper Oxide Nanoparticle Removal in Vertical Flow Constructed Wetlands. *Microbial Ecology* 81 (4): 922–931.

49 Janeeshma, E. and Puthur, J.T. (2020). Direct and indirect influence of arbuscular mycorrhizae on enhancing metal tolerance of plants. *Archives of Microbiology* 202 (1): 1–16.

50 Ferrol, N., Tamayo, E., and Vargas, P. (2016). The heavy metal paradox in arbuscular mycorrhizas: From mechanisms to biotechnological applications. *Journal of Experimental Botany* erw403.

51 Wu, S., Zhang, X., Chen, B., Wu, Z., Li, T., Hu, Y., and Wang, Y. (2016). Chromium immobilization by extraradical mycelium of arbuscular mycorrhiza contributes to plant chromium tolerance. *Environmental and Experimental Botany* 122: 10–18.

52 Bei, L., Xiao-Dong, T., Yu-Fang, G., Jian-Ping, S., and Zhao-Yu, Y. (2015). DNA repair gene XRCC3 Thr241Met polymorphisms and lung cancer risk: A meta-analysis. *Bulletin Du Cancer* 102 (4): 332–339.

53 Wu, S., Zhang, X., Huang, L., and Chen, B. (2019). Arbuscular mycorrhiza and plant chromium tolerance. *Soil Ecology Letters* 1 (3): 94–104.

54 Singh, A., Srivastava, N., Shah, M., Hashem, A., Abd Allah, E.F., and Pal, D.B. (2021). Investigation on hexavalent chromium removal from simulated wastewater using royal poinciana pods-derived bioadsorbent. *Biomass Conversion and Biorefinery* 1–12.

55 Jankong, P. and Visoottiviseth, P. (2008). Effects of arbuscular mycorrhizal inoculation on plants growing on arsenic contaminated soil. *Chemosphere* 72 (7): 1092–1097.

56 Hu, S., Hu, B., Chen, Z., Vosátka, M., and Vymazal, J. (2021). Arbuscular mycorrhizal fungi modulate the chromium distribution and bioavailability in semi-aquatic habitats. *Chemical Engineering Journal* 420: 129925.

57 Wu, S., Chen, B., Sun, Y., Ren, B., Zhang, X., and Wang, Y. (2014). Chromium resistance of dandelion (Taraxacum platypecidum Diels.) and bermudagrass (Cynodon dactylon [Linn.] Pers.) is enhanced by arbuscular mycorrhiza in Cr (VI)-contaminated soils. *Environmental Toxicology and Chemistry* 33 (9): 2105–2113.

58 Subramanian, K.S., Tenshia, V., Jayalakshmi, K., and Ramachandran, V. (2009). Biochemical changes and zinc fractions in arbuscular mycorrhizal fungus (Glomus intraradices) inoculated and uninoculated soils under differential zinc fertilization. *Applied Soil Ecology* 43 (1): 32–39.

59 Zhang, X.H., Lin, A.J., Zhang, X., and Guo, L.P. (2012). The effects of arbuscular mycorrhizal fungi (AMF) on forms of Pb in the upland rice rhizosphere. *Chinese Agricultural Science Bulletin* 28: 24–29.

60 Gonzalez-Chavez, M.C., Carrillo-Gonzalez, R., Wright, S.F., and Nichols, K.A. (2004). The role of glomalin, a protein produced by arbuscular mycorrhizal fungi, in sequestering potentially toxic elements. *Environmental Pollution* 130 (3): 317–323.

61 Gil-Cardeza, M.L., Ferri, A., Cornejo, P., and Gomez, E. (2014). Distribution of chromium species in a Cr-polluted soil: Presence of Cr (III) in glomalin related protein fraction. *Science of the Total Environment* 493: 828–833.

62 Cornejo, P., Meier, S., Borie, G., Rillig, M.C., and Borie, F. (2008). Glomalin-related soil protein in a Mediterranean ecosystem affected by a copper smelter and its contribution to Cu and Zn sequestration. *Science of the Total Environment* 406 (1-2): 154–160.

63 Nakatani, A.S., Mescolotti, D.L., Nogueira, M.A., Martines, A.M., Miyauchi, M.Y., Stürmer, S.L., and Cardoso, E.J. (2011). Dosage-dependent shift in the spore community of arbuscular mycorrhizal fungi following application of tannery sludge. *Mycorrhiza* 21 (6): 515–522.

64 Chen, Y.X., Wang, Y.P., Lin, Q., and Luo, Y.M. (2005). Effect of copper-tolerant rhizosphere bacteria on mobility of copper in soil and copper accumulation by Elsholtzia splendens. *Environment International* 31 (6): 861–866.

65 Estaún, V., Calvet, C., and Camprubí, A. (2010). Effect of differences among crop species and cultivars on the arbuscular mycorrhizal symbiosis. In: *Arbuscular Mycorrhizas: Physiology and Function*, 279–295. Springer.

66 Arias, M.S.B., Peña-Cabriales, J.J., Alarcón, A., and Maldonado Vega, M. (2015). Enhanced Pb absorption by Hordeum vulgare L. and Helianthus annuus L. plants inoculated with an arbuscular mycorrhizal fungi consortium. *International Journal of Phytoremediation* 17 (5): 405–413.

67 Davies, F.T., Jr, Puryear, J.D., Newton, R.J., Egilla, J.N., and Grossi, J.A.S. (2001). *Mycorrhizal Fungi Enhance Accumulation and Tolerance of Chromium in Sunflower (Helianthus Annuus). Journal of Plant Physiology*. 158 (6): 777–786.

68 Davies, P., Walker, A.E., and Grimshaw, J.M. (2010). A systematic review of the use of theory in the design of guideline dissemination and implementation strategies and interpretation of the results of rigorous evaluations. *Implementation Science* 5 (1): 1–6.

69 Shanker, A.K., Cervantes, C., Loza-Tavera, H., and Avudainayagam, S. (2005). Chromium toxicity in plants. *Environment International* 31 (5): 739–753.

70 Weissenhorn, I., Leyval, C., Belgy, G., and Berthelin, J. (1995). Arbuscular mycorrhizal contribution to heavy metal uptake by maize (Zea mays L.) in pot culture with contaminated soil. *Mycorrhiza* 5 (4): 245–251.

71 Ali, S., Rizwan, M., Waqas, A., Hussain, M.B., Hussain, A., Liu, S., ... Abd_Allah, E.F. (2018). Fulvic acid prevents chromium-induced morphological, photosynthetic, and oxidative alterations in wheat irrigated with tannery waste water. *Journal of Plant Growth Regulation* 37 (4): 1357–1367.

72 Citterio, S., Prato, N., Fumagalli, P., Aina, R., Massa, N., Santagostino, A., ... Berta, G. (2005). The arbuscular mycorrhizal fungus Glomus mosseae induces growth and metal accumulation changes in Cannabis sativa L. *Chemosphere* 59 (1): 21–29.

73 Cao, M.A., Wang, P., Hashem, A., Wirth, S., Abd_Allah, E.F., and Wu, Q.S. (2021). Field Inoculation of arbuscular mycorrhizal fungi improves fruit quality and root physiological activity of citrus. *Agriculture* 11: 1297. https://doi.org/10.3390/agriculture11121297.

74 Rahmaty, R. and Khara, J. (2011). Effects of vesicular arbuscular mycorrhiza Glomus intraradices on photosynthetic pigments, antioxidant enzymes, lipid peroxidation, and chromium accumulation in maize plants treated with chromium. *Turkish Journal of Biology* 35 (1): 51–58.

75 Singh, S., Srivastava, P.K., Kumar, D., Tripathi, D.K., Chauhan, D.K., and Prasad, S.M. (2015). Morpho-anatomical and biochemical adapting strategies of maize (Zea mays L.) seedlings against lead and chromium stresses. *Biocatalysis and Agricultural Biotechnology* 4 (3): 286–295.

76 Dong, -W.-W., Zhao, J., and Xu, L. (2008). Remarkable structural transformation of [Zn (tza) 2] during recrystallization, syntheses and crystal structures of [M (tza) 2](M= Zn, Cd, Mn, Co; Htza= tetrazole-1-acetic acid). *Crystal Growth and Design* 8 (8): 2882–2886.

77 Bothe, H., Schmitz, O., Yates, M.G., and Newton, W.E. (2010). Nitrogen fixation and hydrogen metabolism in cyanobacteria. *Microbiology and Molecular Biology Reviews* 74 (4): 529–551.

78 Carbonnel, S. and Gutjahr, C. (2014). Control of arbuscular mycorrhiza development by nutrient signals. *Frontiers in Plant Science* 5: 462.

79 Chu, Q., Wang, X., Yang, Y., Chen, F., Zhang, F., and Feng, G. (2013). Mycorrhizal responsiveness of maize (Zea mays L.) genotypes as related to releasing date and available P content in soil. *Mycorrhiza* 23 (6): 497–505.

80 Wu, C., Chen, L., Lou, X., Ding, M., and Jia, C. (2018a). Fabrication of cobalt-nickel-zinc ternary oxide nanosheet and applications for supercapacitor electrode. *Frontiers in Chemistry* 6: 597.

81 Zhang, X., Ren, B.-H., Wu, S.-L., Sun, Y.-Q., Lin, G., and Chen, B.-D. (2015). Arbuscular mycorrhizal symbiosis influences arsenic accumulation and speciation in Medicago truncatula L. in arsenic-contaminated soil. *Chemosphere* 119: 224–230.

82 Audet, P. and Charest, C. (2007). Heavy metal phytoremediation from a meta-analytical perspective. *Environmental Pollution* 147 (1): 231–237.

83 Yu, Y., Liu, Y., Zhen, S.J., and Huang, C.Z. (2013). A graphene oxide enhanced fluorescence anisotropy strategy for DNAzyme-based assay of metal ions. *Chemical Communications* 49 (19): 1942–1944.

84 Biro, I., Nemeth, T., and Takács, T. (2009). Changes of parameters of infectivity and efficiency of different glomus mosseae arbuscular mycorrhizal fungi strains in cadmium-loaded soils. *Communications in Soil Science and Plant Analysis* 40 (1-6): 227–239.

85 Coughlan, P.J. (2000). In defense of unanimous jury verdicts: Mistrials, communication, and strategic voting. *American Political Science Review* 94 (2): 375–393.

86 Skeffington, R., Shewry, P., and Peterson, P. (1976). Chromium uptake and transport in barley seedlings (Hordeum vulgare L.). *Planta* 132 (3): 209–214.

87 Kuga, S., Yang, J.-H., Takahashi, H., Hirama, K., Iwasaki, T., and Kawarada, H. (2008). Detection of mismatched DNA on partially negatively charged diamond surfaces by optical and potentiometric methods. *Journal of the American Chemical Society* 130 (40): 13251–13263.

88 Nayuki, K., Chen, B., Ohtomo, R., and Kuga, Y. (2014). Cellular imaging of cadmium in resin sections of arbuscular mycorrhizas using synchrotron micro X-ray fluorescence. *Microbes and Environments*, ME13093.

89 Chen, S., Zheng, J., Mei, D., Han, K.S., Engelhard, M.H., Zhao, W., ... Zhang, J.G. (2018). High-voltage lithium-metal batteries enabled by localized high-concentration electrolytes. *Advanced Materials* 30 (21): 1706102.

90 Weiersbye, I., Straker, C., and Przybylowicz, W. (1999). Micro-PIXE mapping of elemental distribution in arbuscular mycorrhizal roots of the grass, Cynodon dactylon, from gold and uranium mine tailings. *Nuclear Instruments and Methods in Physics Research Section B: Beam Interactions with Materials and Atoms* 158 (1-4): 335–343.

91 Orłowska, E., Przybyłowicz, W., Orlowski, D., Turnau, K., and Mesjasz-Przybyłowicz, J. (2011). The effect of mycorrhiza on the growth and elemental composition of Ni-hyperaccumulating plant Berkheya coddii Roessler. *Environmental Pollution* 159 (12): 3730–3738.

92 Wu, S., Hu, Y., Zhang, X., Sun, Y., Wu, Z., Li, T., ... Zheng, L. (2018b). Chromium detoxification in arbuscular mycorrhizal symbiosis mediated by sulfur uptake and metabolism. *Environmental and Experimental Botany* 147: 43–52.

93 Aldrich, M., Gardea-Torresdey, J., Peralta-Videa, J., and Parsons, J. (2003). Uptake and reduction of Cr (VI) to Cr (III) by mesquite (Prosopis spp.): Chromate– plant interaction in hydroponics and solid media studied using XAS. *Environmental Science & Technology* 37 (9): 1859–1864.

94 Chang, Y.W. and Singh, K.P. (2019). Arsenic induces fibrogenic changes in human kidney epithelial cells potentially through epigenetic alterations in DNA methylation. *Journal of Cellular Physiology* 234 (4): 4713–4725.

95 Mittler, R. (2006). Abiotic stress, the field environment and stress combination. *Trends in Plant Science* 11 (1): 15–19.

96 Liu, R.-C., Xiao, Z.-Y., Hashem, A., Abd_Allah, E.F., Xu, Y.-J., and Wu, Q.-S. (2021b). Unraveling the Interaction between Arbuscular Mycorrhizal Fungi and Camellia Plants. *Horticulturae* 7 (9): 322.

97 Feng, J., Liu, R., Chen, P., Yuan, S., Zhao, D., Zhang, J., and Zheng, Z. (2015). Degradation of aqueous 3, 4-dichloroaniline by a novel dielectric barrier discharge plasma reactor. *Environmental Science and Pollution Research* 22 (6): 4447–4459.

98 Yao, Z., Li, J., Xie, H., and Yu, C. (2012). Review on remediation technologies of soil contaminated by heavy metals. *Procedia Environmental Sciences* 16: 722–729.

99 Bai-Hui, R.E.N., Song-Lin, W.U., Bao-Dong, C.H.E.N., Zhao-Xiang, W.U., and Zhang, X. (2015). Cr stable isotope fractionation in arbuscular mycorrhizal dandelion and Cr uptake by extraradical mycelium. *Pedosphere* 25 (2): 186–191.

100 Zhang, X.-F., Hu, Z.-H., Yan, T.-X., Lu, -R.-R., Peng, C.-L., Li, S.-S., and Jing, Y.-X. (2019b). Arbuscular mycorrhizal fungi alleviate Cd phytotoxicity by altering Cd subcellular distribution and chemical forms in Zea mays. *Ecotoxicology and Environmental Safety* 171: 352–360.

101 Zhao, B., Ye, X., Yu, J., Li, L., Li, W., Li, S., ... Chinnaiyan, A.M. (2008). TEAD mediates YAP-dependent gene induction and growth control. *Genes & Development* 22 (14): 1962–1971.

102 Emamverdian, A., Ding, Y., Mokhberdoran, F., & Xie, Y (2015). Heavy metal stress and some mechanisms of plant defense response. *The Scientific World Journal, 2015.*

103 Hashem, A., Abd_Allah, E.F., Alqarawi, A.A., Malik, J.A., Wirth, S., and Egamberdieva, D. (2019). Role of calcium in AMF-mediated alleviation of the adverse impacts of cadmium stress in Bassia indica [Wight] AJ Scott. *Saudi Journal of Biological Sciences* 26 (4): 828–838.

104 Raza, S.H., Athar, H.R., Ashraf, M., and Hameed, A. (2007). Glycinebetaine-induced modulation of antioxidant enzymes activities and ion accumulation in two wheat cultivars differing in salt tolerance. *Environmental and Experimental Botany* 60 (3): 368–376.

105 Reddy, A.M., Kumar, S.G., Jyothsnakumari, G., Thimmanaik, S., and Sudhakar, C. (2005). Lead induced changes in antioxidant metabolism of horsegram (Macrotyloma uniflorum (Lam.) Verdc.) and bengalgram (Cicer arietinum L.). *Chemosphere* 60 (1): 97–104.

Index

Note: Page numbers followed by "f" refers to figures and "t" refers to tables.

Core Microbiome: Improving Crop Quality and Productivity, First Edition. Edited by
Javid A. Parray, Nowsheen Shameem, Elsayed Fathi Abd-Allah, and Mohammad Yaseen Mir.

p

t